# Postcolonialism, Decoloniality and Development

*Postcolonialism, Decoloniality and Development* is a comprehensive revision of *Postcolonialism and Development* (2009) that explains, reviews and critically evaluates recent debates about postcolonial and decolonial approaches and their implications for development studies. By outlining contemporary theoretical debates and examining their implications for how the developing world is thought about, written about and engaged with in policy terms, this book unpacks the difficult, complex and important aspects of the relationships between postcolonial theory, decoloniality and development studies.

The book focuses on the importance of development discourses, the relationship between development knowledge and power, and agency within development. It includes significant new material exploring the significance of postcolonial approaches to understanding development in the context of rapid global change and the dissonances and interconnections between postcolonial theory and decolonial politics. It includes a new chapter on postcolonial theory, development and the Anthropocene that considers the challenges posed by the current global environmental crisis to both postcolonial theory and ideas of development. The book sets out an original and timely agenda for exploring the intersections between postcolonialism, decolonialism and development and provides an outline for a coherent and reinvigorated project of postcolonial development studies.

Engaging with new and emerging debates in the fields of postcolonialism and development, and illustrating these through current issues, the book continues to set agendas for diverse scholars working in the fields of development studies, geography, anthropology, politics, cultural studies and history.

**Cheryl McEwan** is Professor of Human Geography at Durham University, UK. Her main research interests include ethical consumption in the global South, the geographies of transformation in South Africa and the role of postcolonial theory and decolonial politics in shaping responses to global challenges. She is author of *Gender, Geography and Empire* (2000) and *Postcolonialism and Development* (2009), and co-editor of *Postcolonial Geographies* (2002) and *Postcolonial Economies* (2011).

# Routledge Perspectives on Development

**Series Editor:** Professor Tony Binns, *University of Otago*

Since it was established in 2000, the same year as the Millennium Development Goals were set by the United Nations, the Routledge Perspectives on Development series has become the pre-eminent international textbook series on key development issues. Written by leading authors in their fields, the books have been popular with academics and students working in disciplines such as anthropology, economics, geography, international relations, politics and sociology. The series has also proved to be of particular interest to those working in interdisciplinary fields, such as area studies (African, Asian and Latin American studies), development studies, environmental studies, peace and conflict studies, rural and urban studies, travel and tourism.

If you would like to submit a book proposal for the series, please contact the Series Editor, Tony Binns, on: *jab@geography.otago.ac.nz*

For a full list of titles in this series, please visit www.routledge.com/ Routledge-Perspectives-on-Development/book-series/SE0684

# Postcolonialism, Decoloniality and Development

### Second edition

## Cheryl McEwan

Routledge
Taylor & Francis Group

LONDON AND NEW YORK

Second edition published 2019
by Routledge
2 Park Square, Milton Park, Abingdon, Oxon, OX14 4RN

and by Routledge
711 Third Avenue, New York, NY 10017

*Routledge is an imprint of the Taylor & Francis Group, an informa business*

© 2019 Cheryl McEwan

First edition published by Routledge 2009

*British Library Cataloguing-in-Publication Data*
A catalogue record for this book is available from the British Library

*Library of Congress Cataloging-in-Publication Data*
Names: McEwan, Cheryl, author.
Title: Postcolonialism, decoloniality and development / Cheryl McEwan.
Other titles: Postcolonialism and development | Routledge perspectives on
development.
Description: Second edition. | New York : Routledge, 2019. | Series:
Routledge perspectives on development | Includes bibliographical
references and index.
Identifiers: LCCN 2018030336 | ISBN 9781138036710 (hardback : alk. paper) |
ISBN 9781138036727 (paperback : alk. paper) | ISBN 9781315178363 (eBook)
Subjects: LCSH: Postcolonialism--Textbooks. | Developing countries--Textbooks.
Classification: LCC JV51 .M38 2019 | DDC 325/.3--dc23
LC record available at https://lccn.loc.gov/2018030336

ISBN: 978-1-138-03671-0 (hbk)
ISBN: 978-1-138-03672-7 (pbk)
ISBN: 978-1-315-17836-3 (ebk)

Typeset in Times New Roman
by Apex CoVantage LLC

This book is dedicated to my maternal grandparents, Annie (1917–1969) and George Moore (1915–1954) and to my mother. I never had the opportunity to meet my grandparents, but their gift to me is a wonderful mother, whom I love dearly and to whom I owe so much.

# Contents

# Figures

# Tables

# Boxes

# Acknowledgements

This book is a culmination of almost ten years of reading, discussing, writing and mutual learning around the issues first outlined in *Postcolonialism and Development*. The debates about postcolonialism and the terrain of international development have shifted markedly since the first edition, but the fact that there is still a need to bring to the fore the epistemic and material violence and injustice in global relations owes much to the persistent and pernicious legacies of European colonialism. I have countless people to thank for their intellectual generosity and collegiality that have played a role in rethinking many of the arguments in the second edition. Many of the themes have been presented in workshops, seminars and conferences and I am grateful to organizers, presenters and audiences for sharing thoughts and ideas. These include events at the Aotearoa/New Zealand International Development Studies Network; African Studies Association; Association of American Geographers; Development Studies Association; Postcolonial Studies Association; Royal Geographical Society with the Institute of British Geographers; the Developing Areas Research Group of the RGS-IBG; the Centre for Development Research, University of Bonn; and Geography and Development Studies departments at the Universities of Bremen, Cambridge, Glasgow, Manchester and Newcastle.

I have been the grateful beneficiary of the time, funding and visiting positions that have provided invaluable opportunities to learn and think from other places. I owe thanks to the Geography Department at Durham University for maintaining a relatively generous research leave scheme and to the collegiality of my colleagues, whose commitment to the idea of collective endeavour makes this possible in spite of escalating workloads. The Leverhulme Trust, the Economic and Social Research Council and N8 Research Partnership have funded my research in South Africa, a place that has always played a significant role in providing the intellectual stimulus that comes from my ambivalent pleasure and discomfiture at being there. My longstanding research partnerships in South Africa have yielded genuine friendships, especially with Shari Daya, Zaitun Rosenberg and everyone at Flower Valley Conservation Trust. Alex Hughes and David Bek have been and remain constant friends and collaborators. British Academy and Commonwealth Fellowships have enabled me to extend a welcome to scholars from the global South who have also taught me much, especially Lilian Nabulime who is one of Uganda's most talented and soulful artists. Some of the ideas that are included in this book were a product of two Matariki Network travel grants, which enabled me to take up Visiting Fellowships at the University of Otago, and a Massey University International Research Fellowship, which allowed me to spend three very happy months back in Aotearoa/New Zealand. I am grateful to the enormous personal and professional warmth and generosity of colleagues at Otago and Massey Universities, especially Glenn Banks, Tony Binns, Doug Hill, Etienne Nel, Regina Scheyvens, Carolyn Morris and Polly Stupples. Collaboration with Regina and Glenn at Massey gave me the opportunity to learn about decolonial politics from the Pacific, as well as learning through experimentation the optimum temperature at which to drink the many varieties of New Zealand wines. Emma Mawdsley was the finest and funniest fellow traveller, and has been a generous, knowledgeable and appreciated sounding board over the years.

I am enormously grateful to work with Marcin Stanek, whose passion and enthusiasm for decolonizing the classroom is infectious and who brings much to my own endeavours through his generous sharing of knowledge and fieldwork experiences in Latin America. I am also fortunate to have the opportunity to collaborate with like-minded colleagues and first-generation scholars at Durham University on work aimed at tackling some of the embedded social, cultural and economic exclusions within our own institution. The gains may be marginal, but they are meaningful. I am privileged to have the opportunity to supervise, share ideas with and learn from an incredible group of postgraduate researchers at Durham (and a couple at Newcastle) including, most recently, Diana Martin, Rushil Ranchod, Tamlynn Fleetwood, Jonathan Silver, Roman Belete, Ankit Kumar, Lara Bezzina, Lucy Szablewska, Walaa Alqaisiya, Maddy Thompson, Lieya Sahdan, Marcin Stanek, Grace Garside, Cynthia Kamwengo, Matilda Fitzmaurice and Zara Babakordi. At the time of his death, Paul Johnson was developing some fascinating ideas from postcolonial and Marxist theory and a deeply ethical approach to his research on organic activism in Tamil Nadu. He would have been a thoughtful interlocutor on some of the new material in this edition. I continue to feel his loss and remember him with great affection. To receive time and attention from colleagues is a precious gift, and my colleagues in Geography at Durham University are among the most generous and inspirational. I owe a debt of gratitude to members of the Geographies of Life and Culture & Economy research clusters, who are a constant source of dynamism and creativity. I would also like to thank my colleagues with whom I stood and froze in good-humoured solidarity on the picket lines when I should have been finishing this book.

I would like to say a special thank you to the team at Routledge: Andrew Mould, who invited me to write a second edition, Egle Zigaite, who walked me through the submission requirements and was endlessly patient in dealing with repeated delays, and Alaina Christensen, who oversaw the production editing with similar patience. A special thank you to Cynthia Kamwengo, who not only assisted with the index, but has offered me sound advice on what ethical collaboration means in the context of east African universities. I owe heartfelt thanks to dear friends who have kept me going and ensured I got over the finish line: Alex Hughes, Mike Crang, Jo Sharp, Val McDermid, Dorothea Kleine, Alex Norrish, Jackie Underhill and Jo Bradley who all kept me laughing; the Durham and Northumberland guardians of the spirit of 'Grandma Gatewood' who kept my legs going; and the devotees of Omara Portuondo and Celia Cruz who got me dancing (kind of) again. Finally, I wish to thank everyone who ever said anything positive to me or taught me something – I always heard and absorbed it, and it always meant a great deal.

# Copyright acknowledgements

The author and publishers would like to thank the following for granting permission to reproduce images in this work: Royal Albert Museum, Exeter for Figure 1.3 and Indymedia for Figure 6.1. I am also grateful to David Bek (Figure 7.4), Lara Bezzina (Figure 7.5) and Brian Cook (7.1 and 7.2) for granting permission to use their photographs.

# Abbreviations and acronyms

| | |
|---|---|
| AGRA | Alliance for a Green Revolution in Africa |
| ANC | African National Congress |
| ASI | Adam Smith Institute |
| BBC | British Broadcasting Corporation |
| BRICS | Brazil. Russia, India, China, South Africa |
| CBO | Community-Based Organization |
| CIA | Central Intelligence Agency |
| DAC | Development Assistance Committee (of the OECD) |
| DCF | Development Cooperation Forum (of the UN) |
| DfID | Department for International Development (UK Government) |
| FAO | Food and Agriculture Organization |
| FDI | Foreign Direct Investment |
| GAD | Gender and Development |
| GATS | General Agreement on Trade in Services |
| GDP | Gross Domestic Product |
| GEAR | Growth, Employment and Redistribution (South Africa) |
| GNP | Gross National Product |
| GRO | Grassroots Organization |
| HDI | Human Development Index |
| HIPC | Heavily Indebted Poor Countries (IMF initiative) |
| IFI | International Financial Institutions (usually WB, IMF, WTO) |
| IMF | International Monetary Fund |

| | |
|---|---|
| INGO | International Non-Governmental Organization |
| MDGs | Millennium Development Goals (of the United Nations) |
| NAFTA | North American Free Trade Agreement |
| NEPAD | New Economic Partnership for Africa's Development |
| NGO | Non-Governmental Organization |
| NICs | Newly Industrializing Countries |
| ODA | Official Development Assistance |
| OECD | Organization for Economic Co-operation and Development |
| OPEC | Organization of Petroleum Exporting Countries |
| PAC | Pan African Congress |
| PRA | Participatory Rural Appraisal |
| RDP | Reconstruction and Development Programme (South Africa) |
| SDGs | Sustainable Development Goals (of the United Nations) |
| TJM | Trade Justice Movement |
| UN | United Nations |
| UNDP | Uniited Nations Development Programme |
| UNICEF | United Nations Children's Fund |
| USDA | United States Department for Agriculture |
| USAID | United States Agency for International Development |
| VSO | Voluntary Service Overseas |
| WB | World Bank |
| WID | Women in Development |
| WSF | World Social Forum |
| WTO | World Trade Organization |

# **1** Introduction

The possibility of producing a de-colonized, postcolonial knowledge in development studies became a subject of considerable debate in the 1990s. Despite this, however, it is only in the past decade or so that there has been any meaningful dialogue between postcolonial and development studies. Until relatively recently, the two fields 'ignore[d] each other's missions and writings' and were 'giant islands of analysis and enterprise [that] stake[d] out a large part of the world and operate within it – or with respect to it – as if the other had a bad smell' (Sylvester 1999: 703–4). This reflects differences in disciplinary traditions, politics, wariness over motives and divergences in the languages and concepts used to articulate core issues (Table 1.1 summarizes some of the essential differences between the two approaches). In addition, the two fields have often been adversarial and mutually critical. On the one hand, development is one of the dominant western discourses that postcolonial approaches seek to thoroughly challenge and destabilize. On the other hand, there have been criticisms of postcolonialism from those in more applied fields such as development, accusing it of being too abstract and of little relevance. Postcolonialism has been criticized for its alleged failure to connect critiques of discourse and representation to the realities of people's lives, and its apparent inability to define a specific political and ethical project to deal with material problems, such as poverty and inequality, that demand urgent and clear solutions (McEwan 2003).

This book is based on the proposition that both fields are enriched by responding to these mutual criticisms. In so doing, they contribute towards a productive engagement between postcolonial criticisms of how we speak and write about the world, on the one hand, and development concerns with the material realities of global inequalities, on the other. As we shall see, specialists in development studies have more recently begun to view postcolonial approaches as constructive

Table 1.1  *Core differences between development studies and postcolonial studies*

|  | Development studies | Postcolonial studies |
|---|---|---|
| *Nature of field* | An applied field of social science, with origins in economics; managerial in thrust, practical in orientation, and in thrall historically to economic theories and technologies (Sylvester 2006: 66). | Not an applied field, with origins in literary criticism; associated with the humanities and often based in academic departments of language, history, and cultural studies (Sylvester 2006: 66). |
| *Aims* | To develop theory and practice that can assist poorer countries in achieving economic targets and higher, sustainable standards of living. | To re-examine the long historical, cultural and spatial record that has depicted colonies and postcolonies as the problematic children of European history. |
| *Focus* | Measuring 'poverty'; gauging development needs; finding solutions. Concerned with socio-economic issues rather than cultural ones; ignores the question of identity (Baaz 2005). | Analyzing discourses (the social practices through which the world is made meaningful and intelligible), including narratives, concepts, ideologies and signifying practices; critiquing modes of representation (techniques of writing and speaking about the world). Question of identity and culture is at the centre of postcolonial studies; tends to ignore socioeconomic inequality. |
| *Perspective* | Creates spaces and distinctions between places deemed to be either developed or less developed. | Examines interconnections and hybridities created by the world-historical experience of colonialism. |
| *Sources* | Generally conceives of the 'Third World' as a problematic of progress that can be understood and dealt with in statistical and/or technical terms. | Explores everyday lives, archives, and discursive representations through various texts including novels, films, testimonials, official reports, media sources. |

and instructive. This does not mean simply focusing on research carried out in 'post-colonial' contexts (in other words, in countries that were formerly colonized, that are now politically independent and are usually referred to as 'developing'). Rather, it means engagement with postcolonial theories and critiques to create studies of development that are postcolonial in both theory and in practice (Radcliffe 2005).

This book is concerned with the implications of bringing together postcolonial approaches with theories and debates in development studies. It focuses on four areas. First, it provides an account of postcolonial approaches, broadly conceived, as they relate to development studies. It charts the origins of both sets of theories, exploring the ramifications of postcolonialism as a critique of development theory and suggesting that, despite historical divergences, there is scope for increasing intersection and dialogue between the two bodies of theory. Second, it provides a comprehensive review of debates about postcolonialism and development, exploring in more detail the key implications of postcolonial critiques on the writing and doing of development studies. Third, it explores the intersections of postcolonialism with other contemporary re-workings of development theory and practice, such as de-colonial approaches, grassroots and participatory development, indigenous knowledge and global resistance movements. Finally, it explores the possibilities of developing a framework for researchers, practitioners and students of development studies that is postcolonial in theory and in practice.

This chapter introduces both postcolonialism and development, explains the key terminology and highlights the core issues that are explored throughout the book, beginning with some illustrations of why postcolonialism is of contemporary political and cultural significance.

## Why postcolonialism?

At the time of writing, there are two ongoing and unresolved political issues involving former European colonial powers that help explain the importance of postcolonialism, as well as its imbrications with development. The first concerns Germany's announcement in July 2016 that it would formally recognize and apologize for the systematic murder of Namibia's Herero and Nama peoples between 1904 and 1907. Now recognized as the first genocide of the twentieth century, German troops oversaw the extermination of 85 per cent of the Herero and over 50 per cent of the Nama populations of what is now Namibia, expropriated their land and seized their source of wealth – their cattle. While German colonialism in Africa was not on the scale of that of Britain, France or Belgium, its legacy in Namibia is similarly profound. Following Germany's loss of its African colonies during the First World War, Namibia was until 1990 controlled by South Africa's white-minority government, which suppressed recognition of the genocide. Following independence, the Ovambo-dominated government has also shown little interest in acknowledging the near extermination of other ethnic groups. Today, the Herero and Nama make up about 10 per cent and 7 per cent of Namibia's population respectively and live in some of the country's poorest and most underdeveloped regions, while farmers descended from original German settlers still own land seized from local people.

The proposed German apology is striking because other former European colonial powers have been profoundly reluctant to acknowledge the violence associated with and legacies of their imperial history. There has been much public debate in Belgium concerning acknowledgement of its shameful past, including an official apology in 2002 for its role in the assassination of Congo's first Prime Minister, Patrice Lumumba, in 1961 (Kerstens 2008). It is thought that 10 million people – roughly half the country's population – died under the rule of King Leopold II of Belgium between 1885 and 1908 (Hochschild 1998) and many more suffered the consequences of slavery, violence and destruction. Global outrage at Leopold's spectacularly brutal exploitation of a land and peoples he considered his own personal property led eventually to the annexing of Congo by the Belgian state in 1908. It remained a Belgian colony until 1960, during which time the exploitation of Congo's natural and human resources continued on a grand scale. Belgium has never officially recognized or apologized for the cost of invasion and decades of exploitation of Congo. Today, the Democratic Republic of Congo has one of the world's lowest levels of development, ranking 176th out of 187 countries on the Human Development Index (UNDP 2016), despite its enormous wealth of natural resources. It is a country with vast economic resources, but is riven by civil war and corruption and, in recent years, an estimated 6 million lives have been lost to fighting, disease and malnutrition.

A similar case involves the UK government, which in 2013 reluctantly expressed 'sincere regret' and compensated 5,000 Kenyans imprisoned and tortured during the Mau Mau rebellion in its former colony in the 1950s. However, as discussed below, Britain has largely failed to come to terms with its colonial past. The official German apology for the Herero genocide is significant within this broader context of wilful amnesia. However, while the apology was due in June 2017, negotiations are still ongoing at the time of writing and a formal apology is still forthcoming. One sticking point is that the German government is clear that recognition will not involve reparations (Chutel 2016). There is thus no acknowledgement of the profound consequences that colonialism has had for the economic, political, social and cultural development of Namibia. Instead, German negotiators have proposed setting up a foundation for youth exchanges with Namibia and funding various infrastructure projects, such as training centres, housing developments and solar power stations. This will involve bilateral negotiations between the Namibian government and Berlin, without Herero or Nama participation, whose representatives are concerned at being further marginalized as development aid is directed elsewhere. This may explain why, in January 2017, descendants of the genocide issued a lawsuit against Germany in the United States and the prospects of a resolution still seem far off.

The second issue involving a former European colonial power that helps explain the importance of postcolonialism erupted in March 2015, with the formation of a protest movement in Cape Town in response to the lack of transformation in post-apartheid South Africa. The Rhodes Must Fall movement began with demands to remove the statue of Cecil Rhodes – a major architect of British imperialism in southern Africa – from the University of Cape Town (UCT) campus. At the heart of these demands is a much deeper concern with the failure in South Africa to acknowledge the legacies of the past in perpetuating institutionalized racism, to decolonize education and to transform racialized inequality. It is also a reaction against the fact that most South African universities still emulate the colonial morals of Oxbridge at a time when many black South Africans want to see the objectives of higher education more closely aligned with the requirements of alleviating mass poverty and upliftment (Mbembe 2015). The Rhodes statue was removed in April 2016. The Rhodes Memorial remains on land adjacent to UCT, but this has also been vandalized (Figure 1.1). The protests spread across South Africa and elsewhere, including to Britain where similar demands have been made to remove the statue of Rhodes at Oriel College, Oxford University.

The increasingly violent nature of the protests at UCT has drawn much criticism. There is a danger that simply removing the symbols of empire risks erasing and forgetting the past. Marina Warner (2016), for example, argues that rather than tearing down statues,

**Figure 1.1** *The Rhodes Memorial below Devil's Peak, Cape Town, showing signs of vandalism from 2015*

(Source: Author)

South Africa might learn from other former colonies, such as India, where 'On Delhi's old parade ground . . ., a site of much extravagant, ornamentalist imperial pomp and ceremony, the viceroys and heroes of the Raj, who once proudly dominated the city squares and streets up and down the sub-continent, have been put out to pasture'. Warner argues for the importance of keeping history and its makers in our sights, but in contexts in which their deeds are not trumpeted as glorious (as are often those of Rhodes) and that provide the basis for open and critical discussion about the failures and crimes of the past. In the case of Delhi, this has involved removing the statues of former British rulers, governors and officials of the British Raj from various public locations to the Coronation Park and old parade ground. There has been a strong reaction against Rhodes Must Fall at Oxford University, in part because benefactors threatened to withdraw significant bequests if the statue is removed. However, the Rhodes Must Fall movement has been successful in articulating a demand that the past be acknowledged and prompting some reflection on the enduring legacies of colonialism. In South Africa, it has become a focal point for grievances about the persistence of racialized structural inequalities and the role of education in perpetuating these. In Britain, it has shone a light, for a moment at least, on the country's failure to deal with its problematic colonial past. While the politics of colonial pasts in countries like Germany, Belgium, France and Portugal remain fraught, the level of popular and academic debate in recent years contrasts sharply with the 'self-conscious forgetting' (Larmer 2016) in Britain. The collective amnesia about the levels of violence, exploitation and racism in many aspects of British imperialism reflects broader cultural forces that focus attention on more positive and benevolent aspects of history. For example, school children are taught about Britain's role in ending the slave trade, but not its role in orchestrating its abuses. This may explain why a survey in 2014 found that most Britons (59 per cent) think the British Empire is more something to be proud than ashamed of (19 per cent) (Dahlgreen 2014). Rhodes Must Fall attempts to create a more open public debate about Britain's colonial past, and a better understanding of how this has shaped the contemporary world and how Britain is perceived by others.

These recent events remind us of the ways in which history has shaped and continues to shape the present, of the enormity of European imperialism and of the entanglements between development in Europe and underdevelopment in Africa, Latin America and other parts of the world. As Karl Marx (1977: 915) wrote, imperialism and colonialism were essential components of the origins and globalization of capitalism:

> The discovery of gold and silver in America, the extirpation, enslavement and entombment in mines of the aboriginal population, the beginning of the conquest and looting of India, the turning of Africa into a warren for the commercial hunting of black-skins, signalled the rosy dawn of the era of capitalist production.

The economies and cultures of the world we live in today have been undeniably shaped by imperialism and colonialism. Consider, for example, the consequences of the African slave trade. Those individuals and nations that carried out the trade profited on an enormous scale and remain to this day among the most powerful economic nations. Britain's ports (e.g. Bristol and Liverpool) and other large cities developed out of the slave trade. It is estimated, for example, that 80 per cent of Manchester's industrial growth was a consequence (direct or indirect) of the trade, while Birmingham became the 'City of a Thousand Trades' supplying guns, military equipment, chains and shackles to Britain's colonies. Estimates by historians suggest that around 12 million Africans were sent forcibly to the Americas and the Caribbean islands and sold into slavery between 1450 and 1807 (the numbers removed from their homes and dying in captivity were considerably higher). This began with the Spanish Conquistadors (Figure 1.2) and culminated in Britain shipping more than 300,000 slaves a year. Vast areas of the African continent were emptied of their strong, young people, producing an economic decline that blighted generations to come while providing the free labour that underpinned the accumulation of enormous wealth by white industrialists and landowners overseas. As we shall see, postcolonial approaches require that we both recognize these connections – between past and present and between different parts of the world – and acknowledge their consequences, not only in how they have shaped contemporary cultures and economies, but particularly as they relate to power relations.

These issues raise questions about how we tell stories about the past – a concern of critical importance to postcolonial approaches. For example, following criticisms of mainstream histories of the abolition of slavery, the 2007 bicentennial commemorations attempted to challenge popular perceptions that this is simply a story of British

**Figure 1.2**   *Manacas Iznaga Tower and slave bell on a former Spanish-owned sugar plantation, Valle de Los Ingenios Sancti Spiritus, Cuba*

(Source: Author)

enlightenment and heroic achievement. Traditional histories have focused on the roles of the likes of William Wilberforce, a lone abolitionist voice in Parliament who persisted in introducing anti-slavery motions over a period of 18 years until his campaign was successful. However, the formal British commemorations in 2007 acknowledged for the first time that there were also black leaders of the abolition movement, such as Olaudah Equiano (Figure 1.3) and Ignatius Sancho, who inspired thousands of ordinary Britons in a broad popular movement concerned with universal equal rights. Equiano – a freed slave – published his autobiography, *The Interesting Narrative of the Life of Olaudah Equiano, or Gustavas Vassa, the African* in 1789, which became an instant best seller and inspiration to the abolitionists. Ignatius Sancho – a freed slave, 'man of letters' and composer of music – came to symbolize the humanity and intelligence of Africans at a time when this was both denied and disputed. The

**Figure 1.3** *Olaudah Equiano (Reproduced with permission from Royal Albert Museum, Exeter)*

stories of black leaders, however, are rarely at the forefront of the history of abolition, and even the significant role played by the slave rebellions in the Caribbean islands (e.g. Haiti, the largest and most successful rebellion that ended both slavery and French control of the colony) in the late eighteenth century are too often forgotten in the celebrations of the triumph of British parliamentary civility. Similarly, the extraordinary and tenacious efforts by E.D. Morel, a young British shipping company official, to expose the appalling abuses in Leopold's Congo at the beginning of the twentieth century are quite rightly lauded. However, the first to write about Congo as a land of coherent societies, each with its own culture and history that was being systematically destroyed by violent imperialism, and the first to voice criticism of Leopold's regime as theft of land and freedom was George Washington Williams, a black American. By the time of his travels in 1890, over 1,000 Europeans and Americans had visited or worked in the Congo, but as Hochschild (1998) argues, Williams was the only one to speak out passionately about what others denied or ignored. Williams died in 1891; his cry of outrage went unheeded for several more years, and his name is largely forgotten in the telling of the moral crusade against Leopold's excesses.

The wilful erasure of the violence of colonialism remains a source of deep controversy for indigenous peoples within former colonies. In the United States, for example, the teaching in schools of 'American exceptionalism' – the idea that the US was a new and uniquely free country, borne out of revolution and based on the principles of democratic and personal freedom – remains a source of profound insult for Native Americans. Not only does this dismiss thousands of years of Native American histories on the continent, but it erases the genocide, conquest and dispossession its pre-colonial inhabitants. Speaking about the 400th anniversary of the first permanent European settlement at Jamestown, Virginia in 2007, Chief Ken Adams writes:

> the story for me and many other Virginia Indians is a story of sorrow and pain; a story of growing up in a society where Indian culture had almost been completely destroyed . . . When the settlers reached Jamestown, conflict with Indians began almost immediately . . . Today, we have private citizens patrolling the United States border with Mexico to help keep out illegal aliens. To the Virginia Indians, the first British settlers were illegal aliens. The British government finally sent enough people to take over all the land, which the Indians owned, and in the process of the wars that followed, 90 percent

of an entire human race of people died. Our leaders talk about our
nation being founded on Judeo-Christian principles. And yet the loss
of life, liberty and land, experienced by many Native Americans, was
in direct violation of these principles. In 1607, there were hundreds
of Indian villages along the waterways of Virginia and now, sadly,
on those same waterways there are only two: the Pamunkey and the
Mattaponi.

(www.co-ppliving.com/coopliving/issues/2005/June
per cent202005/food.htm)

The erasure of the violence of colonialism remains a feature of many
white settler societies. For example, a tourist brochure in Hermanus,
a wealthy resort in the Western Cape of South Africa, recounts the
history of the area as follows:

Long before European farmers set foot in the area, indigenous Khoi
and San peoples grazed their cattle and sheep on the sweet limestone
veld on the coastal lowlands . . . Many died of smallpox and slowly
disappeared and the remainder joined tribes to the East. In the early
1700s new inhabitants arrived in the area. Cape Colony farmers made
use of the pastures . . .

(Hermanus Cliff Path tourist brochure, 2017)

This account creates a fictional *terra nullius* – a supposed empty
land occupied and owned by no one – into which the Europeans
settled after the original inhabitants had either died of disease or
wandered off elsewhere. It fails to acknowledge that Khoi and
San peoples thrived in this area for at least 20,000 years until the
arrival of Dutch settlers in the eighteenth century, who introduced
smallpox and violent conflict. Indigenous peoples died in large
numbers trying to resist the destruction of their livelihoods by
the Dutch East India Company's enclosure of traditional grazing
land for farms. Throughout the eighteenth century, the Khoi and
San were steadily driven off their land by colonial expansion,
which destroyed their social structures and traditional life and
reduced those who survived to slaves, bondservants or labourers
on farms (Guelke and Shell 1992). Contrary to the sanitized history
presented to tourists, the Khoi and San did not disappear quietly
prior to the arrival of Europeans, but were subject to what some
scholars view as genocide (Adhikari 2010). This is not the story
told to tourists and the erasure of the truth of colonialism continues
to do violence to the descendants of those indigenous peoples who
survived.

The history of European slavery, colonialism and imperialism reminds us that global relations remain shaped by power dynamics that have existed since the fifteenth century. These power dynamics are not only economic and political but are also deeply cultural. They relate to questions about who has the power to write histories and to represent other peoples and places in the present. Postcolonial approaches arise from the fact that this power still resides overwhelmingly in the West and, in former settler societies, with the descendants of European colonizers. These representations are intertwined with political and economic interventions by wealthy countries in other parts of the world, which historically have taken various forms: slavery, imperialism, colonialism, decolonization and 'development'. This book considers these intertwinings as they relate specifically to development. It explores the implications of postcolonial approaches, which seek to reveal and challenge these long-entrenched power dynamics, for understanding development. Rather like the counter-story about Jamestown told by Chief Ken Adams, postcolonialism enables counter-stories to be told about the history of interventions in the name of 'development' by wealthy countries in poorer ones.

The relationships between colonialism, decolonization and development are not simply about globally unequal power relations, but are also played out in national contexts. As we have seen, white settler societies, many of which are considered 'developed' countries, are still dealing with the consequences of colonialism. Indigenous groups have invariably fared extremely poorly in economic and political terms, either being completely marginalized in development processes (as is the case in much of North America) or falling victim to targeted state interventions in the name of development. For example, post-independence nation-building in Latin American countries placed great emphasis upon *mestizaje*, the process of racial and cultural miscegenation (the mixing of 'races' by interbreeding). The implicit aim was to erase cultural and ethnic identities by creating a single, mixed-race, national identity. Mexican nationalism, for example, prioritized *mestizaje* such that 90 per cent of the population is now considered *mestizo* (Gutiérrez 1995). Consequently, indigenous groups throughout Latin America, such as the Quechua of the central Andes and the Mapuche of Chile and Argentina, have been marginalized and discriminated against. In Australia, generations of young Aboriginal children were taken from their parents in a state-sanctioned policy of forced assimilation, which lasted from the nineteenth century into the late 1960s. These children are known as the 'stolen generations' and their story was the subject of the films *Rabbit-Proof Fence* (2002) and *Australia* (2008).

Settler societies have also had to negotiate the increasing visibility and activism of indigenous groups following independence, often articulated around human rights in relation to forms of economic development. In Canada, for example, resistance by Inuit and Cree groups to the encroachment of hydroelectric projects into indigenous lands in the early 1970s initiated a protracted process of land claims settlements (finally concluded in 2005), specifically in Inuvialuit and Nunavut (Northwest Territories), Nunavik (northern Quebec), and Nunatsiavut (northern Labrador). By far the largest land claim agreement in Canadian history is the Nunavut Land Claims Agreement signed in 1993. This provided title to the Inuit to 352,240 square kilometres of land in the eastern half of the former Northwest Territories. The agreement provided clear rules of ownership, rights and obligations towards the land, water and resources of Nunavut, as well as $1.14 billion and a Training Trust Fund to ensure that the Inuit have access to sufficient training to enable them to meet their responsibilities under the claim. The Final Agreement also included an undertaking by Canada to recommend legislation to Parliament to establish a Nunavut territory. On 1 April 1999 Nunavut became Canada's newest and, at one-fifth the size of the country, largest territory. As discussed subsequently, Canada's partial recognition of the territorial claims of indigenous peoples does not necessarily protect them from further exploitation, particularly where they include rich mineral resources such as oil.

In Australia, most Aboriginal peoples continue to live in communities with some of the highest rates of poverty, ill-health and deprivation in the 'developed' world, and where there is a 17-year gap in life expectancy in comparison to other Australians. The plight of Aboriginal peoples has become a source of embarrassment to liberal Australians. The formal apology on 13 February 2008, issued by Prime Minister Rudd on behalf of the Australian government, was thus of great symbolic importance. This was issued to all Aboriginal peoples for laws and policies that 'inflicted profound grief, suffering and loss'. The 'stolen generations' of thousands of children forcibly removed from their families were referred to specifically. Although some Aboriginal groups are critical that the apology was not accompanied by compensation, others celebrated the fact that this was the first time the Australian government had publicly acknowledged past mistreatment of Aboriginal peoples. Australia, like Canada, has engaged in a state-sanctioned process of reconciliation, which involves establishing and enacting new relationships between settler and indigenous groups. In Canada, postcolonial politics have been framed around justice and human rights in relation to indigenous peoples and their development. The apology by the Australian government emerged more from the needs of liberal white Australians than it did from those of Aboriginal peoples (see Gooder and Jacobs 2002) – reconciliation eases white settler guilt without returning land and sovereignty to Aboriginal peoples, or even acknowledging this return as the only ethical standpoint. It will only be of more than symbolic significance if it facilitates the emergence of a similar kind of postcolonial, socio-economic justice.

Some traditional accounts of development may work to encourage students in wealthy countries to imagine that 'development' does not really concern them; it happens in other places (places called the 'Third World', the 'developing world', the 'South' or the 'non-West') that are not really connected to 'us'. Alternatively, these accounts may foster an unacknowledged assumption that the answer to development problems lies exclusively in the 'developed world'. This book takes a rather different approach, encouraging critical thinking about the ways in which we are all implicated in development and about the nature of this involvement. Development is thus an issue about which we should all be concerned, even passionate, because we are all potentially implicated in and affected by it. The history of the slave trade, for example, reveals connections between development in one part of the world and underdevelopment in others. There is a need to think more deeply about these connections – between 'us' and 'them', 'here' and 'there' – both in the past and in the present. Before unpacking the complexities of postcolonialism and its implications for development, however, it is expedient to explore what is meant by 'development' and other terms of importance throughout this book.

## Terminology

As we shall see, problematizing language is a fundamental aspect of postcolonial approaches. It is important, however, to delineate some of the key terms used throughout this book.

## Development

Development is a term that creates numerous difficulties in defining its meaning:

- it is one of the most complex words in the English language
- it means different things at different times and in different places
- it means different things within countries (e.g. between different social groups)
- it has no clear meaning and there is little international consensus, but many nation states and international organizations claim to pursue it
- it is the stuff of myth, mystique and mirage (Power 2003)
- it is assumed by development organizations and governments to be possible, a natural process of evolution, progressive and implying 'good change', but some critics see it as problematic and even destructive

In a generic sense, development has been defined, not unproblematically, as 'evolution', 'progress' or 'modernity' (see Chapter 3). Western modernity has been fixated on progress through industrialization, which is understood as development. Cowen and Shenton (1995) first referred to this as 'little d' development: the idea of a natural, immanent, evolutionary process without intentionality, such as the shift from agrarian to capitalist economies. They contrast this with 'big D' development: an intentional practice, often performed by governments, to deal with the problems created by 'little d' development. 'Big D' development originated in capitalist economies in the nineteenth century as 'a means to create order out of the social disorder of rapid urbanization, poverty and unemployment' (*ibid.*: 32) or as a 'will to improve' (Li 2007). In development studies, development has been defined as the use of resources to relieve poverty and improve the standard of living of a nation. It is often thought of as the means through which a traditional, low-technology society is changed into a modern, high-technology society, with a corresponding increase in incomes. This is a notion of intentional practice in developing countries, but as we shall see, it is also underpinned by the idea of an immanent (evolutionary) process, rooted problematically in western understandings of progress and modernization. Narrowly economic definitions of development, based on indicators such as per capita Gross National Product (GNP), production, consumption and investment, have also been criticized. Consequently, many contemporary definitions define it in a more holistic sense to include improvements in social justice, such as a more equitable distribution of income, or improvements in women's and minority rights.

One thing that can be said is that development is a powerful term; it has the power to fascinate, seduce and create dreams and expectations. As discussed in subsequent chapters, there are important questions to be asked about how and why development has become *normative* (defining what should be done) and *instrumental* (determining how it should be done), usually pointing to things that are lacking or deficient (e.g. economic advancement, knowledge) or things that need to be intensified (e.g. democracy).

## Differentiating between countries

There are many ways of ranking and categorizing countries in terms of their levels of development. As we shall see, the way in which parts of the world are described tells us a great deal about who has the power to decide which qualities and indicators are valued and which are denigrated. Rather than using terms that suggest both a hierarchy and a value judgement ('First' and 'Third World', 'developed' and 'developing'), the terms 'global North' or 'North' and 'global South' or 'South' are generally used throughout this book. This is, of course, geographically inaccurate and too generalized to encompass the complexities within and between nations, but it is perhaps the least problematic means of distinguishing between relatively wealthy countries and continents (Europe, Japan, Australia, New Zealand, USA and Canada) and relatively poorer ones (Africa, Asia, Latin America, the Caribbean and the Pacific), since it also implies an interconnected world and a global context in which to consider inequality and poverty (Chapter 4 explores further the problematic of labelling different parts of the world). As a political designation that distinguishes between the 'haves' and the 'have nots', the terms have a certain political value. It is perhaps most useful to think of North/South as a *metaphorical* rather than a *geographical* distinction, where North refers to the pathways of transnational capital and South to the marginalized poor of the world regardless of geographical location (Dirlik 1997; see also Mohanty 2002). Thus, impoverished Australian Aboriginal communities might still be thought of as part of the global South despite being positioned within a wealthy country; conversely, the super-rich of Brazil, India and China might be considered part of the global North. Therefore, as a metaphorical concept, the North/South binary is a useful shorthand for referring to relatively wealthier and poorer parts of the world, but also has scope to recognize the analogous effects in both North and South of neoliberal globalization and of what critics view as 'a savage restructuring of class and social relations worldwide . . . in the interests of capital': the privatization of social provision, the dismantling of welfare states, the erosion of security in wide aspects of peoples' lives, the driving of millions into increasingly precarious work or into structural unemployment, and the simultaneous erosion of workers' and civil rights (Lazarus 2011: 7).

The terms 'North' and 'South' are used throughout the book in preference to others that have achieved popular parlance within development studies. Although the term 'Third World' did not originally imply a sense of hierarchy, it has come to be used as shorthand for countries considered to be least developed. It was originally used during the Cold War to refer to those countries, mostly newly independent, that were non-aligned to either the industrialized, capitalist countries of the 'First World' (the USA and its allies) or the communist/state socialist countries of the 'Second World' (the USSR and its allies), and that sought a 'third way' through which to develop their economies and societies. The collapse of state socialism from the late 1980s and the ending of the Cold War have effectively removed the distinction between different regions of the world. The fact that 'Third World' has come to be used in a hierarchical sense is considered in more detail later in this chapter. The 'developed'/'developing' binary is often used to distinguish between richer and poorer countries, but this implies the existence of a magical cut-off point where countries stop being 'developing' and become 'developed'. It fails to recognize the dynamism of all societies ('developed' countries are also always 'developing' and perhaps sometimes 'de-developing') and it fails to consider inequality and poverty within supposedly 'developed' countries (Willis 2005: 16). Terms such as 'More Economically Developed Countries' (MEDCs) and 'Less Economically Developed Countries' (LEDCs) are popular, but refer exclusively to economic development, neglecting other factors of human well-being that are also important aspects of 'development'. The terms global North and South capture more than simply economic indices of development and can be used to challenge ideas about 'developed' and 'developing' nations. They capture a relational dualism: both global North and global South are products of the same global historical processes and their uneven effects. Today these processes are disrupting traditional geographies of core and periphery and the driving forces of global capitalism are shifting to the South (Ong 2006). As we shall see, the last two decades have witnessed many countries of the global South, particularly some of the big Asian economies of China, India and Indonesia, but also Brazil and South Africa, experience high economic growth rates. This is, to some extent, rebalancing global consumption and the global economy (World Bank 2011; UNDP 2013) as growth in much of the global North remains slow, giving rise to new geographies of development (Sidaway, 2012). South–South trade, which bypasses Europe and the USA, has become increasingly significant and continues to grow (World Bank, 2011). The growing economic significance of some global South countries is also enhancing their political influence (Wang 2008; Stephen 2012). As Raghuram *et al.* (2014) argue, the North–South relations that inform postcolonial thinking about development have been replaced by a much more complex and variegated spatial matrix of power relations. And as wealthier countries in the North experience economic crisis, political corruption and racial tension, they are increasing having to respond to similar challenges faced in societies normally associated with the global South concerning the nature of democracy, national borders, labour and capital and multiculturalism (Comaroff and Comaroff 2012). As we shall see, scholars and practitioners in the global North might seek to better understand these issues by engaging with theories developed in the global South.

Some political activists working for global justice use the terms 'Majority World' to refer to countries of the global South and 'Minority World' to refer to countries of the global North. According to the Human Development Index, in 2015 85 per cent of the world's population lived in developing countries; most of the world's population thus live in the poorest countries, a stark indicator of increasing global inequality in the twenty-first century. Esteva and Prakash (1998) refer to the One-Third versus Two-Thirds division, wherein social majorities (the Two Thirds, or metaphorical South) have a poorer quality of life than social minorities (the One Third, or metaphorical North). However, simple binaries between Minority and Majority worlds do not capture the presence and influence of elites within the Majority world – the super-rich who are part of the global 1 per cent with the greatest wealth and power. The term 'Majority World' does highlight the fact that *Eurocentrism* underpins many of the terms used to refer to most of the world's population. In other words, European or western ideas – representative of a small minority of the world's population – are seen to have the greatest importance and validity. A western perspective is seen to be the only way of comprehending the world. Eurocentrism relates to power over knowledge and, as we shall see, has been central to the formulation and deployment of development theories throughout history. Comaroff and Comaroff (2012: 113–4) summarize it thus: western Enlightenment thought (see below) has 'posited itself as the wellspring of universal learning' and knowledge, while the non-West (the Third World, the Underdeveloped World, the Global South) is seen 'primarily as a place of parochial wisdom, of antiquarian traditions, or exotic ways and means' and as a place of 'unprocessed data'. In other words, the non-West is a 'reservoir of raw fact' from which only western scholars are equipped to develop their 'testable theories and transcendent truths' (*ibid.*: 114). On occasion, therefore, the terms western and 'the West' are used interchangeably with global North. The West is less a geographical term, however, but refers more explicitly to ideologies, cultures and systems of thought, originally centred in Europe and exported through colonialism to the Americas, Australia and New Zealand.

Western thought is considered to originate in Greco-Roman and Judeo-Christian cultures, evolving through the Renaissance (the European revival of learning and advancement of sciences from the fourteenth century), European imperialism from the fifteenth century onwards and the Enlightenment (the 'Age of Reason' that shaped systems of governance in Europe and North America during the eighteenth century; see Chapter 3). As with all definitions, it is important to acknowledge that 'western' cannot be reduced to a singular set of beliefs, but refers to cultures and philosophies that are also complex and sometimes contradictory. What we define as western, for example, includes Christian moral traditions and religious values alongside secular values that often have a rationalist anti-clerical tradition. In the modern sense, western is usually used to refer to those countries that have chosen democracy as a form of governance, favour capitalist modes of economic organization, espouse the values of free international trade and co-operate both economically and militarily.

The primary problem with dividing the world into 'West' and 'non-West' is that this implies a false separation between two distinct entities and the valorizing of one set of values (those that are identified as western – Christianity, rationalism, civilization, democracy, capitalism) over the other (non-western – those seen as non-Christian, non-rational, undemocratic and non-capitalist). In short, it encourages a binary understanding of the world that is both hierarchical and over-simplified. It carries with it other assumptions, such as modernity and development being inseparable from the rise of Enlightenment in the West and shaping a 'distinctively European mission to emancipate humankind from a prehistory of bare necessity, enchantment and entropy' (Comaroff and Comaroff 2012: 114). It also encourages forgetfulness about the deeply intertwined histories and cultures of the West and 'non-West' and, particularly, the scientific and cultural debts that western countries owe to the 'non-western' world. As we shall see, postcolonial approaches attempt to counter this forgetfulness and to problematize binary views of histories and cultures that tend to perpetuate the dominance of western modes of thought. They also allow for the inversion of the Eurocentric spatiality of theory – the notion that theory originates in and is dispersed from the West – to consider how theorizing from elsewhere might provide insights into the workings of the world. In other words, development is theorized from the global South and countries in the global North can be the repositories of that knowledge.

## Subaltern

A key term used throughout this book is *subaltern*, the meaning and usage of which is complex (McEwan 2008b). There are two distinct but related ways in which the term has been understood and deployed (Box 1.1). Both meanings of subaltern – the military analogy and the general meaning of subordinate – have been particularly significant within postcolonial theory. As discussed in Chapter 2, postcolonial theorists have examined the processes by which elites in colonized countries internalized Eurocentric ideas of the inferiority of their own cultures and ensured the dominance of the European powers. This dominance was achieved largely through processes of colonialism, whereby imperial power was grafted onto existing mechanisms of power. Westernization in India and the African colonies, for example, was a tactic for forming an administration and military from local elites, whose access to power would be limited to a subaltern role. Rather like the subaltern officers in the British army, these elites were essential to the chain of command and in maintaining order amongst the masses; the hegemony of the British colonial power was maintained through the consent of these subaltern administrative and military ranks. Other, more subordinated, groups (e.g. proletarian and peasant groups) were not drawn into these structures but were left to re-negotiate their position relative to these local elites. For this reason, the supposed superiority of the colonizers was accepted and left unchallenged.

---

## Box 1.1 The meanings and usage of 'subaltern'

1. As a military expression. Originally used to refer to junior officers deriving from its literal meaning – 'subordinate'. In the British army, for example, subalterns are commissioned officers below the rank of captain. These ranks are subordinate to the superior officers, but serve a role in connecting the lowest ranks with the commanding ranks and thus ensuring that the chain of command is maintained.
2. As a broader term to refer to any person of inferior station or rank. This also draws on the literal meaning of 'subordinate'. Since the 1980s, this meaning has become widely used, primarily within postcolonial theory, to mark the subordinate positions of groups of formerly colonized peoples who remain marginalized primarily by power relations that work through race and class.

The creation of subaltern groups or individuals is achieved through a process known as *hegemony*. This term comes from the writings of Italian theorist, Antonio Gramsci, who used it to refer to the dominance of one group or class in society over others, achieved not through force but through the consent of those other groups (see Chapter 2). Consent is achieved through the dominant group associating itself with moral and intellectual leadership in a society. The term subaltern is now more popularly used, particularly within postcolonial theory, to describe groups who are excluded and do not have a position from which to speak, for example peasants and women in postcolonial societies. This usage was inspired by the Subaltern Studies collective who applied Gramscian perspectives to explorations of Indian history in the 1980s (see Chapter 2). Thus, in current philosophical and critical usage subaltern describes specifically a person or groups of people rendered voiceless and without agency by their social status. Gayatri Spivak's 1988 essay 'Can the subaltern speak?' has had considerable influence on this broader deployment, especially in its deconstruction of gender and representation in India.

Since subaltern can refer to any person or group of inferior rank and/or status, it might be employed in discussions of race, ethnicity, class, gender, sexuality, religion and so on. However, care needs to be exercised in its usage. Spivak, for example, objects to careless use of the term and its appropriation by marginalized groups who are not specifically subaltern. She asserts that subaltern is not just another word for the oppressed or marginalized, nor a general claim to disenfranchisement within a system of hegemonic discourse. Rather, it signifies very specifically a group of people whose voices cannot be heard or that are wilfully ignored in dominant modes of narrative production. Thus, as it is used in postcolonial theory, subaltern refers to those who have no access to processes of cultural imperialism, which locates them in a space of difference (or 'alterity'). In this sense, then, subaltern is not simply the oppressed. For example, working classes in the global North can be said to have been oppressed, but they are not subaltern by the fact that they have historically had access to cultural imperialism and are positioned favourably within hegemonic discourse. They might be oppressed by class structures, but they remain privileged by skin colour. Postcolonial studies have attempted to create intellectual space to allow the subaltern to speak, rather than always being spoken for by either elites or colonizing Northern representatives.

Postcolonial approaches have revealed and unravelled these processes of colonization of minds as well as territories. Of importance are the particular resonances that understanding how power works have for development studies, especially for reflecting on how Eurocentric ideas about development theory and practice continue to be accepted by and deployed within the former colonies, and in explaining why alternatives very often have not emerged. Attempts within postcolonial approaches to create intellectual space to allow the subaltern to speak, rather than always being spoken for by local elites or western representatives, also have implications for the nature of development studies.

## Contested definitions of postcolonialism

Like development, postcolonialism is a difficult and contested term. As long ago as 1995, cultural theorists expressed concern at the proliferating usages of the term: 'the increasingly unfocused use of the term "post-colonial" over the last ten years to describe an astonishing variety of cultural, economic and political practices has meant that there is a danger of it losing effective meaning altogether' (Ashcroft *et al.* 1995: 2) However, others have suggested that rather than fearing excess or loss, perhaps we should celebrate the 'open constellation of meanings associated with a term that crops up in academic writings, journalism and literature' (Sidaway 2000: 593). Box 1.2 summarizes five main understandings of postcolonialism, which are of relevance throughout this book and explained in further detail below.

## Box 1.2 Summary of the meanings of postcolonialism

1) Postcolonialism as 'after-colonialism' – written as post-colonialism to signify the notion of time or a new epoch
2) Postcolonialism as a 'condition', related to the state of 'after-colonialism'
3) Postcolonialism as a metaphysical, ethical and political theory – dealing with issues such as identity, race, ethnicity and gender, the challenges of developing post-colonial national identities, and relationships between power and knowledge
4) Postcolonialism as literary theory – critiquing the perpetuation of representations of colonized and formerly colonized people as inferior, and countering these with alternative representations from writers in (de-)colonized countries
5) Postcolonialism as anti-colonialism – a critique of all forms of colonial power (cultural, political and economic, past and present)

Thus the 'post' of 'postcolonialism' has two meanings, referring to:

- a *temporal* aftermath (1 and 2 above) – a period *after* colonialism
- a *critical* aftermath (3 to 5 above) – cultures, discourses and critiques that lie *beyond*, but remain closely influenced by, colonialism (Blunt and McEwan 2002). Importantly, this meaning does not consign colonialism to the past, but explores and critiques the ways in which it endures into the present. For this reason, Ann Laura Stoler (2016: ix) uses (post)colonialism to emphasize both colonial 'presence' in tangible and intangible forms and the persistence of colonial 'presents'.

These two meanings do not necessarily coincide and it is, in part, their problematic interaction that has made postcolonialism such a contested term.

## Postcolonialism as 'after-colonialism' or temporal period

Postcolonialism is sometimes assumed to refer to 'after-colonialism' or 'after-independence' (Ashcroft *et al*, 1995: 2), describing the wide range of social, cultural and political events arising specifically from the decline and fall of European colonialism that took place after World War Two. However, while European colonialism may no longer exist as before, it is far from clear that colonialism has been relegated to the past. As Table 1.2 illustrates, colonies and colonial powers still exist. Britain and France, for example, still possess scattered territories around the world. Indonesia – in relation to disputed territories in East Timor, Irian Jaya, Sipadan, Ligitan and Batam – and China – in relation to Tibet – for example, are putative colonial powers. The devastation of Puerto Rico by Hurricane Maria on 20 September 2017 and the subsequent humanitarian crisis exposed the profound colonial condition of the island. An

*Table 1.2 Examples of twenty-first century colonies (or 'overseas territories')*

| Governing state | Overseas territories (date of first formal colonization in brackets) |
|---|---|
| *Britain* | Anguilla (1650–1967, 1969–), Ascension (1815), Bermuda (1609), British Indian Ocean Territory (1810), British Virgin Islands (1666), Cayman Islands (1670), Falkland Islands (1833), Gibraltar (1704), Montserrat (1632–1667, 1668–1782, 1784–), Pitcairn (1838), Saint Helena (1651), South Georgia and South Sandwich Islands (1775), Tristan da Cunha (1816), Turks and Caicos (1679). |
| *France* | Bassas da India (1897), Clipperton Island (1855), Europa Island (1897), French Guiana (1644–1809, 1817–), French Oceania (1842), French Polynesia (1842), Glorioso Island (1897), Guadaloupe (1635–1813, 1814–), Juan de Nova Island (1897), Martinique (1625), Mayotte (1841), New Caledonia (1853), Réunion (1642–1810, 1815–), Saint-Barthélemy (1648–1784, 1878–), Saint Martin (1648), Saint-Pierre and Miquelon (1604–1713, 1763–), Tromelin Island (1814), Wallis and Futuna (1887). |
| *The Netherlands* | Aruba (1634), Bonaire (1642), Curaçao (1634), Dutch Antilles (1848), Saba (1640), Saint Eustatius (1636), Saint Maarten (1648). |
| *Portugal* | Azores (1432), Madeira (1420). |
| *Spain* | Ceuta (1640), Isla de Alborán, Isla Perejil (1668), Islas Chafarinas (1847), Melilla (1497), Peñón de Alhucemas (1559), Peñón de Velez de la Gomera (1508). |
| *China* | Aksai Chin (1962), Tibet (1951). |
| *Indonesia* | Moluccas Islands (1950), West New Guinea (1962). |
| *USA* | American Samoa (1900), Bajo Nuevo Bank (1856), Baker Island (1856), Guam (1898), Howland Island (1856), Jarvis Island (1856), Johnston Atoll (1858), Kingman Reef (1856), Midway Island (1867), Navassa Island (1858), Palmayra Atoll (1898), Puerto Rico (1898), Seranilla Bank (1856), Swains Island (1925), US Virgin Islands (1917), Wake Island (1898), Guantanamo Bay concession (*sui generis*), Cuba (1898). |

unincorporated territory of the USA, its residents are US citizens, but have no representation in Congress and thus no sovereignty. The US government controls its ports and maritime waters. It confiscated aid delivered to Puerto Rico from other countries while simultaneously refusing assistance, delaying recovery after the hurricane and worsening a public health crisis and threat of epidemics (Sotomayor *et al.* 2017). The relationships between colonial power and colonial territory differ in each case, but the fact is that these territories are still colonies (Box 1.3). In addition to actual colonies still existing, critics also argue that wealthy nations continue to enjoy economic and political advantages that accrue to them because of the alleged abuses of globalized capitalism (Box 1.4). Thus, neo-colonialism is viewed as a continuation of the domination and exploitation of formerly colonized countries but through different means, primarily inequitable international trade and geopolitical relations.

## Box 1.3  Examples of present-day colonies: American Samoa and Tibet

In the late nineteenth century, Britain, Germany, and the United States were competing for influence in Samoa. In 1899, the three powers agreed to divide the islands between Germany and the US, with Britain withdrawing all claims in return for acknowledgement of its rights in other Pacific territories. The 1899 line of division remains the international boundary today between American Samoa and the independent Republic of Samoa. In 1951, control of the islands was shifted from the US Navy to the Department of the Interior (Gray 1960). The Samoans gained a measure of self-government when American Samoa approved its first constitution in 1966. This constitution is still in effect. It provides a tripartite system of government similar to the standard US model, with some unique concessions to local custom. The islands' chief executive continued to be a Washington-appointed governor until 1977, when the position was made elective. Since then, the islands have had considerable autonomy, particularly in local affairs, although certain powers remain reserved to the Secretary of the Interior. Samoans are American nationals, but not American citizens. They owe allegiance to the US and have American diplomatic and military protection, but are not entitled to a representative in Congress. Samoa is an 'unincorporated' territory, meaning that not all provisions and protections of the US Constitution apply there. Samoans can travel freely to, and reside in, the US and Samoa's economy remains partly dependent upon US aid. For this reason, American Samoa is on the east side of the International Dateline (three hours behind California), while Samoa shifted back to the west side in 2012 because Australia and New Zealand have become its biggest trading partners.

The status of Tibet has a long and complicated history and is deeply contested. International attention was focused on the territory in 2008 during the run-up to the Beijing Olympics. It is a mainly-Buddhist territory, governed as an autonomous region of China, but the allegiances of many Tibetans lie with the exiled spiritual leader, the Dalai Lama, who is seen by his followers as a living god, but by China as a separatist threat. Tibet was unified as kingdom in the seventh century. It became a part of the Mongol-ruled Chinese empire in the thirteenth century and was further incorporated into the Chinese empire in the seventeenth century under the Qing Dynasty (Beckwith 1987). The Dalai Lama lineage was established in 1578, and rose to political power under the fifth Dalai Lama (1617–1682), administering religious and administrative authority from the traditional capital, Lhasa, until 1951. In 1912, the thirteenth Dalai Lama unilaterally declared separation from China. Until 1951, Tibet possessed *de facto* independence, although no nation has ever recognized its independence. Under Chinese military pressure, the Tibetan and Chinese governments signed an agreement reintegrating Tibet into China in 1951. The current Dalai Lama later repudiated this agreement and in 1959, following a failed uprising and the collapse of the Tibetan resistance movement, he became the head of the Tibetan Government in exile in India.

Tibet's status is currently an issue of intense political debate. In terms of sovereignty, Tibetan nationalists believe Tibet was independent and has been colonized, but China claims its sovereignty dates back over centuries. China has been accused of repression, with 1.2 million Tibetans believed to have been killed under Chinese rule, which China disputes. China is also accused of actively suppressing Tibetan identity and, while it acknowledges some abuses, China now claims to be supporting the revival of Tibetan culture. China claims to have brought improvements in health and the economy, but many Tibetans believe that development has favoured Han Chinese immigrants. No country openly disputes China's claim to sovereignty, although the Free Tibet movement campaigns for an independent Tibet (see www.freetibet.org/). The Dalai Lama believes that for it to modernize, Tibet must remain within the People's Republic of China, but he also wants China to give a full guarantee to preserve Tibetan culture. The issues in Tibet are thus not necessarily about nominal independence from China, but more about the autonomy of the Tibetan people, and the impacts of Chinese rule on their human rights, their freedom to express cultural and religious beliefs, and their abilities to share in Tibet's development.

## Box 1.4 Neo-colonialism: Britain's twenty-first century scramble for Africa and oil extraction in North America

Critics argue that a determined attempt by powerful countries and global corporations to plunder the earth's natural resources, especially its strategic energy and mineral resources, represents a new and devastating form of colonialism. This includes what has been described as a new 'Scramble for Africa' where powerful external actors (including former European colonial powers and 'Rising Powers' such as China and India) compete to take control of the continent's mineral resources; it also includes resource extraction in former settler societies, such as those in North America.

According to a report by War On Want (Curtis 2016), British companies aided and abetted by the British government, now control vast proportions of Africa's key mineral resources: gold, platinum, diamonds, copper, oil, gas and coal. The report demonstrates that 101 mostly British companies listed on the London Stock Exchange have mining operations in 37 sub-Saharan African countries and collectively control over $1 trillion worth of Africa's most valuable resources. Five British government officials have taken up seats on the boards of mining companies operating in Africa, which the report suggests demonstrates complicity by the UK government. Augmented by World Trade

Organization rules, Britain's leverage over Africa's political and economic systems has resulted corporations being able to generate vast revenues (e.g. Glencore's revenues are ten times the gross domestic product of Zambia). Echoing colonial discourses (see Chapter 4) Britain claims to be helping to deliver economic development in Africa and $134 billion has flowed into the continent each year in the form of loans, foreign investment and aid. However, $192 billion has been simultaneously extracted from African countries mainly in profits by foreign companies. The report claims that there are other disturbing echoes of nineteenth century colonialism: the displacement of people, killings, labour rights violations, environmental degradation and tax dodging. As in the late nineteenth century, Africa appears to have become a free-for-all, with African governments having none, or at best only very small (3–20 per cent) shareholdings in projects.

In the former settler colonies of North America, oil extraction and its impacts on indigenous peoples has become a lightning rod for protest. For example, while the tar sands development in Canada's northern Alberta has brought income to some people, and wealth to a few, it has been criticized for its profound impact on the environment and on the lives of many indigenous groups. Indigenous peoples' ability to hunt, trap and fish has been severely curtailed and they are often too fearful of toxins to drink water and eat fish from waterways polluted by the 'externalities' of tar sands production. Some indigenous spokespersons talk in terms of a slow industrial genocide being perpetrated against them (Huseman and Short 2012). In January 2017, President Trump signed executive orders to complete the Keystone XL pipeline across the Great Plains and the Dakota Access pipeline in the northern plains, which will connect the Alberta tar sands to refineries in the southern United States. President Obama had previously halted both projects because of concerns about the environmental and cultural impacts. The Dakota Access pipeline project has encountered fierce opposition from the Standing Rock Sioux peoples, whose reservation is immediately downstream of the point where the pipeline will cross the Missouri River. They fear the threat of an oil spill and poisoned water sources and are concerned about the disturbance of sacred burial grounds during construction of the pipeline. The Oceti Sakowin peoples of Dakota have formed alliances with the Cree and Dene First Nations in Alberta to fight against the Keystone XL pipeline and protect their inherent right to self-determination as indigenous peoples. As one indigenous leader (Goldtooth 2015) explains:

Our acts of resistance to the Keystone XL pipeline are a perfect example of us wising up to the ongoing modern colonialist game, and a proactive step toward protecting future generations from the worst impacts of climate change. As indigenous peoples . . . we were handed down the original teachings on how to live in balance with Mother Earth. We must see all aspects of life as related, to respect the feminine principle of creation and to maintain a sustainable relationship with the land. These tenets are antithetical to the extractive economy we are faced with today. The land, air, and water are commodified . . . The lack of *proper consultation with tribal nations* along the proposed route of the Keystone XL pipeline violates *basic tenets of US Federal Indian Law* and the principle of *free, prior and informed consent* recognized in international law.

Focusing on the temporal difference between a colonial past and post-colonial present not only obscures colonial and neo-colonial inequalities that persist today, but can also obscure the power relations between colonizer and colonized. Moreover, defining the world purely in terms of western expansion, and studying people and places only because they have been colonized, means that 'colonialism returns at the moment of its disappearance' (McClintock 1995: 11). Given these problems with this temporal notion of postcolonialism, where the term is taken in this book to mean 'after-colonialism' it is written as post-colonialism. However, postcolonialism is not often used simply to refer to 'after-colonialism'. Rather, it refers either to a condition, or a set of approaches and theories that have become ways of criticizing the material and discursive legacies of colonialism still apparent in the world today and still shaping geopolitical and economic relations between the global North and South (Radcliffe 1999: 84). As Loomba suggests, 'it is more helpful to think of postcolonialism not just as coming literally after colonialism and signifying its demise, but more flexibly as the *contestation* of colonial domination and the legacies of colonialism' (Loomba 1998: 12, emphasis added). Similarly, 'Postcolonialism may be better conceptualized as an historically dispersed set of formations which negotiate the ideological, social and material structures of power established under colonialism' (Jacobs 1996: 25).

## Postcolonialism as a condition

Postcolonialism as a condition refers to the political, cultural and economic realities of societies living with the legacies and in the aftermath of colonialism. Several writers have insisted on the need to be careful *not* to use the term postcolonial as though it describes a single condition (e.g. McClintock 1992; Loomba 1998). The plurality of experiences of colonialism makes it such that virtually every place on earth might be claimed to be postcolonial (Myers 2006), but they are not postcolonial in similar ways. In this respect, McClintock (1992: 87) describes postcolonialism as 'unevenly developed' globally: 'Argentina, formally independent of imperial Spain for over a century and a half, is not 'postcolonial' in the same way as Hong Kong (. . . independent of Britain only in 1997). Nor is Brazil 'postcolonial' in the same way as Zimbabwe'.

In addition, while it is important to challenge a temporal binary between a colonial past and postcolonial present, it is also important to challenge a spatial binary between colonial centres and postcolonial margins. The effects of colonialism were, and are, not just one-way, transported from metropolis to colony. Rather,

> Postcolonial studies have shown that both the 'metropolis' and the 'colony' were deeply altered by the colonial process. *Both* of them are, accordingly, also restructured by decolonisation. This of course does not mean that both are postcolonial *in the same way.* Postcoloniality . . . is articulated alongside other economic, social, cultural and historical factors, and therefore, in practice, it works quite differently in various parts of the world.
>
> (Loomba 1998: 19)

It is also the case that not only are formerly colonized countries living with the political, cultural and economic legacies of colonialism, so too are the former colonizers. Thus, while cities like Lusaka, Singapore and Sydney can be said to be postcolonial, so too can London, Birmingham, Paris and Madrid. The economies, peoples and cultures of such cities have been and remain thoroughly shaped by colonialism and its aftermath (see, for example, Jacobs 1996; Henry *et al.* 2002; McEwan *et al.* 2005). Moreover, the experience of postcoloniality is very different in diverse regions of the world.

Issues also arise about how to classify certain countries. As Sylvester (1999: 713) asks: 'Are Australia, the USA and Canada colonial, owing to their unresolved native issues? Are they post-colonial because some among them broke from Europe, or postcolonial because all are working out an ongoing European genealogy?' Achille Mbembe (2001: 102) uses the term 'the postcolony' to refer to the shared experiences of places that were formerly colonized and had a 'given historical trajectory . . . recently emerging from the experience of colonization and the violence which the colonial relationship involves'. This is a useful term, therefore, for distinguishing between former imperial metropoles and places that were formerly colonized, both of which can be said to be postcolonial. Mbembe's understanding of postcolonialism as a condition is based largely on the experience of Africa, where continued dependence on the North and its political and economic domination render it relatively easy to make the case that colonial relationships have not ended. For Mbembe, the postcolonial condition is one in which the neo-colonial practices of Northern powers are virtually indistinguishable from the imperial practices of a century ago. Postcolonial conditions for many countries of the South are thus not actually postcolonial in too many senses, except in the temporal sense of 'after-colonialism'.

## Postcolonialism as theoretical approach

### 1. Metaphysical, ethical and political theory

As a metaphysical, ethical and political theory, postcolonialism can be thought of as an approach that addresses issues such as identity, race, ethnicity and gender, the challenges of developing post-colonial national identities, and relationships between power and knowledge. It does so in terms of how colonial powers produced and used knowledge of colonized peoples in their own interests and in how this knowledge continues to structure inequitable relations between formerly colonized and colonizers.

As we shall see throughout this book, postcolonial approaches seek to interrogate this knowledge and representations of the South through critical analysis of the relationship between power and knowledge. In other words, they examine relationships of power that determine who creates 'knowledge' about other places and peoples and the consequences of this knowledge, be that in the form of colonialism in the past or development and geopolitical interventions in the present. They also seek to demonstrate how the language of colonialism still shapes western ideas about other parts of the world. In challenging this knowledge and the processes that produce it (research, fieldwork and so on), postcolonialism highlights the significance of ethics in development. As discussed in Chapter 7, it helps create new possibilities for learning between development scholars and practitioners working in both North and South. It also helps pluralize the production of knowledge – in other words, knowledge is not simply produced in the North but in a variety of places. Postcolonialism demands more globally informed, rather than western-centric, knowledge.

### 2. Postcolonialism as literary theory

As a literary theory, postcolonialism examines literature produced both by authors in colonial countries and by colonized peoples responding to colonial legacies by 'writing back', or challenging colonial cultural attitudes through literature. Postcolonial literary critics re-examine classic literature focusing on the social discourse that shaped it. For instance, in *Orientalism*, Said (1985, first publ. 1978) analyzes the works of Balzac, Baudelaire and Austen, exploring how they were influenced by and helped to shape a societal fantasy of European racial superiority. Some postcolonial fictional writers interact with traditional colonial discourses, but modify or subvert it, perhaps by retelling a familiar story from the perspective of an oppressed minor character in the story. Jean Rhys's *Wide Sargasso Sea* (1966) is one example.

Rhys was a 'Creole' writer, a West Indian-born European, culturally marked as an inferior colonial but racially and institutionally privileged in relation to Africans in the Caribbean. She wrote *Wide Sargasso Sea* as a pseudo-prequel to Charlotte Brontë's *Jane Eyre*. She tells the story of the first Mrs Rochester – Bertha Mason, also a Creole woman – from a different perspective. Whereas Brontë draws on standard Victorian representations of Creole women in depicting Bertha as the mad woman in the attic, Rhys focuses on Rochester's rejection of her because of her Creole heritage. This, and the sense of trauma that she feels when she is taken away from the Caribbean to London, provides a more sympathetic account of her descent into madness, which went unexplained in *Jane Eyre*. Rhys's novel is a sophisticated example of examining critically European perceptions of the Caribbean Creole community.

Postcolonial literary theory uses a wide range of terms, like 'writing back', re-writing and re-reading, which describe the interpretation of well-known literature from the perspective of the formerly colonized. The desire to rewrite the master narratives of European discourse is a common postcolonial practice, with texts like *The Tempest, Robinson Crusoe* and *Great Expectations* also re-written from different perspectives. This rewriting offers a critical commentary on power, which is racialized, patriarchal and cultural. The telling of a story from another point of view is an extension of the deconstructive project. Deconstruction is central to postcolonial literary theory. Rather than focusing on what is written, a deconstructive reading explores the gaps and silences in a text (e.g. the story behind Bertha Mason's madness). Since writing has long been recognized as one of the strongest forms of cultural power, the rewriting of key narratives of colonial superiority is liberating for writers from the former colonies. Postcolonial literary theory thus addresses the ways in which the literature of the colonial powers has been used to justify colonialism through the perpetuation of representations of colonized people as inferior. It also explores the ways in which writers from formerly colonized countries have attempted to articulate, reclaim and celebrate their cultural identities. As we shall see, this kind of 'writing back' has significant implications for challenging the dominant discourses of development.

In summary, postcolonialism as ethical, political and literary theory is used with reference to a set of theoretical perspectives or conceptual strategies aimed at 'hearing or recovering the experiences of the colonized' (Sidaway 2000: 591). Postcolonial theory aims at 'getting past the post', or not simply thinking of postcolonialism as 'after-colonialism'. It also aims at contesting the 'lingering and debilitating modes of thought and action that comprise postcolonial conditions' (Myers 2006: 290). These specific meanings of postcolonialism have profound implications for development studies and form the focus of this book.

## Postcolonialism as anti-colonialism

As we have seen, there is no single definition of postcolonialism. Indeed, some critics reject the use of the term on grounds that it presupposes a continued socio-cultural dominance by former colonial powers. However, broadly speaking, postcolonial approaches and perspectives can be said to be *anti-colonial*:

> Definitions of the post-colonial, of course, vary widely, but for me the concept proves most useful when it is not used synonymously with a post-independence historical period in once-colonized nations, but rather when it locates a specifically anti- or *post*-colonial *discursive* purchase in culture, one which begins in the moment that the colonizing power inscribes itself onto the body and space of its Others and which continues as an often occluded tradition into the modern theatre of neocolonialist international relations.
>
> (Sleman 1991: 3)

Here, postcolonial is used to signify aesthetic, political and theoretical perspectives elaborated in literary and cultural theory, and is committed to critique, expose, deconstruct, counter and transcend the cultural and broader ideological legacies and presences of imperialism. Postcolonial approaches have become increasingly important across a range of disciplines over the last 20 years, and are now beginning to influence critiques of and ideas about development. Postcolonialism draws together both metaphysical, ethical and political theory and literary theory in diverse strategies that are broadly anti-colonial. These strategies are summarized below.

## Core strategies of postcolonial critiques and the emergence of decolonial theory

Four core strategies form the fabric of the complex field of inquiry of postcolonial studies, based in the 'historical fact' of European colonialism and the diverse material effects to which this phenomenon has given rise:

1.  *Destabilizing the dominant discourses of imperial Europe, such as history, philosophy, linguistics and 'development'.* These discourses are unconsciously ethnocentric, rooted in European cultures and reflective of a dominant western world-view. Postcolonial strategies problematize the very ways in which the world is known, challenging the unacknowledged and unexamined assumptions at the heart of European and American disciplines that are profoundly insensitive to the meanings, values and practices of other cultures.
2.  *Challenging the experiences of speaking and writing by which dominant discourses come into being.* For example, a term such as 'the Third World' homogenizes peoples and countries and carries other associations – economic backwardness, the failure to develop economic and political order, and connotations of a binary contest between 'us' and 'them', 'self' and 'other' (Darby 1997: 2–3) – which are often inscribed in development writings. These practices of naming are not innocent. Rather they are part of the process of 'worlding' (Spivak 1990), or setting apart certain parts of the world from others. Edward Said (1985, first publ. 1978) has shown how knowledge is a form of power, and by implication violence; it gives authority to the possessor of knowledge. Knowledge has been, and to large extent still is, controlled and produced in the North. The power to name, represent and theorize is still located here, a fact which postcolonialism seeks to disrupt.
3.  *Invoking an explicit critique of the spatial metaphors and temporality employed in western discourses.* Whereas previous

designations of the Third World signalled both spatial and temporal distance – 'out there' and 'back there' – a postcolonial perspective insists that the 'other' world is 'in here' (Chambers 1996: 209). The South is integral to what western discourses refer to as 'modernity' and 'progress', contributing directly to the economic wealth of the North through its labour and raw materials. In addition, the modalities and aesthetics of countries in the South have partially constituted languages and cultures in the North. Postcolonialism, therefore, attempts to re-write the hegemonic accounting of time (history) and the spatial distribution of knowledge (power) that constructs the South as the 'Third World'.

4. *Attempting to recover the lost historical and contemporary voices of the marginalized, the oppressed and the dominated, through a radical reconstruction of history and knowledge production* (Guha 1982). Postcolonial theory has developed this radical edge through the works of political and literary critics such as Spivak, Said and Bhabha (see Chapter 2) who, in different ways, have sought to recover the agency and resistance of peoples subjugated by both colonialism and neo-colonialism.

These core strategies are explored in further detail throughout this book, specifically as they relate to development. However, the book also recognizes that these core concerns, which one critic calls 'metropolitan postcolonialism' (Kaiwar 2015: x), have been criticized for eschewing structural relationships of power and the polarizing impact of capitalism on a global scale. Critics argue that they represent a set of ideas present in Anglophone universities and are a shift from an idea of the 'postcolonial' as:

> the agendas of development and distributive justice that informed the concerns of the first two generations or so of progressive thinkers and public figures with the end of the formal colonisation of a number of countries of what was once called the 'Third World'.
>
> (Kaiwar 2015: ix)

These criticisms have crystallized in recent years in the emergence of decolonial theory. As we will see, while they share some common ground, there are significant differences in postcolonial and decolonial thinking that have emerged in different contexts. Decolonization as a material concern (e.g. the return to indigenous peoples of sovereignty over the lands from which they have been dispossessed) does not equate with – and some critics would argue is incommensurable with – decolonization as invoked metaphorically within postcolonial theory (e.g. decolonizing knowledge). Despite this, as Tuck and Yang (2012: 28) argue, 'opportunities for solidarity [still] lie in what is incommensurable rather than what is common across these efforts'. This book argues that engaging 'the postcolonial' with ideas of development and with emerging anti- and de-colonial theory from global South contexts is important in ensuring that postcolonialism is not simply a metropolitan concern, but remains relevant in critical theories that seek alternatives to social, economic and political injustice. Postcolonial theory enables those marginalized by relationships of power the means to narrate from their situation, and it demands that those positioned more favourably within relationships of power be prepared to listen and learn. This provides new imaginings of what has previously been taken for granted, including development. It is a different political project to decolonization, but may provide a foundation for a deeper engagement with decolonization. As Asher (2013) suggests, those concerned with race, gender and other inequalities, representation, and the future of socio-natural world, particularly development scholars, would deepen their learning from engaging closely with both postcolonial and decolonial theory (explored further in Chapter 2).

## Postcolonialism and development

The aim of this book is two-fold. First, it attempts to bring postcolonial, decolonial and development theory into critical dialogue. Second, it attempts to examine how this critical dialogue can inform both postcolonial and development theory and explore the possibilities of producing a truly decolonized, postcolonial knowledge in development studies. In theoretical terms, postcolonialism has been greatly influenced by Marxism and poststructuralism (Blunt and Wills 2000). For example, in critiquing modernity it draws on the political-economy approaches of Marxism to explore how dominant groups in society come to exercise power and authority over less powerful or subjugated groups. Significantly, postcolonial theories address Marx's 'blind spot' (Castro-Gómez 2008): the racialized injustices of colonialism. Whereas Marxist approaches prioritize class differences, postcolonial approaches explore how these class differences are also racialized, particularly in colonial/post-colonial contexts. Poststructuralism emerged as a body of critique of structuralist approaches, like Marxism, which it accused of ignoring the importance of culture in shaping social relations. Postcolonial approaches have thus also been shaped by the cultural and linguistic analyses of poststructuralism (discussed in more detail in Chapter 2). However, the focus on postcolonialism as literary theory has tended to downplay the fact that postcolonial approaches are inspired by *both* Marxism and poststructuralism. And yet it is precisely this combination that makes postcolonialism relevant to development studies. The politics of postcolonialism diverge sharply from other discourses and, although it shares similarities with some radical critiques of development (see Chapter 3), its radicalism rejects established agendas and accustomed ways of seeing. This means that postcolonialism is a powerful critique of 'development' and an increasingly important challenge to dominant ways of apprehending global relations.

Postcolonial theory is deeply critical and suspicious of the 'development project', since this is part of what postcolonial theorists see as the dominant, universalizing and arrogant discourses of the North. In particular, the extent to which Northern 'development' agendas have assumed that they alone can define and solve development 'problems' is seen as profoundly problematic (an example of this can be seen in Box 1.5). Vigorous environmental and aboriginal politics have also emerged to critique western modernity and its fixation on progress through industrialization (Larsen 2006). Conversely, development studies has also criticized postcolonialism, precisely because of its tendency to focus on issues of culture and representation and a purported tendency to neglect the lived experiences and material realities of postcoloniality. Some writers have expressed frustration with postcolonial studies for its disconnection from actual studies of development (Simon 1998). It would be easy to imagine, therefore, that postcolonialism and development are too divergent to have any meaningful dialogue, and this has tended to be the case until relatively recently. However, these mutual criticisms, far from widening the historical divergences of the two approaches, have the potential to facilitate greater convergence. Moving towards a development studies that is postcolonial in theory and practice is now seen as desirable by many development scholars.

## Box 1.5 Development as the preserve of Northern experts: UN Report on the UK Housing Crisis 2014

The idea that development issues are the preserve of Northern experts and delivered to poorer countries in the global South is apparent within the development sphere itself (Chapter 5), but also within governments and news media in developed countries. In 2013, the United Nations special investigator on housing, Raquel Rolnik, made a research trip to Britain investigate housing provision; her report was published by the UN in February 2014. The report was critical of the UK government's bedroom tax policy (also known as the Spare Room Subsidy), which reduces Housing Benefit Entitlements to people living in social housing if they have one or more spare bedrooms. The report argued this negatively 'impacts on the right to adequate housing and general wellbeing of many vulnerable individuals and households'. The report also said that lack of investment in housing over several decades meant Britain now faces a crisis of housing affordability and availability; it called for a significant increase in housing stock, increased protections for tenants in the rapidly growing private rented sector who find themselves with 'very few rights and little security', and a series of welfare reforms to be re-assessed to ensure they do not impact disproportionately on the most vulnerable individuals.

The reaction by the UK government and sections of the media reveal that while Britain is assumed to be able to dictate social and economic development and human rights policies to developing countries, the reverse does not apply. Conservative party chairman, Grant Shapps, dismissed Rolnik as 'a woman from Brazil' and wrote a letter of complaint to the UN Secretary-General, Ban Ki-moon, calling her proposals 'an absolute disgrace' (*The Guardian* 3 February 2014). The government's then Housing Minister dismissed the report as 'a misleading Marxist diatribe'. The (consistently xenophobic) *Daily Mail* newspaper ran the story of her briefing report under the headlines: 'Outrage as "loopy" UN inspector lectures Britain: she's from violent, slum-ridden Brazil, yet still attacks us on housing and human rights' (11 September 2013) and 'Raquel Rolnik: a dabbler in witchcraft who offered an animal sacrifice to Marx' (12 September 2013).

The dismissal of Rolnik because she is both Brazilian and a woman says a great deal about who right-wing commentators in the UK – even those in the government – view as able to claim expertise in dealing with development challenges. Significantly, Rolnik's report was written in response to an invitation from the UK government. On 7 February 2017, Housing Minister Sajid Javid made a statement to the House of Commons acknowledging that 'Our housing market is broken' and launched a White Paper purporting to deal with the needs of 'the couple in the private rented sector handing half their combined income straight to their landlord' and the fact that 'for far too long, we have not built enough houses' (www.gov.uk/government/speeches/housing-white-paper-statement). He made no mention of the Rolnik Report, which had previously made precisely these points.

In tracing the interconnections between postcolonialism and development and the ways in which they might mutually inform each other, key points need to be borne in mind. First, place and space matter and understanding the spatiality of development is important.

This means understanding the relationships between people and places at a global level, but also that these relationships are rooted in localities. Development has always been about spatial imaginaries that operate at local, national and international scales; underpinning many development interventions are the ways in which the South is perceived and represented in the North. Postcolonialism seeks to bring these spatial imaginaries into question. For example, the North is perceived to be the centre, the originator of development ideas and policies. The South is a separate entity, the periphery to which development ideas and policies are exported. Postcolonialism challenges this idea of two separate entities, each with distinct histories and trajectories. It demonstrates how the centre and periphery – the 'here' and the 'there' – have always been interconnected and mutually constituted, if often in highly unequal ways.

Second, representations and what postcolonial theorist Edward Said (1978) refers to as 'imagined geographies' are significant. Representations of the 'Third World' and 'the West' are imbued with moral, cultural and socio-political attributes (e.g. civilization or lack of, modernity or backwardness, development or underdevelopment). These representations often shape development policies (see Chapter 4). Popular images in the North of the 'Third World' are formed through mass media – television, radio, cinema, written press, school and university education, friends and family, tourism. These are not necessarily false or wrong, but they are partial because they are often not based on first-hand experience, or they are mediated (e.g. through tourism) and are thus in many ways 'fictions'. They are often simplified and distorted images, mental maps of the world or imagined geographies. They represent individual and collective imaginations – they are a product of how the geography of Africa, Asia or Latin America is imagined and how the 'development' of these regions is understood. A key concern of postcolonial approaches is to think critically about how these maps are drawn, and how they affect perceptions of political, socio-economic or cultural differences between world regions, peoples and places. Of significance in this book is the notion of 'development' as a way of representing and defining the South, which has led to countries outside of the North being viewed as lacking or lagging behind. However, such representations ignore the diversity, dynamism and complexities of such countries and the lives of the people who live in them. Reflecting on how defining the South through development erases the richness and diversity of lived experience in popular imaginations in the North is thus of significance.

Third, Eurocentrism/western-centrism has been central to 'development'. Many people will have remarkably similar mental maps because they are formed using the same sources, but these fail to acknowledge that the world might be viewed differently by, for example, Chinese, Russian, African or Muslim peoples. Eurocentrism/western-centrism creates specific, imagined geographies of the world in which Europe and North America are seen as the highest stages of civilization and global progress and as the pivotal axes in global development.

Fourth, 'development' is about power. This is revealed in questions such as: 'By whom is development being done? To whom?' . . . These should be asked particularly when 'solutions' are put forward that begin 'We should . . .' without making clear who 'We' are and what interests 'We' represent (Allen and Thomas 2001: 4). European and North American financial institutions, such as the World Bank (WB) and International Monetary Fund (IMF), the World Trade Organization (WTO), and global development organizations, such as the United Nations (UN), have enormous power and capacity to put forward dominant spatial imaginations of other peoples and places to provide 'truths' about successful growth, miracle economies and economic recovery. As Rist (1997) argues, development in some conceptions might be understood as a dogmatic belief, almost a religion: 'From their pulpits, global development agencies preach a number of creeds and doctrines of good change while forcing countries to seek their blessing and baptism at the altar of development' (in Power 2003: 7). This book is not about a simple 'pro' or 'anti' debate concerning such organizations and the states in the North that dominate them. Rather it aims to understand the power of these states, ideas and institutions in specific places, at certain times in world history, and in the present.

Fifth, development is a kind of global 'industry' and we might think of the WB and the IMF, based respectively in New York and Washington, as the headquarters. As we shall see in Chapter 3, the global development industry is currently dominated by neoliberal ideology, which preaches restraint on state intervention and social spending and the prominence of the market. Ulrich Beck (2000) has called it an ideological 'thought virus' devised in and spreading from Europe and North America. It is problematic because it promotes an economic growth-first strategy where social and welfare concerns come later. It promotes a false image of international markets as fair and efficient, and privileges 'lean government', economic deregulation and the removal of state subsidies to marginal and disadvantaged communities. It also forecloses alternative paths of development; the economy should dictate its rules to society rather than vice versa. Consequently, the global development industry is part of a global capitalist economy that continues to produce inequality and uneven development:

> A big part of 'economic development', i.e. the wealth, of the rich countries *is* wealth imported from the poor countries. The world economic system *generates* inequality and it runs on inequality . . . It is a fraud to hold up the image of the world's rich as a condition available to all.
>
> (Lummis 1992: 46–7)

Finally, a postcolonial approach reminds us that 'development' is not simply a European and American invention because it is also shaped by agency and resistance in the South. Development theories and ideas are formed and contested in the South (see Chapter 3). Postcolonialism also demands that development is properly opened to the presence and significance of voices, knowledge and agency in the South (see Chapter 6). Moreover, development is not simply a problem of the South. The early twenty-first century shows evidence of widening and deepening patterns of global inequality. An Oxfam report (Hardoon *et al.* 2016) calculates that the richest 1 per cent of the world's people now have more wealth than the rest of the world combined. In 2015, just 62 individuals (mainly from the USA and Europe, but including individuals from China, India and Mexico) had the same wealth as 3.6 billion people – the bottom half of humanity, with their wealth rising by 44 per cent between 2010 and 2015. Meanwhile, the wealth of the bottom half fell by 41 per cent over the same period. By 2017, with better data from India and China, Oxfam calculated that just eight men own the same wealth as the 3.6 billion people who make up the poorest half of humanity (www.oxfam.org/en/pressroom/pressreleases/2017-01-16/just-8-men-own-same-wealth-half-world; accessed 06/04/17). What is striking about these patterns are the growing disparities of wealth within countries, including within developed countries, the fact that most of world's poor (72 per cent) live not in poor countries, but in middle income countries (e.g. India, China, Nigeria, Bangladesh, Pakistan, Indonesia, South Africa) (Sumner 2012) and the gendered nature of wealth and poverty. Growing economic inequality undermines growth and social cohesion everywhere, but the consequences for the world's poorest people are particularly severe. And yet, as far back as 1998, a high-ranking US official claimed that a child born in New York had less chance of reaching the age of five, or of learning to read, than one born in Shanghai, China. As we have seen, the North and South, developed and developing, First and Third Worlds are not as separate and distinct as is sometimes imagined, but the divide between the world's richest and poorest is growing ever wider.

Bringing together development and postcolonialism in meaningful dialogue, and attempting to produce a development studies that is postcolonial in theory and practice, means acknowledging the significance of issues of language and representation, and understanding the power of development discourse and its material effects on the lives of people subject to development policies. It also means acknowledging the already postcolonial world of development in which contemporary re-workings of development theory and practice, such as decolonial thinking, grassroots and participatory development, indigenous knowledge and global resistance movements, inform postcolonial theory (Chapter 6). As discussed in Chapter 8, the idea of the Anthropocene, the relationship between climate change and carbon-fuelled development and the dilemmas of tackling this without exacerbating existing global inequalities, present perhaps the most pressing challenges for theorizing development. They also raise the question of whether we are 'beyond development' as conceived of in the West. While the Northern powers remain largely deaf to ideas emerging in other parts of the world and are becoming increasingly parochial in outlook, planetary futures are arguably already being played out in Africa, Asia, Latin America (Mbembe 2013; 2015; 2016). In places where ecological crisis and precarity of human life are most acute, new theories of exchange, democracy, human and non-human rights – development in its widest sense – are already being formed. Failing to acknowledge and engage with these contemporary reworkings of development in other parts of the world means previously universal and omnipotent western development risks becoming increasingly anachronistic and irrelevant.

# Summary

- Until recently, there has been little dialogue between postcolonialism and development studies because of differences in disciplinary traditions, politics, wariness over motives and divergences in the languages and concepts used to articulate core issues.
- Both fields are enriched by a productive engagement between postcolonial criticisms of how we speak and write about the world and development concerns with the material realities of global inequalities.
- Specialists in development studies have begun to view postcolonial approaches as constructive and instructive.
- Postcolonialism has several meanings. However, it is as political, ethical and literary theory and as an anti-colonial approach that its uses are most pertinent to development.
- Postcolonial critiques of development aim to understand the power of development ideas, knowledge and institutions and their consequences in specific places at particular times.
- Producing a development study that is postcolonial in theory and practice means acknowledging the significance of language and representation and understanding the power of development discourse and its material effects on the lives of people subject to development policies.
- Postcolonial development studies require acknowledging and engaging with the already postcolonial world of development in which contemporary re-workings of development are emerging.

# Discussion questions

1  What are the major differences between postcolonial and development studies?
2  How do postcolonial approaches challenge established histories and why is this important?
3  Why should we be concerned about the terminology we use to describe regions of the world?
4  What are the problems of defining postcolonialism as 'after-colonialism'?
5  Of what relevance are the core strategies of postcolonial critiques in challenging the field of development?

# Further reading

Asher, K. (2013) 'Latin American decolonial thought, or making the subaltern speak' *Geography Compass*, 7, 12, 832–42. Succinct and clearly written article that traces some of the differences and commonalities between postcolonial and decolonial theory.

Hall, S. (1996) 'What was "the post-colonial"? Thinking at the limit'. In I. Chambers and L. Curti (eds), *The Postcolonial Question: Common Skies, Divided Horizons* London, Routledge, pp. 242–60. Useful book chapter that problematizes the meaning of the term 'postcolonial'.

Loomba, A. (1998) *Colonialism/Postcolonialism* London, Routledge. Clearly written overview of the meanings of postcolonialism and the principal elements of postcolonial approaches.

McClintock, A. (1992), 'The angel of progress: pitfalls of the term "postcolonialism"', *Social Text*, 31/32, pp. 84–98. Useful article that, like Hall (1996), outlines some of the main problems with the term 'postcolonialism'.

McEwan, C. (2002) 'Postcolonialism'. In V. Desai and R. Potter (eds) *The Companion to Development Studies*, London, Arnold, pp. 127–31. Summary of the key elements of postcolonialism and its implications for development studies.

Sumner, A. (2012) 'Global poverty and the "new bottom billion" revisited: exploring the paradox that most of the world's extreme poor no longer live in the world's poorest countries' *IDS Working Paper* www.ids.ac.uk/project/the-new-bottom-billion. An interesting article that explores the changing geography of global poverty and unsettles traditional binaries about rich and poor countries.

Sylvester, C. (1999) 'Development studies and postcolonial studies: disparate tales of the "Third World"' *Third World Quarterly*, 20, 4, pp. 703–21. Succinct account of the major differences and divergences between postcolonialism and development studies.

# Useful websites

Www.antislavery.org Website of Anti-Slavery International (UK charity in existence since 1839) containing information of the history of the slave trade as well as slavery in the present-day.

www.ipcs.org.au/ The Australian-based Institute of Postcolonial Studies; contains links to various sites connected with the study of postcolonialism, journals, and reading lists.

www.itk.ca/ Website of the Inuit Tapiriit Kanatami, which has accomplished major settlements of comprehensive land claim agreements in the four Inuit regions of Canada and campaigns for the protection of Inuit rights under the Canadian Constitution.

http://www.ienearth.org Indigenous Environmental Network, an alliance of indigenous peoples whose shared mission is to protect the sacredness of the earth from contamination and exploitation by respecting and adhering to indigenous knowledge.

news.bbc.co.uk/1/hi/world/asia-pacific/7242057.stm Full text of the Australian government's historic apology to Aboriginal peoples (13 February 2008).

policy-practice.oxfam.org.uk/publications/an-economy-for-the-1-how-privilege-and-power-in-the-economy-drive-extreme-inequ-592643 Oxfam International's 2016 report on increasing global inequality.

media.waronwant.org/sites/default/files/TheNewColonialism.pdf?_ga=2.1843059.39998409.1498639734–1814348306.1498639734 War On Want's 2016 report, authored by Mark Curtis, on the 'new colonialism' and Britain's twenty-first century scramble for African energy and mineral resources.

# 2 Histories and geographies of postcolonialism

## Introduction

There is no single origin of postcolonialism. Indeed, what is referred to as postcolonialism is, as Chapter 1 illustrates, a diverse set of approaches, theories and strategies. This chapter takes a broad approach to postcolonialism to understand it as a variety of responses to colonialism and decolonization that have inspired both liberation struggles and academic inquiry. Many of these responses predate the period when the term postcolonial began to gain currency and are now claimed retrospectively as continuous or contiguous with postcolonial politics and modes of cultural practice.

Many scholars trace the origins of postcolonialism to the emergence of postcolonial literary and cultural criticism and theory in the 1980s and 1990s. For example, Leela Gandhi (1998) traces postcolonialism back to the Subaltern Studies group in South Asian studies; Phillip Darby (1997) suggests that postcolonialism grew out of the study of fiction written in ex-colonial countries. Others argue that postcolonial theory was already established in Anglophone social sciences, including in geography and development studies, well before the Subaltern Studies group found a wide readership and it was, in fact, through postcolonial theory (and especially Said's endorsement in a foreword to the *Selected Subaltern Studies* in 1988) that the latter's work to critique and 'unthink' Eurocentrism gained traction (Brennan 2014; Lazarus 2016). So, does postcolonialism, or postcolonial theory, exist in any coherent way? What we know now as postcolonial theory is essentially concerned with revealing the situatedness of knowledge, and particularly the universalizing knowledge produced in imperial Europe (Said 1993, 1999) and the West more broadly. The idea of situated knowledge has a long intellectual pedigree in feminist and postcolonial scholarship. Knowledge is never impartial, removed, or objective, but always *situated*, produced by actors who are positioned in specific locations and shaped by numerous cultural and other influences. This determines what counts as knowledge, who creates it, where it is generated and how and for whom it is disseminated. Not only is there a politics of knowledge, therefore, but knowledge is always partial. Postcolonialism reveals and problematizes the power relations through which what counts as knowledge and how it is produced has tended to be determined in the West. In addition, rather than being thought of as situated, western knowledge forms are assumed to be universal.

Postcolonialism is also conditioned by its places of formation (Clayton 2000; Lester 2003). It is thus a *geographically* dispersed contestation of colonial and neo-imperial power and knowledge. It is important not to erase the differences – in context, motivation, approaches, types of critique and politics – between the varieties of writings that we now call postcolonialism. What holds these writings together are similarities in their commitment to challenging cultural hegemony (be that from the West or from post-colonial elites), their commitment to anti-racism and anti-colonial politics and their focus on matters of culture. Broad similarities can be teased out, but it is important to remember that very few scholars refer to themselves as postcolonial critics or theorists. Even those we now think of as the leading postcolonial theorists (e.g. Said and Spivak) have rarely referred to themselves as such.

Anti-colonial and anti-racist resistance could surface in any number of ways, in any number of contexts, through any number of practices. However, as Robert Young (2001: 165–6) argues, anti-colonialism as a theorized political position took a relatively restricted number of forms. As Table 2.1 illustrates, historically, postcolonial theory is the product of many of these, but often in very different ways.

Young (2001) outlines the diverse geographies and histories of anti-colonialism and how these movements have in turn inspired postcolonial theory:

- the role of Marxist internationalism in liberation movements in China, Vietnam, Cuba and later in some African countries (e.g. the Mau Mau in Kenya)
- Mao's socialist revolution in China in 1949
- Nasser's anti-colonialism in Egypt (culminating in the Suez Crisis in 1956)
- the Bandung Conference in 1955 (see Chapter 3)

*Table 2.1  Relationship between anti-colonial politics and postcolonial theory*

| Political, ideological and military resistance to colonialism | How postcolonial theory relates to this resistance |
|---|---|
| European moral and humanitarian objection (enlightenment, anti-slavery campaigns, rights theory). | Affiliated with in broad terms. |
| European liberal economic objections (Smith, Cobden, Bright, Bastiat). | Does not affiliate with. |
| European/non-European rivalry between imperial powers (e.g. Britain–France–Germany–Russia, USA–Japan–Russia). | Does not affiliate with. |
| Assertion of political rights to self-determination in settler colonies (e.g. USA, Canada, Bolivia, Transvaal/Orange Free State). | One constituency of postcolonial theory, associated particularly with indigenous peoples, situates itself in this. |
| Colonial nationalism (bourgeois, cultural, religious) (e.g. India, Pakistan, Ireland, Scotland, Poland, Turkey, Egypt, Kenya, China). | Those still linked to the legacy of this have often been most keen to distance themselves from postcolonial theory, and to characterize it as 'western'. |
| Anti-colonial internationalism (e.g. pan-Africanism, pan-Arabism, pan-Islamism, the *Khalifat* movement, the *négritude* movement, African Socialism). | Finds few supporters today, despite being central to the thinking of many liberation ideologies; probably because it was not incompatible with Marxist internationalism. |
| Industrial strikes, agitation over land reform, communalism, protest-migration, peasant revolts (displaced forms of resistance against colonial power or transformed into anti-colonial activities by communism and/or nationalism). | Much of postcolonial theory bears historical traces of this. |
| Marxist internationalism and the armed national liberation movements (e.g. China, Vietnam, Cuba, Angola, Mozambique). | Renewed interest in internationalism is emerging in some postcolonial thought, especially that with a strong association with humanism; some political synergies with anti-globalization movements. |

*Source*: adapted from Young 2001: 165–6

- the postcolonial movements in Latin America that gave rise to dependency theory
- the influence of Che Guevara, Fidel Castro and the Tricontinental in Latin America

- the rise of Anglophone and Francophone African Socialism (e.g. C.L.R. James and Senghor)
- the emergence of pan-Africanism (e.g. Nkrumah in Ghana)
- the writings of Fanon (on French imperialism) and Amilcar Cabral (on Portuguese imperialism)
- the violent resistance to colonialism in Algeria and Ireland
- the emergence of Marxism in India
- Gandhi's counter-modernity in India and South Africa

Young gives a more complete account of these diverse movements than is possible here. However, the following attempts to further expand on some of these historically and geographically specific movements that have inspired postcolonial theory. It should also be noted, as Young (2001: 165) suggests, that some tensions within postcolonial theory continue to erupt from traditional sources of conflict within the liberation movements themselves. These are often between bourgeois (and the bourgeois diaspora) nationalisms and more engaged, often more localized, politics concerned with social justice and equality that remain inspired by the memory of resistance.

This chapter provides a review of some of the key political, cultural and literary movements that might be considered as influencing the development of contemporary postcolonial politics and ethics. It then outlines the major influences on postcolonial theory, arguing that these can be traced to key writers in specific colonial and postcolonial contexts. First is the influence of writers involved in anti-colonial politics, especially in Africa. Second is the emergence of women's anti-colonial movements and the critiques of western feminism, particularly from the 1970s onwards. Third is the emergence in the 1980s of the Subaltern Studies collective in postcolonial India. Of significance is the way in which these scholars inspired debates concerning issues such as knowledge production, the histories of colonialism and liberation and the (ir)retrievability of subaltern voices, which in turn has inspired much of subsequent postcolonial theory. The chapter then outlines the impact and influence of key postcolonial theorists, focusing specifically on the writings of Edward Said, Homi Bhabha and Gayatri Chakravorty Spivak, before tracing some of the differences between postcolonial and decolonial theory: what is incommensurable between them, where there are possible solidarities, and how both matter in the context of development.

## Cultural and political movements, 1890s–1970s

Different strands of thought have contributed to what we now understand as postcolonialism. Although there are no direct links between the political and cultural movements of the late nineteenth and twentieth centuries and the postcolonial theory that emerged after the 1980s, there are some similarities in the politics and ethics that inspired them. These similarities are focused on struggles against racism, liberationist politics, challenges to white political and cultural domination and the assertion of black identities and cultures. While there were many differences between and within these movements, reflecting both the temporal and spatial contexts in which they emerged and the different views of the key personalities involved, there is little doubt that their influence travelled and percolated, inspiring in different ways and in different contexts around the world.

### Pan-Africanism from the 1890s

Pan-Africanism is usually seen as a political movement that emerged out of the Atlantic slave trade and the removal and dispersal of people from the continent to disparate parts of the world. Although African slaves and their descendants had diverse origins and cultures, one shared experience was of their location in systems of exploitation where African origin and skin colour became a sign of their subaltern status. Modern pan-Africanism became prominent in the late nineteenth century, aiming to set aside cultural differences (for which it has subsequently been criticized) and to assert this shared experience to foster solidarity and resistance to exploitation. The intention was to unify and uplift both native Africans and those in the African diaspora. Key figures included Trinidadian Henry Sylvester Williams, who established the African Association in 1897 that later became the Pan African Association, and W.E.B. Du Bois, who became one of the most prominent intellectual leaders and political activists on behalf of African Americans in the first half of the twentieth century. Du Bois founded the National Association for the Advancement of Colored People (NAACP) in 1909, became Editor-in-Chief of its major publications through which the political cause of pan-Africanism was debated and, through his own writings, became an important advocate of African American civil rights (Lewis 1995; 1997). In turn, the NAACP inspired black political activists in Africa, notably the South African Sol Plaatje (Box 2.1).

## Box 2.1 Sol Plaatje and Marcus Garvey

Solomon Tshekisho Plaatje (1876–1932) was an accomplished black South African intellectual, journalist, linguist, politician, translator and writer. As an activist and politician he spent much of his life campaigning for the enfranchisement and liberation of African people. He was a founder member and first General Secretary of the South African Native National Congress (SANNC), which would become the African National Congress (ANC) in 1926, and a key figure in the struggle against the dispossession of black South Africans in the late nineteenth and early twentieth centuries. His *Native Life in South Africa* (1914) was a scathing indictment of the Native Land Act of 1913, which effectively dispossessed black South Africans of their lands, and is still considered by many to be one of South Africa's greatest political texts. As the writer of the first novel in English by a black South African (*Mhudi* 1913), Plaatje was concerned to refute the widely held fallacy of the 'uncivilized' character of black people, while simultaneously attacking the white South African government's policies of segregation and land redistribution. While he wrote in English to convey his political message, he was also concerned with preserving his own Setswana language and folklore. His political activism took him to Britain, where he continued his campaign against the Native Land Act in meetings with prominent politicians. He also travelled to the US, where he met and interacted with prominent black leaders such as Marcus Garvey and W.E.B. Du Bois.

Marcus Mosiah Garvey, Jr. (1887–1940) was a Jamaican publisher, journalist and political commentator. In 1914 he founded the Universal Negro Improvement Association and African Communities League (UNIA-ACL). He is best remembered as a proponent of the Back-to-Africa movement that sought to encourage the African diaspora to return 'home' to Africa. Central to Garvey's Pan-African philosophy was black separatism and a notion that Africa could only be redeemed by those of African ancestry. He also advocated anti-colonialism, suggesting that the European colonial powers should leave Africa. His speech of 1927, in which he urged diasporic Africans to look towards Africa for inspiration because a king would one day be crowned there, was considered prophetic by many Rastafarians, coming just three years before the coronation of Emperor Haile Selassie. During the 1920s, under Garvey's leadership, UNIA claimed to have over a million members and thus was the largest single pan-African movement (Winston, 1998).

Pan-Africanism influenced liberation struggles in Africa and some of the first rulers of post-independent African nations were also advocates of pan-African politics (e.g. Jomo Kenyatta, the first President of Kenya, and Kwame Nkrumah, the first President of Ghana, who advocated a United States of Africa). Haile Selassie, the Emperor of Ethiopia between 1930 and 1974, was also a pan-Africanist and advocate of greater unity among African nations. In turn, Haile Selassie became a key symbolic figure for the Rastafarian movement, which was founded in Jamaica in 1930 among working class and peasant black people. Selassie's status as the only African monarch of a fully independent state was interpreted by the Rastafarians as fulfilling a Biblical prophecy of the return of the messiah. Central to the Rastafarian movement is a focus on Afrocentric social and political aspirations, of which Marcus Garvey has been a significant proponent (Box 2.1). Pan-Africanism also played a significant role in the South African liberation movement, with the founding of the Pan-Africanist Congress (PAC) in 1959. This was a rival to the African National Congress (ANC) and objected to the ANC's non-racial policies in the fight against apartheid. The PAC wished to take a bolder approach based more on mass action and a notion that South Africa was exclusively an African country. Although poorly organized and never having the popular support of the ANC, the PAC inspired the Black Consciousness Movement and its leader Steve Biko (discussed subsequently).

## The Harlem renaissance in the 1920s and 1930s

The Harlem Renaissance refers to the emergence of African American literature, art, music and theatre during the 1920s and 1930s. It was borne out of the migration of hundreds of thousands of African Americans to cities in the northern United States and the emergence of a new urban African American middle class. Although centred in the Harlem district of New York City (Hutchinson 1997), which became a centre of social and cultural change, the movement had a major impact in cities throughout the USA and elsewhere. Coinciding with the political undercurrents of pan-Africanism and drawing inspiration from the likes of Du Bois and Garvey, African American artists, novelists, poets, playwrights and musicians began creating works rooted in their own culture instead of imitating the styles of Europeans and white Americans. Perhaps most notable was the popularizing of jazz and blues – black music associated with the South of the USA that migrants brought with them to the nightclubs of Harlem. An African American literary movement coincided with this, whose participants were often the grandchildren of slaves who had left the South after the Civil War to seek a better standard of living and escape deep-rooted racial prejudices. The Harlem literary scene was also augmented by migrants from the Caribbean.

Characterizing the Harlem Renaissance was an overt racial pride that sought to challenge the pervading racism and stereotypes of African

Americans as incapable of possessing intellect, or of producing great literature, art, music, poetry and theatre. This assertion of intellect, creativity and cultural presence was bound up with contemporaneous political movements to promote progressive politics, civil rights and racial and social integration. There was no singular Harlem Renaissance style; rather Harlem became the centre of cultural foment and celebration of a wide range of cultural expressions including traditional and experimental styles. Some conservative elements of the black intelligentsia took issue with certain depictions of black life, but for many African Americans the Harlem Renaissance was about the assertion of humanity in defiance of racism and a means of demanding civil rights.

Despite civil rights being some time in coming, this period witnessed a shift in the attitudes of many white people. Some offered overt support through patronage, publication and the opening of job opportunities for the most talented artists, musicians and writers. Others undoubtedly wanted to capitalize upon and exploit the popularity of black American culture at the time. However, in addition to the network of support provided by the black middle classes through business links and publications (e.g. Du Bois published the works of Harlem Renaissance writers), the support of white Americans helped open doors that led beyond the African American community. The Harlem Renaissance thus reflected a significant era of social and intellectual change in twentieth century American life. Although it began to decline with the worldwide economic downturn and the Great Depression of the late 1920s, it laid the foundation of the Civil Rights Movement that came to fruition after World War Two and continued to inspire black artists and writers throughout the twentieth century. It also provided significant inspiration for early anti-colonial movements.

## Négritude movements of the 1940s and 1950s

The Harlem Renaissance inspired an anti-colonial literary and political movement that became known as négritude – loosely translated from French as 'blackness'. This emerged in the 1930s among a group of black writers in the French colonies, including the future Senegalese president Léopold Senghor, who wrote poetic portraits of African civilizations, Martinican poet Aimé Césaire and Guianan poet Léon Damas. The term itself was probably first used by Césaire, whose first published work 'Negreries' was notable for its reclamation of the word 'nègre' as a positive term, when this had previously been almost exclusively used in a pejorative sense (Césaire 1997; 2000). As with pan-Africanism, the négritude writers promoted solidarity in a common black identity and rejected, specifically, French colonial racism. Shared black heritage was considered the best weapon with which to fight against French political and intellectual hegemony and domination. They drew particularly on the works of Harlem Renaissance writers such as Langston Hughes and Richard Wright who addressed the themes of 'blackness' and racism. Indeed, several négritude writers had been introduced to Harlem Renaissance writers while studying in Paris. (The Harlem connection was also shared by the closely parallel development of *negrismo* in the Spanish-speaking Caribbean; despite a difference in language the movements drew on similar influences and shared a common purpose.) They were also inspired by events on Haiti, which had declared its independence from France in 1804 following a successful slave rebellion and which, at the beginning of the twentieth century, witnessed the flourishing of black culture (James 1963).

While it gained support among some sections of the French intelligentsia (most notably from Jean-Paul Sartre), négritude was criticized by some black writers in the 1960s as insufficiently militant. For example, neither Césaire in Martinique nor Senghor in Senegal (see Senghor 1998) envisaged political independence from France but promoted the idea that blacks could participate as equals with the French rulers. Négritude did not promote radical black liberation in the form of independence, it did not persuade the French colonialists of the equality of blacks, nor did it define a new kind of aesthetic that would free black people and black culture from dominant white conceptualizations. Despite this, the critique of humanism and its occlusion of race and racism developed by négritude scholars remains powerful today. As discussed in Chapter 8, scholars seeking to problematize the erasure of race from posthumanist philosophies of Anthropocene, and to refocus attention on topics all too often sidestepped by posthumanism such as race, colonialism and slavery, are returning to decolonial theorists like Césaire (see Todd 2015), as well as Senghor, Du Bois and Fanon (Gilroy 2015). Négritude scholars thus remain of importance in the continuing struggle to decentre Eurocentric and heteropatriarchal theories that claim to diagnose planetary problems and author their solutions.

## Anti-colonial literature

Postcolonial literary criticism came to the fore in the 1980s but had antecedents in the anti-colonial literature that emerged in the mid-twentieth century, particularly in Africa. One notable example is the Nigerian novelist and poet, Chinua Achebe, an esteemed and controversial literary critic and one of the most widely read African authors of the twentieth century. Achebe's fictional and non-fictional writing is primarily concerned with African politics, western representation of Africa and Africans, the intricacies of pre-colonial African culture and civilization, and the deleterious effects of colonialism on African societies. His defining work is the novel *Things Fall Apart* (1958), which explores the devastating effects of British colonialism on Ibgo society and which received renewed publicity on its 50th anniversary. The book has sold over 10 million copies and has been translated into over 50 languages. Achebe's essay, 'An image of Africa: racism in Conrad's *Heart of Darkness*' (1975) has become one of the most influential, controversial, widely studied and debated essays of its kind around the world. It countered traditional readings of Conrad's novel as anti-imperialist by arguing that *Heart of Darkness* dehumanizes Africans, rendering Africa as 'a metaphysical battlefield devoid of all recognizable humanity, into which the wandering European enters at his peril'. Despite being overlooked by the western academy, Achebe is widely considered a literary champion of Africans and an advocate for the dignity of voiceless and dispossessed peoples. He was an early exponent of postcolonial literary criticism and inspired a whole swathe of critical writings from former colonial countries in the 1970s and 1980s (sometimes referred to as 'Commonwealth' writing).

Perhaps less well known in Anglophone contexts, but no less significant in post-colonial Africa, is Sembène Ousmane. Born in Senegal to a Muslim Wolof family in 1923, Sembène is considered one of the greatest African writers and film directors. His early novels and films represent a powerful critique of economic and racist oppression of Africans under European colonialism. His later novels are notable for their excoriating critique of corrupt African elites in post-independence contexts. Sembène's participation in the 1947–1948 Niger–Dakar railway strike inspired his seminal novel *Les Bouts de Bois de Dieu* (*God's Bits of Wood*, 1960). The book deals with the diverse ways in which Senegalese and Malians responded to colonialism. It was notable for its account of the strikers' resistance to the mistreatment of the Senegalese people and their fight to assert their rights, as well as for its depiction of African women as equals. It is widely considered a masterpiece of postcolonial French literature, as is Sembène's 1973 novel *Xala*. The latter focuses on the corruption and moral collapse of post-colonial Senegal through the figure of El Hadji Abdou Kader Beye, a rich but morally bankrupt businessman. Sembène also used film to expand his audience beyond the literate African elite. His first, *La Noire de* (1960), based on one of his own short stories, was the first feature-length film ever released by a sub-Saharan African director. Recurrent themes of Sembène's films are the history of colonialism, the failings of religion, critique of the new African elites and the strength of African women. He was enormously influential in providing an alternative view of Africa (Murphy 2001).

Other colonies witnessed the emergence of anti-colonial literature during this period. For example, Mahatma Gandhi wrote *Hind Swaraj* or 'Indian Home Rule' in 1908, in which he advocated freedom from British colonial rule. In the Caribbean, colonial subjects like Trinidadian C.L.R. James explored the paradox of European colonialism by juxtaposing their own subjugation under colonialism with European discourses on rights and sovereignty. James had begun to campaign for the independence of the West Indies while in Trinidad in the 1930s, and his *Life of Captain Cipriani* (1932) and the pamphlet *The Case for West-Indian Self Government* (1933) were his first important published works. He was also a leading champion of pan-Africanism. His book *The Black Jacobins: Toussaint L'Ouverture and the San Domingo Revolution* (1938) was a widely acclaimed and inspirational history of the Haitian slave uprising and revolution. The later work of Guyanan Walter Rodney was notable for its contribution to pan-Africanism and the Black Power movement in the Caribbean and North America. His most influential book, *How Europe Underdeveloped Africa* (1972), portrayed Africa as consciously exploited by European imperialists, leading directly to the modern underdevelopment of most of the continent.

Imaginative literatures by 'Third World' authors were seen in the aftermath of colonialism as the keys to recovering the memories of colonial experience that had been rendered abject by imperialism. In addition, they were important avenues for recovering writing styles that imperialism had violently erased through the imposition of European languages and education. An example of an early postcolonial writer is Ngũgĩ wa Thiong'o, a Kenyan author and the first East African to publish a novel in English. Ngũgĩ embraced Fanonist Marxism (see below) in his third novel, *A Grain of Wheat* (1967). He subsequently renounced English, Christianity and the name James Ngugi as colonialist. He changed his name to Ngũgĩ wa Thiong'o and began to write solely in Gĩkũyũ and Swahili. Whilst in exile following a period of imprisonment, he wrote his most influential book, *Decolonizing the Mind* (1986), in which he argues for Africans to write in their native languages to overcome the suppressions of colonialism and to build an authentic African literature. Contemporary writers have drawn on Ngũgĩ's notions of the power of language. Some appropriate the language of the colonizers to 'write back' from the margins and reclaim postcolonial cultural identities. Salman Rushdie's novel *Midnight's Children* (1981) is one example, its liberal mix of English and Indian languages being a departure from conventional Indian English writing. More recently, postcolonial studies have been less concerned with gathering narratives of experience through literature and more concerned with theorizing postcoloniality. Again, there are multiple origins of these kinds of approaches, but one important influence was the work of Frantz Fanon.

## Fanon's psychology of colonialism and racism

Frantz Fanon was perhaps the pre-eminent thinker of the twentieth century on the issue of decolonization and the psychopathology of colonization. He was born in Martinique in 1925, was taught by Césaire and became active in anti-colonial activism in French North Africa. His works inspired both anti-colonial liberation movements throughout the world and postcolonial theorists, notably Edward Said and Homi Bhabha (see below). Fanon's ideas changed over time (see Box 2.2). His first book, *Black Skin, White Masks* (1952) analyzed the impact of colonial subjugation on the black psyche. Based on Fanon's personal experience as a black intellectual in a white-dominated world, the book elaborates the ways in which the colonizer–colonized relationship is normalized. As a colonized subject and recipient of French education, Fanon conceived of himself as French, but felt profound disorientation and disillusionment after encountering French racism, which decisively shaped his psychological theories about culture. *Black Skin, White Masks* charts Fanon's disillusionment with a culture in which he was raised, which he perceived as his own, but in which he was discriminated against (Haddour 2006). Fanon develops an understanding of racism as generating harmful psychological constructs that both blind the black man to his subjection to a universalized white norm and alienate his consciousness. (Fanon's account has been criticized by feminists for failing to consider the experience of subjugation of black women.)

Fanon was perhaps the first postcolonial writer to capture the compromise of mimicry that arose out of the central contradiction at the heart of the colonial enterprise – that, on the one hand, the colonized were irrevocably different from the colonizers; and, on the other, the colonized could become like the colonizers through being civilized. This notion, based on the idea of universal humanism, legitimated European colonial expansion. Drawing on his own experience, he argued (1986: 11, 117) that the colonized 'is sealed in his blackness', never able to attain the culture of the colonizers, which instead he is urged to mimic and desire:

> The white world, the only honorable one, barred me from all participation. A man was expected to behave like a man. I was expected to behave like a black man – or at least like a nigger. I shouted a greeting to the world and the world slashed away my joy. I was told to stay within bounds, to go back where I belong . . . I tell you, I was walled in: No exception was made for my refined manners, or my knowledge of literature, or my understanding of the quantum theory.

## Box 2.2  A summary of Fanon's changing ideas

1  As a member of the colonized elite, Fanon initially disavowed his West Indian identity and identified with the cultural models of the French colonizers.
2  After fighting for his 'mother country' (France) in World War Two, in which he was a decorated soldier, Fanon became aware of racism amongst his comrades. In France, he was treated not as a 'Frenchman', but as a 'Negro'. His encounter with racism intensified while he studied in Lyon after the war, and this led him to renounce his Frenchness. This shift in Fanon's thinking provided the inspiration for his first book, *Black Skin, White Masks* (1952), which reflected his concern with psychoanalysis and his ambivalent relationship with France and Martinique. This ambivalence never left Fanon, but his writing and politics became more overtly concerned with liberation.

While *Black Skin* subscribes to a notion of assimilation into the dominant white, colonial culture, his later work rejects this as impossible.

3   Unable to resolve his unattainable identification with Martinique as the place that was simultaneously home but in which he could never be anything but a colonized and, therefore, inferior subject, Fanon began to identify with Algeria. Following the outbreak of the Algerian War of liberation from France in 1954, Fanon became increasingly anti-colonial. In 1956 he became an active militant in the cause of both the Algerian War and African freedom more broadly. He became the intellectual leader and political spokesperson of the Algerian War and understood it in revolutionary terms, espousing violent conflict as the only means through which to defeat European imperialism and racism.

4   Internal conflicts at the heart of the Algerian Revolution inspired Fanon to abandon the narrow political concerns of violent revolution and to champion a universalist conception of politics based in humanism. Fanon identified with the Algerian elite, who revolted against French culture and the notion of assimilation. However, he also rejected the mythic rhetoric of négritude (including his former teacher Césaire's ideas), because it ignored the cultural, national and ethnic differences that traversed the notion of blackness. Fanon's new humanism was based on an ethics that acknowledged differences in identity and history. For him, decolonization was the only way out of the absolute dehumanization in which the colonized lived; the doctrine of assimilation only succeeded in denying the colonized historical agency. This further shift in Fanon's thinking forms the basis of *The Wretched of the Earth* (1961), which reflected more Fanon's interest in revolutionary praxis and humanist politics that acknowledges difference. It also warned against the negative effects of reactionary and bigoted nationalism that anticipated 'the ethnonationalist "switchbacks" of our times, the charnel houses of ethnic cleansing: Bosnia, Rwanda, Kosovo, Gujarat, Sudan' (Bhabha 2004: xvi).

Sources: Memmi 1973 and Haddour 2006

Fanon exposed the colonial logic in which the colonized were never quite able to internalize properly the European cultural codes, nor be allowed access to universal humanism and its associated rights. The colonized could be civilized but would never be quite the same. This contradiction underpinned the maintenance of the power and privilege of the colonizer. It maintained a border of difference between colonized and colonizer, whilst simultaneously legitimating colonial conquest and the 'white man's burden' to civilize the Other (discussed in more detail in Chapter 4).

Fanon was also a most incisive writer on the production of alienation, destruction of self-esteem and the cementing of inferiority complexes wrought by colonialism. He wrote in direct criticism of Octave Mannoni, a French psychologist whose famous work *Prospero and*

*Caliban* dealt with the psychology of the colonizer and the colonized. Mannoni argued that colonialism must be understood as produced by psychic differences between the colonized, who suffered from a dependency complex that made them revere their ancestors, and the colonizers, who feared their own inferiority and sought to assert their dominance. In contrast, Fanon argued that the inferiority complex was a product of the colonized internalizing the colonizer's image of them as inferior. The colonized come to look at themselves through the eyes of the colonizer. To gain esteem from the colonizer, the colonized try to escape their blackness, to 'turn white' by adopting the culture and language of the colonizer. After Fanon's death in 1961, such readings of the psychology of racism exerted considerable influence on black consciousness movements around the world, both in the US Civil Rights Movement and in anti-apartheid resistance (Box 2.3).

---

## Box 2.3 Steve Biko: black consciousness in apartheid South Africa

In a nation where blacks were the clear majority (20 million to four million), apartheid (the system of spatial, political and economic separation of ethnic groups from 'white' South Africa that was inscribed in law between 1948 and 1994) existed because it possessed an overpowering psychological grasp on the minds of black people. For centuries, Africans were led to believe that they were an inferior people incapable of development, and that whites intrinsically possessed all that was good and all that was superior. Biko perceived the problem as a twofold psychological phenomenon. On the one side, following generations of exploitation, whites believed in the inferiority of blacks. On the other, blacks had developed an ingrained dependence fostered by white domination. Biko advocated solidarity among black people to break with the chains of oppression.

In 1968, Biko and his confederates formed SASO (the South African Students Organization), with Biko's philosophy of Black Consciousness at its core. The new organization and its ideological underpinnings appealed to young black South Africans, and the movement spread rapidly. Black Consciousness stands as Biko's most lasting contribution to the liberation struggle in South Africa. It uplifted a mass of people, inspired hope and gave direction and purpose to the lives of black South Africans. Black Consciousness was then a battle for the mind-war waged in the subconscious. In 1972, Biko emphasized this point:

> [Black Consciousness] is more than just a reactionary rejection of Whites by Blacks. The quintessence of it is the realization by the Blacks that, in order to feature well in this game of power politics, they have to use the concept of group power and to build a strong foundation

for this. Being an historically, politically, socially and economically disinherited and dispossessed group, they have the strongest foundation from which to operate. The philosophy of Black Consciousness, therefore, expresses group pride and the determination by the Blacks to rise and attain the envisaged self. At the heart of this kind of thinking is the realization by the Blacks that the most potent weapon in the hands of the oppressor is the mind of the oppressed.

Source: Arnold 1978: xvii, xix

Steve Biko died from severe head injuries in South African police custody on 12 September 1977 and is seen by many as a martyr in the anti-apartheid struggle. His radical politics were shaped more directly by Fanon and Césaire, in contrast to ANC leaders, such as Nelson Mandela and Albert Lutuli, who came to subscribe to a more Gandhian line of peaceful civil disobedience as a form of resistance. However, like the ANC, Biko was willing to hold discussions with white South Africans and concede them a place in a future black-ruled South Africa. The latter proved ultimately to be a successful route in bringing about a largely peaceful end to apartheid, with the election of the ANC government in April 1994. Throughout the 1970s and 1980s, Black Consciousness played a significant role in mobilizing black people in South Africa; Biko's contribution was acknowledged in the ANC's adoption of his image on publicity posters for the 1994 election.

Fanon's explanation in *Black Skin* of the effects of being colonized by language was also influential: speaking the language of the colonizer meant that the colonized accepted, or were coerced into accepting, the collective consciousness of the colonial power, which identified blackness with inferiority, evil and sin. To escape the association of blackness with evil, the black man wears a white mask and internalizes white cultural values, creating a fundamental disjuncture between consciousness and body. This focus on alienation reveals the Marxist roots of Fanon's writings. Fanon's arguments have been central in postcolonial critiques of colonialism and racism. Of significance is his insistence that the category 'white' depends for its stability on its negation, 'black', that notions of white superiority cannot exist without an opposite (black inferiority), and that the origins of these notions were formed through imperial conquest. *The Wretched of the Earth* (1961) proposed the overcoming of the binary system in which black is bad and white is good through peasant revolution. This faith in the African peasantry and emphasis on language anticipates the work of Ngũgĩ. Fanon was also a major influence on Brazilian philosopher of education Paulo Freire (who in turn was an influence on Biko). Freire's *Pedagogy of the Oppressed* (1972) may be best read as an extension of *The Wretched of the Earth* in that it emphasized the need to provide native populations with an education that was simultaneously new and modern (rather than traditional) and anti-colonial (not simply an extension of the culture of the colonizer).

Fanon emerged as a global figure in the 1980s, primarily in English, Cultural Studies and Postcolonial Studies programmes (Haddour 2006: xiv). Renewed interest, particularly in *Black Skin*, positioned him as a leading exponent of postcolonial theorizing. This was undoubtedly inspired by Bhabha's foreword to the 1986 edition that inspired new readings. However, *The Wretched of the Earth* was also an attack on the insidious neocolonialism that shapes relations between former colonized and the West and has inspired those working within critical development theory. In this book, Fanon wrote that the end of direct colonial rule did not necessarily mean the end of colonialism, and that there was a need for a cultural and political revolution that would guarantee the independence of the Third World. He anticipated the embroiling of the former colonies in the emerging Cold War (see Chapter 3) and that aid, rather being reparation for the damage wrought by colonialism, would become a tool of ideological control and manipulation. Instead, he called for the Third World 'not to want to catch up with anyone' (Fanon 1967: 254–5).

The spirit of Fanon's politics lives on in contemporary global justice movements, which are arguably the anti-colonial movements of the twenty-first century (see Chapter 6). His ethics of decolonization calls for a different kind of globalization, 'a new world order in which segregated economies must be abolished' (Haddour 2006: xxi). He also attacks the 'long history of domination, of exploitation and pillage' that creates and perpetuates poverty (Fanon 1967: 39). As he puts it: 'Europe is literally the creation of the Third World. The wealth which smothers her is that which was stolen from the under-developed people'. His work thus shares similarities with other critics of 'underdevelopment' during the 1960s, such as Andre Gunder Frank in Latin America and Walter Rodney in Africa. Underdevelopment was defined as deliberately created poverty by imperialism, including lack of access to health care, drinkable water, food, education and housing. Industrialized countries were accused of actively blocking or deforming the development of agrarian countries in the South, by means of policies and military interventions intended to protect their global power and superior position in world trade. They argued that very little had changed in the lives of formerly colonized peoples with the formal ending of colonialism. In prophetic language that conjures up the abject poverty of parts of contemporary Africa, Fanon (1967: 76) writes:

> The mass of the people struggle against the same poverty, flounder about making the same gestures and with their shrunken bellies outline what has been called the geography of hunger. It is an underdeveloped world, a world inhuman in its poverty; but it is a world without doctors, without engineers and without administrators. Confronting this world the European opulence is literally scandalous, for it has been founded on slavery, it has been nourished with the blood of slaves and it comes from the soil and subsoil of that underdeveloped world.

That this 'geography of hunger' persists in the twenty-first century confirms Fanon's continued relevance as a critic of neocolonial economic relations. Despite this, his call for a redirection of military spending in the West towards poverty alleviation, and for liberation of the South through development, remains unheeded. Moreover, post-colonial governments in the South did not heed Fanon's warning against the pitfalls of nationalism, particularly in Africa. As Haddour (2006: xxv) argues, Fanon's work sadly remains as relevant today as it was when it was written: 'It is important to revisit *The Wretched of the Earth* and to re-read it as we enter into the new age of globalized terror and violence'. Equally, postcolonialism could be more engaged with the problems that globalization poses for emergent post-colonial nations – the neocolonialism that shapes international development and geopolitical relations and the failure of the international community to help the 'wretched of the earth' lift themselves out of poverty.

## Latin American freedom struggles

Latin America occupies a special place in the history of anti-colonialism and its relation to postcolonial theory (Young 2001). Much of it has been independent from European rule for over 200 years. Indeed, it was post-colonial before many European states became imperial powers and before some (e.g. Germany and Italy) had even become nation states. Its history of liberation is complex, with resistance to Spanish and Portuguese imperialism arguably dating back over 500 years, giving rise to the anti-colonial movements of the early nineteenth century and anti-imperialism movements of the late nineteenth and twentieth centuries (Figure 2.1). Much of the continent became independent of Spanish and Portuguese rule between 1808 and 1825, but colonies facing the Caribbean (Honduras, British and Dutch Guiana) only became independent after 1945 (and, as Table 1.2 illustrates, French Guiana remains French). As with the USA in relation to Britain, Brazil went on to become more of an economic powerhouse than its former colonial master, Portugal. To add to the complexity, political and cultural times that would be chronological or linear in many contexts are 'folded' in Latin America. For example, European countries often progressed in a linear fashion from feudalism to industrialization and modernity, with feudal organization of societies and livelihood patterns disappearing as modern governance and economic systems evolved to take their place. However, in Latin America this transition is not linear and, rather like a folded piece of paper, feudalism and modernity are juxtaposed simultaneously. Feudal land ownership patterns and livelihoods persist alongside democratic governance and industrial modernity. Enormous GNPs exist alongside extreme poverty.

Since the early nineteenth century, Latin America has been subject more than any other region in the world to neocolonialism in the form of US economic, political and cultural imperialism (Young 2001: 194). This gave rise to Latin American dependency theory (see Chapter 3), but rarely to nationalism as a form of resistance. For example, the United States' CIA was able to topple the Goulart and Allende governments in Brazil and Chile respectively and enable the imposing right-wing dictatorships in their place. Rarely has the general population of many Latin American countries been able to exercise self-determination. Independence movements, like that in the USA, were essentially bourgeois (led by *criollos* – white European settlers). In some countries, such as Argentina, a US-style policy of European immigration and extermination of indigenous peoples facilitated the establishment of a predominantly European society. Young (2001) argues that for most people in these countries, little changed with independence. The conditions of the peasantry and of local, indigenous peoples have, if anything, become worse with urbanization and social division; landlessness and extreme poverty have become acute.

**Figure 2.1** *José Martí, nineteenth-century martyr of the Cuban independence struggle with Spain, Plaza De Revolution, Havana*

Although Zapata's Mexican rebellion of 1910 was the true precursor of insurrections against exploitative power, the Cuban revolution of 1959 (led by Che Guevara and Fidel Castro) was the first to successfully achieve land redistribution and, because of its international perspective, to articulate anti-imperialism in Latin America. It also inspired similar movements in South East Asia and Africa. Dependency theory has its origins in and takes its inspiration from these political movements, providing a theoretical innovation that has become part of mainstream Marxism and a major critique of development theory (see Chapter 3). The writings of Guevara (see Figure 2.2) were an important influence; like Fanon, 'Guevara's sights were set on the global reach of injustice' (Young 2001: 211). The Zapatista rebellion against the neoliberal North American Free Trade Agreement (NAFTA) in Chiapas, Mexico in 1994, led by Subcomandante Marcos, suggests that there is negligible difference in the principles for which modern-day resistance movements fight, and for which Guevara fought and was executed by CIA-trained troops in Bolivia in 1967. Latin American anti-imperialism brought together social, theoretical and political thought, personified through Guevara (in Latin America), Cabral and Fanon (in Africa) and Ho Chi Minh

**Figure 2.2** *Che Guevara Ironwork on the Interior Ministry Building, Plaza De Revolution, Havana*

Source: Author

(in Asia). As Young (2001: 213) puts it: 'Postcolonialism was born with the *Tricontinental*' – a journal that established a syncretic body of writing that would provide 'the theoretical and political foundations of postcolonialism' around the world.

## Women, anti-colonialism and the hegemony of western feminism

The history of anti-colonialism, much like the history of colonialism and imperialism, has been dominated by male political theorists, anti-colonial activists and leaders. However, this is not to say that women were absent from anti-colonial politics and movements. Feminist historical work from the 1980s revealed the significant roles that women played in anti-colonialism. For example, Jayawardena (1986) provided a comprehensive account of women's participation in nationalist movements in Asia and the Middle East from the late nineteenth century onwards. Similarly, Sangari and Vaid (1989) explored the role of women in liberation movements, despite the masculinist nature of anti-colonial struggles that restricted the participation of women and their subsequent erasure in the writing of the histories of these struggles. As Young (2001: 361) argues, the different profile of women in anti-colonial struggles stems from the gender-based inequalities that formed the origin of feminist politics. As a result, no single figure registers anti-colonial positions like a Gandhi or Nkrumah. Even Celia Sánchez, who played a prominent role in the Cuban revolution, did not have access to the same public profile as Castro or Guevara. However, while gender inequalities may have rendered women less visible, many played important roles in anti-colonial struggles. Examples include:

- Led armed rebellions in India in 1857–1858 (e.g. Rani Lakshmi Bai of Jhansi, the Rani of Ramgarh and Begum Hazrat Mahal) (Young 2001: 362). Also played a prominent role in Civil Disobedience movement of 1920s and 1930s, which marked a major step in the emancipation of women in India.
- Fought in anti-colonial war in Vietnam against the French (e.g. Nguyen Thi Nghia, Nguyen Thi Minh Khai and Minh Khai, who were all executed) (*ibid.*).
- Fought in Malaya, Cuba, Nicaragua and many other liberation struggles in Latin America (Collinson 1990; Jayawardena 1986: 208–9; Yuval-Davies 1997: 98).

- Involved in anti-French resistance in Algerian War of Independence (Fanon 1980).
- Long history of women's involvement in anti-colonialism in Africa. For example, Mbuya Nehanda fought against British imperial presence in Shona, Rhodesia and was executed in 1897; the women warriors of Dahomey fought against French imperialism and were subsequently massacred in 1894; Abo women's revolt in Nigeria against British taxation in 1929; in 1959 the Kon women of Eastern Nigeria instigated a major uprising against the deterioration of their position as farmers and the fear of losing their land to men. Women participated in armed struggle in Angola, Eritrea, Guinea-Bissau, Mozambique, Namibia and Zimbabwe (Urdang 1979). Women played a key role in anti-colonial consciousness-raising and solidarity-building in Ghana during the 1950s (Young 2001: 363) and in the Kenyan independence movement over a long period from 1895. This included Mary Muthoni Nyanjiru, who was shot in 1922 after organizing a workers' strike, the many women involved in the 'Mau Mau' revolt of the 1950s (at one point, 35,000 women were imprisoned by the British) and the many Kikuyu women who were active at local levels (Young 2001: 366).
- Prominent in the anti-apartheid struggle in South Africa throughout the twentieth century (e.g. Mrs Molisapoli, who was active in the resistance against Pass Laws in the early twentieth century, and the Federation of South African Women, which worked closely with the ANC from the 1960s).
- Played a role in anti-colonial nationalist movements in Turkey, Egypt and Iran.

Excavating such stories now forms part of the continual transformation of contemporary feminist postcolonial theory, which presents challenges to dominant western feminisms.

Although women's involvement in anti-colonial struggles took a variety of different forms and emerged from diverse political and cultural motivations, not all of them explicitly feminist, anti-colonial approaches have on occasion converged with feminist politics to produce powerful critiques of western feminism. Until the 1980s, there was a tendency to assume a commonality in the forms of women's oppression and activism worldwide (Morgan, 1984). Western feminists assumed that their political project was universal, and that women globally faced the same universal forms of oppression. However, divisions among women based on nationality, race, class, religion, region, language and sexual orientation have proved more divisive within and across nations than western theorists acknowledged or anticipated.

At an abstract level, assumptions by western feminists about what their political project entails have been called into question by a range of criticisms under the broad rubric of postcolonialism. Encounters with different feminisms and different gender relations have raised issues about what exactly it means to be feminist and have ensured that a western-centric political vision is no longer acceptable. Since the 1980s, black feminists have explored the ways in which feminism is historically located in the dominant discourses of the West, a product of western cultural politics and therefore reflecting western understandings of sexual politics and gender relations. Indeed, in many cultures (particularly in the South) feminism is associated with cultural imperialism. In their influential essay, Amos and Parmar (1984) trace the historical relationship between western feminism and imperial ideologies, institutions and practices. They argue that like gender, the category of feminism emerged from the historical context of modern European colonialism and anti-colonial struggles; histories of feminism must therefore engage with its imperialist origins.

The 1970s and 1980s saw an outpouring of critical work by black feminists (Davis 1982; Dubois 1978; Lorde 1984; Rich 1986), which began to have influence. Black feminism represented a different current of feminist thought that brought together gender domination with racism and capitalism. In the 1970s, this was not restricted to black women but included women from Latina/Chicana origins (see Anzaldua and Moraga 1981), Chinese, or 'Third World' origins. In the USA, the Combahee River Collective, which emerged out of the civil rights movement, had a significant impact on Women's Studies, which had previously been predominantly white. In the British context, careful literary and historical work demonstrated that white British women's historical experience, in all its complexity and variation, was often bound up culturally, economically and politically with imperial concerns and interests (Ferguson 1992; Lewis 1996; Melman 1992; Midgley 1992; Ware 1994). Of course, some British feminist movements have acted politically to oppose such hegemonies (e.g. the Greenham Common women in the context of East–West geopolitics).

As Burton (1999: 218) argues, however, the original intention of Amos and Parmar's essay was 'not to clear the way for a more politically accountable historiography of Euro-American women's movements, but rather to make space for histories of black women, women of colour, and anti-colonialist and nationalist women'. She contends that:

> Before the 1980s, it was possible for even some of the most accomplished feminist historians in the West to express surprise that there had been women's movements and feminist cultures outside Europe and North America before the 1960s, even as they failed to realise the neocolonialist effect this kind of ignorance was having on the production of postcolonial counter-histories.

Chandra Talpade Mohanty has been amongst the most influential postcolonial feminist theorists. Her essay 'Under western eyes: feminist scholarship and colonial discourses' (1988) is a devastating critique of the political project of western feminism. The discursive construction of the category of the 'Third World woman' as a hegemonic entity forms the focus of Mohanty's criticisms. She argues that western feminisms have tended to gloss over the differences between Southern women, ignoring the diversity of experiences of oppression and its contingency on geography, history and culture. Her analysis of the insufficiency of western epistemological frameworks for recovering, let alone understanding, the cultural and historical meanings of women's experiences and structural locations outside the West has been enormously influential. Her criticism of the invisibility of black and Third World women in histories of feminism precipitated an outpouring of publications. These focused especially on Indian and Egyptian women's movements (Badran 1995; Baron 1994; Jayawardena 1995; Southard 1995), but also on countries and cultures having less self-evident (or less well-known) relationships to European empires, such as Iran (Afary 1996; Kandiyoti 1991; Shahidian 1995). The outcome of this feminist and anti-imperialist scholarship has been an attempt to re-orient western feminisms, such that they are no longer perceived as exclusive and dominant, but as part of a plurality of feminisms, each with a specific history and set of political objectives, as well as sharing some common ground. Contrary to the widespread belief that the inspiration, origins and relevance of feminism are bourgeois or western, related to a specific ideology, strategy or approach, it is now recognized that feminism does not simply originate in the West. There were many pre-colonial women's movements around the world and various forms of feminism have existed and continue to exist across cultures.

Black feminist and postcolonial critiques have also offered more profound examinations of the racism and ethnocentrism at the heart of (white) western feminisms. As bell hooks (1981: 8–9) argues:

> All too frequently in the women's movement it was assumed one could be free of sexist thinking by simply adopting the appropriate feminist rhetoric; it was further assumed that identifying oneself as oppressed freed one from being an oppressor. To a grave extent such thinking prevented white feminists from understanding and overcoming their own sexist–racist attitudes toward black women. They could pay lip service to the idea of sisterhood and solidarity between women but at the same time dismiss black women.

Fellows and Razack (1998: 335) refer to this as a 'race to innocence', a 'process through which a woman comes to believe her own claim of subordination is the most urgent, and that she is unimplicated in the subordination of the other women'. The relationship between (white) western and 'other' feminisms has often been adversarial, partly because of the failure of white women to recognize that they stand in a power relationship with black women that is a legacy of imperialism, and partly because the concepts central to feminist theory in the West become problematic when applied to black women (hooks 1984; Mohanty 1988). One example is the explanation given for inequalities in gender relations. Many black and Southern activists object to western feminism that depicts men as the primary source of oppression. For black women there is no sole source of oppression; gender oppression is inextricably bound up with 'race' and class. There is perhaps a tendency in some of this criticism to homogenize 'western feminism' – socialist feminists also identify capital as a source of oppression (Delphy 1984), and lesbian feminists have criticized marginalization by sexuality (Bell and Klein 1996). However, this criticism has also forced recognition that assumptions at the heart of western feminisms do not necessarily reflect the experience of black women (Carby 1983; Nain 1991). Furthermore, in many cultures black women often feel solidarity with black men and do not advocate separatism; they struggle with black men against racism, and against black men over sexism. These debates have generated theories that attempt to explain the interrelationship of multiple forms of oppression, such as race, class, imperialism and gender, without arguing that all oppression derives ultimately from men's oppression of women.

Similar criticisms have been levelled at understandings of the public–private dichotomy. A large part of western feminist literature is dedicated to critiquing the separation of public and private spheres, arguing that it devalues women's contribution to society, and that it has been used to confine women and inhibit their input. A major problem with this idea is that it ignores the contentions of some commentators in other parts of the world that the private realm does indeed exist separately from the public one, but that both domains are needed and political. For example, instead of motherhood being a private occupation forced on some women, which limits their political inputs or contributions, it is reconstructed as a chosen political occupation with important social and economic repercussions. The activities of some Islamist feminists (e.g. Moroccan writer Fatema Mernissi, Egyptian author and activist Nawal El Saadawi), and the Argentinian Mothers of Disappeared (Fisher 1993), are examples where women have sought an empowering 'private' function, challenging western feminist assumptions about the home, family and motherhood as a site of oppression. This is not to argue, of course, that mobilizing around 'private issues' is always empowering in terms of challenging entrenched gender ideologies or strategic interests.

More broadly, many women of the South resent the bourgeois preoccupations of western feminisms, particularly within gender and development debates. Economic exploitation and political oppression, as well as provision of basic needs such as clean water and children's education, are considered more pertinent than issues of sexual politics and gender oppression that often motivate middle-class feminism in both North and South (Schech and Haggis 2000: 88). In Latin America, this class and race divide is exemplified by the distinctions between 'feminine' and 'feminist' politics that emerged during the 1970s and 1980s (Marchand 1995). Feminine politics are articulated around *practical* gender interests, such as health care, nutrition and shelter, potable water and secure livelihoods that are vital to the more immediate survival of poorer women and their families. Feminist politics are organized around longer-term *strategic* gender interests, such as overturning the gender division of labour, gaining control over one's own reproduction, and attaining legal and political equality, which have long been the concern of middle-class feminisms

(Alvarez 1990). These differences created divisions and tensions in Latin American women's activism, as well opportunities for forming new alliances around issues such as domestic violence and sexuality. Similar tensions existed between women in East and West, and a significant body of criticism emanated from women in post-communist countries of Eastern Europe and the former Soviet Union (see Drakulić 1993; Einhorn 1993; Funk 1993; Funk and Mueller 1993). Differences in tradition, culture, personality, beliefs and desires, therefore, demand the interrogation and destabilization of dominant western feminist discourses. Postcolonial approaches seek to 'provincialize' (Chakrabarty 1992) western feminisms rather than see them as a paradigmatic form of feminism *per se*.

As Roberts and Connell (2016: 137) argue, 'There is an enormous wealth of feminist thought in the global periphery', but much of this is ignored in the metropole. Consequently, some of the most powerful thinkers and pioneers in world feminism (e.g. Bina Agarwal, Heleieth Saffioti, Li Ziaojiang, Raden Adjeng Kartini, He-Yin Zhen and Huda Sharawi) remain invisible in the mainstream economy of knowledge. Much contemporary feminist thought from post-colonial societies remains unacknowledged even when it has influenced mainstream thinking. For example, 'The discovery of "intersectionality" in the metropole was long preceded by debates about gender, race, class and caste from in the colonized world from India to Brazil' (*ibid.*). Critiques of western feminisms, such as that by Roberts and Connell, have some parallels in their critique of the knowledge economy with both anti-colonial politics and the postcolonial theories that began to emerge in the 1980s.

## Subaltern studies: postcolonial history, nationalism and identity

In addition to the variety of approaches that we might now call postcolonial and that emerged in anti-colonialism and liberation movements, a further impetus to postcolonial studies, and to postcolonial theory specifically, emerged in India in the 1980s. The Subaltern Studies Group was a collective of scholars including Ranajit Guha, Gyan Prakash, Dipesh Chakrabarty, Gayatri Chakravorty Spivak, Shahid Amin, David Arnold, David Hardiman, Gyanendra Pandey and Partha Chatterjee. They were interested in the post-colonial societies of South Asia in particular, and the developing world in general. Originally, the collective adopted an approach inspired primarily by Gramscian Marxism, of history from below, focusing on the lives of peasants – subalterns – rather than on elites. This was intended to counter colonialist versions of (primarily Indian) history and the elitism of bourgeois-nationalists in the historiography of Indian nationalism, both of which had been apparent in debates in the 1970s between Cambridge University scholars and orthodox Marxist historians in India. Both these traditional Marxist and elitist narratives of Indian history focused on the political consciousness of elites in mobilizing the masses, ending the domination of Britain as the colonial power and building an independent nation state. In contrast, the Subaltern Studies group focused on the role of non-elites as agents of political and social change, concentrating specifically on the role of peasants. Subaltern Studies, therefore, was not simply a critique of European hegemony, but also of elites within post-colonial societies.

The Subaltern Studies Group played a vanguard role in postcolonial theory. Subaltern, both as a term of reference and as an object of enquiry (see Chapter 1), was popularized within academic literature by the collective. Their use of the term subaltern was somewhat broader than the military analogy, referring to wider processes of subordination, particularly in the writing of histories of independence struggles and post-colonial nation-building. The Subaltern Studies scholars appropriated Gramsci's term to locate and re-establish a voice or collective locus of agency in post-colonial India that moved beyond the role of elites. They attempted to formulate a new narrative of the history of India and South Asia.

The collective made numerous influential contributions. The first was in studying the active participation and intellectual creativity of subaltern classes in processes of nation-state formation. The second was in elaborating the discourses and rhetoric of emerging political and social movements, in contrast to highly visible actions like demonstrations and uprisings. The third was in attempting to recover the silenced voices of the formerly colonized without losing sight of the structural inequalities between the dominant and the subjugated. The argument in the early writings of the collective is that subaltern groups were also able to subvert the authority of those who had hegemonic power, and they thus played a more significant role in historical change than had previously been acknowledged. There are clear similarities between subaltern histories and accounts of women's roles in anti-colonialism that were emerging at the same time. Historical narratives thus need to reflect this in contrast to the grand, heroic narratives of independence and nation-building that erase the historical agency of subaltern groups.

Tensions eventually grew within the collective between those who subscribed more closely to Marxist approaches and those who were engaging with poststructuralist critiques. According to Chakrabarty (2005: 468), critics suggest that:

> Subaltern Studies was once 'good' Marxist history in the same way that the English tradition of 'history from below' was, but that it lost its way when it came into contact with Said's Orientalism, Spivak's deconstructionism, or Bhabha's analysis of colonial discourse.

Chakrabarty sees this as a serious misreading of the radical departures of Subaltern Studies from the outset. In addition, it was arguably these tensions that inspired debates concerning issues such as knowledge production, historicism and the (ir)retrievability of subaltern voices that have inspired much of what is now referred to as postcolonial theory since the 1980s, and that has had a significant impact across a range of disciplines.

## Postcolonial theory

While the origins of postcolonialism can be traced to numerous different and disparate sources and while, in theoretical terms, it has been greatly influenced by Marxist political-economy and post-structuralist cultural and linguistic analysis, Edward Said, and subsequently Homi Bhabha and Gayatri Spivak, have arguably had the greatest influence within postcolonial theory.

### Edward Said and colonial discourse

Said is best known for his book *Orientalism* first published in 1978, which is often considered the founding text of contemporary postcolonial theory. Through this, Said introduced the notion that colonialism operated not only as a form of military and economic domination but also as a *discourse* of domination. Central to this is the idea that knowledge is never innocent but is profoundly connected with the operations of power. As Young (2001: 383) argues, Said's critique of the cultural politics of academic knowledge, from the basis of his experience growing up as an 'Oriental' in two British colonies (Palestine and Egypt), founded postcolonial studies as an academic discipline, 'invested in the political commitment and the locational identification of its practitioners'. Like Fanon, Said brought together the political commitments and ideological critiques of anti-colonialism with the theoretical work of poststructuralism Said was notable in shifting the analysis of colonialism, imperialism and struggles against it to questions of discourse. This was not simply a question of language, but of a discursive regime of knowledge, drawing its inspiration from the writings of French theorist Michel Foucault.

*Orientalism* is not about non-western cultures, but about western representations of these cultures. It examines how the formal study of the 'Orient' (what is now known as the Middle East), along with key literary and cultural texts, 'consolidated certain ways of seeing and thinking which in turn contributed to the functioning of colonial power' (Loomba 1998: 43–4). Said critiques the constellation of false assumptions underlying western attitudes towards Asia and the Middle East and decries the subtle and persistent Eurocentric prejudices against Arabs and Islamic peoples and their cultures. He draws implicitly on Fanon's ideas about binaries ('othering') to suggest that a long tradition of false and romanticized images served as an implicit justification for Europe and the USA's colonial and imperial ambitions. However, colonial discourse did not simply justify colonial rule after the fact, as it were, but operated more instrumentally, as 'a kind of Western projection onto and will to govern over the Orient' (Said 1985: 95). In other words, power and knowledge are inseparable in discourse. Orientalism – the study of the Orient – was ultimately a political vision of reality whose structure promoted a binary opposition between the familiar 'self' (Europe, the West, 'us') and the strange 'Other' (the Orient, the East, 'them').

This creation of opposition is crucial to European self-conception:

> [I]f colonised people are irrational, Europeans are rational; if the former are barbaric, sensual, and lazy, Europe is civilisation itself, with its sexual appetites under control and its dominant ethic that of hard work; if the Orient is static, Europe can be seen as developing

and marching ahead; the Orient has to be feminine so that Europe can be masculine.

(Loomba 1998: 47)

Said demonstrates a 'consistent discursive register of particular perceptions, vocabularies and modes of representation common to a wide variety of texts, from literary to racial theory, from economics to autobiography, from philosophy to linguistics' (Young 2001: 387–8). According to Said, discourses about the Orient, and by extension the South, were ideological representations with no corresponding reality; they said more about the West than they did about the real world they purported to represent. Said also denounces Arab elites for internalizing Orientalist ideas of Arabic culture. Despite criticisms (see Box 2.4), his use of Foucauldian theory to argue that colonial discourse has material effects on those being represented has been particularly influential, especially within the fields of literary theory, cultural and development studies and human geography and, to a lesser extent, history and oriental studies. The importance of Said's discussion of Orientalism lies in its exposure of the subtle and persistent Eurocentric prejudice against Arab-Islamic peoples and their culture, and of the links between false images in western culture and its colonial and imperial ambitions. It has inspired numerous examinations of Eurocentric prejudice against other parts of the world. Writing in 1980, Said criticized what he regarded as poor present-day understandings of Arab culture in the West:

> So far as the United States seems to be concerned, it is only a slight overstatement to say that Moslems and Arabs are essentially seen as either oil suppliers or potential terrorists. Very little of the detail, the human density, the passion of Arab-Moslem life has entered the awareness of even those people whose profession it is to report the Arab world. What we have instead is a series of crude, essentialized caricatures of the Islamic world presented in such a way as to make that world vulnerable to military aggression.
>
> (Said 1980)

His account of the workings of discourse and its imbrications with imperialism and global geopolitics is still relevant in the twenty-first century (see Gregory 2004). That *Orientalism* is still debated fervently is indicative of its status within postcolonial theory.

Said's work has been extremely important in demonstrating how the power to represent other places (the power to name, to describe, to publish, to claim and construct knowledge), was instrumental in reinforcing a sense of difference between the West and non-West, which also translated into a sense of superiority and justified various political interventions that underpinned imperialism. The same processes persist in present-day development (see Chapter 4) and geopolitical discourses (see Gregory 2004).

---

## Box 2.4  Criticisms of Said's 'colonial discourse'

- Makes large historical generalizations from the analysis of a limited number of literary texts.
- Colonial discourse analysis dehistoricizes and treats all texts as synchronic, as if they existed in an ahistorical, unchanging, spatialized continuum.
- Colonial discourse analysis explores representation as a representation, while the historian analyzes the representation in terms of what it represents. Colonial discourse analysis does not search for historical 'truths', but assumes that a more truthful version of history could be told.
- Colonial discourse forms a homogeneous totality that overrides the particularity of historical and geographical difference which, given the complexities of colonialism, is problematic. Thus, while Said criticizes the West for its totalizing discourses, he could equally be criticized for creating a homogeneous West and ignoring the differences between European empires. It fixes an East versus West divide and assumes this has existed from classical Greece to the present day. Said addresses this criticism in *Culture and Imperialism* (1993).
- Said employs too determining and univocal a notion of discourse from which he claims no westerner can escape, but this is undermined by his own analysis of the complexity and range of positions taken up by the writers he discusses.
- By focusing on western literary texts, Said does not disrupt the marginalization of 'other' cultures but re-centres the West. He ignores the self-representations of the colonized and focuses on the imposition of colonial power rather than resistances to it.
- Critics and followers of Said have tended to emphasize the textual nature of colonial discourse, which differs from Foucault's notion, as outlined in *The Archaeology of Knowledge* (1969), of discourse as a regime of knowledge and disciplinary practice. Said is thus criticized for not exploring colonial discourse as a material practice.

Source: adapted from Young 2001: 389–92 and Loomba 1998: 48–9

---

## Homi Bhabha: ambivalence, mimicry and hybridity

Like Said, Indian scholar Homi Bhabha challenges the tendency to treat postcolonial countries as a homogeneous category about which stereotypes can be perpetuated, but his ideas are more heavily influenced by poststructuralism, most notably the writings of Derrida, Lacan and Foucault. His work is noteworthy for identifying ambivalence in colonial dominance and thus producing a more nuanced understanding of colonial power. One simple example of ambivalence is that the colonizer wants and needs the colonized to be similar to himself, but not the same. If the colonized continues to behave in his traditional ways, he brings no economic gain to the colonizer. But, if the colonized changes too much and is found to be exactly the same as the colonizer, the colonizer is left with no argument for his supremacy. In revising Said's notion of colonial discourse, Bhabha emphasizes its ambivalence and heterogeneity, rather than its fixed homogeneity, suggesting its outcomes were less certain rather than always successfully realizing its intention.

In *The Location of Culture* (1994), Bhabha draws on Fanon's writings and deploys concepts such as mimicry (imitation) and hybridity (mixing) to challenge the colonial production of binary oppositions (centre/margin, civilized/savage, enlightened/ignorant), suggesting that cultures interact, transgress and transform each other in a much more complex manner than binary oppositions allow. In other words, imperialism was not simply a case of the colonizer transforming the colonized. Rather, the coming together of the two meant that hybrid spaces and cultures were created in the colonies, even if the power relations within these spaces were always unequal. Thus, hybridity has the potential to intervene and dislocate processes of domination through re-interpreting and redeploying dominant discourses; the spaces where differences meet become important. According to Bhabha (1994), colonial power should be understood as a 'production of hybridization rather than the noisy command of colonialist authority or the silent repression of native traditions'. The incorporation of hundreds of words from different Indian languages into English is one simple but obvious example of such hybridization. Examples include: bangle, bazaar, caravan, cot, jungle, juggernaut, pajama, pundit, shampoo, thug (from Hindi); atoll, aubergine, avatar, bandana, candy, cash, dinghy, karma, sugar (from Sanskrit); catamaran, curry, mantra, pariah (from Tamil); bungalow, khaki (from Urdu).

In his well-known essay, 'Signs taken for wonders' (1985), Bhabha uses the example of the dissemination of the Bible in India to explain his understanding of hybridization as produced by colonial power, demonstrating how the supposedly unchangeable 'Word of God' was transformed through Indian interpretations, readings and usages. Colonial power, its message and its authority are revealed as ambivalent and fractured, since the Bible was often not read, but sold or bartered or used as waste or wrapping paper. In addition, the vegetarian Hindu reading of the Christian communion in terms of cannibalism challenged the superiority of the British colonizers. It asked questions such as 'How can the word of God come from the flesh-eating mouths of the English?' and 'How can it be the European Book when we believe it is God's gift to us?' (Bhabha 1994: 102–22). Thus, in the very practice of domination, the language of the colonizing master becomes hybrid – neither one thing nor the other. Colonial authority undermines itself by not being able to replicate its own self perfectly: 'the colonial presence is always ambivalent, split between its appearance as original and authoritative and its articulation as repetition and difference' (Bhabha 1985: 150). This gap marks a failure of colonialism and becomes a site for resistance (Loomba 1998: 177).

Resistance, for Bhabha, takes shape through mimicry. In Fanon's writings, colonial authority works by inviting black subjects to mimic white culture; for Bhabha such an invitation itself undercuts colonial hegemony and undermines authority. In celebrating cultural heterogeneity, Bhabha has been criticized for relying on notions of original and distinct cultures. He has also been accused of universalizing and generalizing the colonial encounter – the split, ambivalent, hybrid colonial subject is in fact 'curiously universal and homogeneous . . . he could exist anywhere in the colonial world' (Loomba 1998: 178). The notion of hybridity has also been criticized for downplaying the bitter tension and clashes between colonizers and colonized, and of misrepresenting the dynamics of anti-colonial struggle. Nationalist struggles and pan-nationalist movements (e.g. négritude) were fuelled by the anger and alienation of the colonized. However, Bhabha's emphasis on the subversive effects of hybridization has been influential. His reworking of Said's notion of discursive domination, by highlighting the formation of colonial subjectivities as a process that is never fully or perfectly achieved, has been particularly important. It has drawn attention to the agency of the colonized in ways that were absent in *Orientalism*. For Bhabha, colonial discourses were neither fixed nor all-powerful, but became diluted and hybridized, so that the identities of both colonizer and colonized were always unstable and intertwined. His work is useful in de-colonizing development because of its concern with shifting theory-making to an 'ex-centric site' (1994: 6), disrupting the idea that theory emerges in the West and exploring the agency in knowledge-making of most of the world's population, whose concepts of development and modernity do not originate with the European Enlightenment. There are thus similarities with Chakrabarty's attempts to 'provincialize Europe' (see below).

## Gayatri Chakravorty Spivak and the subaltern

Gayatri Spivak has perhaps made the most innovative and substantial contributions to postcolonial forms of cultural analysis. She has been referred to as a 'Marxist-feminist-deconstructionist', since she sees each of these approaches as necessary but insufficient by themselves, always in tension, yet productive together. Her overriding ethico-political concern has been the tendency of institutional and cultural discourses and practices to exclude and marginalize the subaltern, especially subaltern women. She thus adds an important corrective to Said and Bhabha's work, which largely occludes the significance of class and gender. Her work is considered difficult, in part because of a dense writing style (as with Bhabha) and, in part, because she transgresses disciplinary boundaries. However, one of her concerns is that the academy is constituted into discrete disciplines so as to be unable to address the most serious of global questions (think of the inter-disciplinary approach required to address the causes and consequences of climate change, for example).

Spivak's response is to relate the micro-politics of the academy to the macro-narrative of imperialism/global capitalism. She sees teaching as a political act and explores the links between the development of the university in the North and forms of exploitation in the South. Like many postcolonial theorists, the notion that power is knowledge and knowledge is power is central to her work. Through deconstruction, Spivak is concerned to expose the global division of labour, the gendering of this with the super-exploitation of women, and the links between the academy and global exploitation. (Deconstruction is a process usually associated with French philosopher Jacques Derrida, by which the texts and languages of western philosophy appear to shift and complicate in meaning when read against the assumptions and absences they reveal within themselves.) As she states in the abstract to 'Can the Subaltern Speak?' (1988a), an essay that encapsulates much of her thinking: 'An understanding of contemporary relations of power, and of the Western intellectual's role within them, requires an examination of the intersection of a theory of representation and the political economy of global capitalism'.

Spivak shares Said's concern with Eurocentric and elite knowledge, but her focus is less on this knowledge per se (in contrast with Said's reading of western literature) and more on its effect on the subaltern. Four elements of Spivak's concern with the subaltern have broader ramifications for how knowledge is constructed about other places and peoples:

1 As a political project – Spivak argues that attempts to understand subaltern classes only in terms of their adequation to European models has been deeply destructive (she refers to this as epistemic violence). Her aim is to allow the subaltern consciousness to find expression, which will then inflect and produce forms of political liberation (also bypassing European models).

2 The politics of teaching – students must confront the problems of the world but should also understand their own position in relation to these problems, which in the case of the relatively privileged and elite involves complicity in the perpetuation of power relations and continued epistemic violence. Spivak suggests that to confront this complicity involves being aware of it and vigilant about its potential effects. Avoiding it requires:

3 'Unlearning one's privilege as one's loss' – this requires a double recognition. First, recognizing that one's race, ethnicity, class, gender and nationality create relative privilege. Second, having recognized one's relative privileges, recognizing one's own prejudice and learned responses that are conditioned by these privileges. In short, our relative privileges have given us a limited knowledge, but have prevented us from gaining other knowledge, which we are not equipped to understand. This is very different from western rationalism which, since the Enlightenment, has deemed that the world is knowable through observation. Spivak argues that there is some knowledge, experiences and modes of existence that are closed off from the privileged view. This is only resolvable by:

4 Learning to learn – having unlearned one's privileges as loss, one should learn to learn anew, which opens up the possibility of

gaining knowledge of others. This involves an ethical relationship with the other that facilitates the opening of spaces where those others can have a voice and where they can be taken seriously. Spivak argues that this is a task for everyone, including herself. She is, after all, also in a position of relative privilege as a Brahmin scholar and member of the Indian elite, educated in the USA and now working in the US academy.

The implication of Spivak's arguments is a need for continual questioning and reflexivity about one's relative position, and vigilance of the potential for epistemic violence in our attempts to create knowledge about and to represent the subaltern.

Spivak's most influential (and controversial) essay, 'Can the subaltern speak?' (1988a) explores whether subalterns can speak themselves or are condemned only to be represented and spoken for in a necessarily distorted fashion. Spivak locates the problem, not with the inability of the subaltern to speak, but with the unwillingness of the culturally dominant to listen. In this case, the culturally dominant are not only the British colonizers, but the Indian elites who also oppress the subaltern subject. Spivak's intention, therefore, is to challenge the simple division between colonizers and colonized by inserting the 'brown woman' as a category oppressed by both (Loomba 1998: 234). Many readers have found this essay difficult to read and understand, which in part explains why Spivak's main arguments have sometimes been misinterpreted. However, it is one of the most influential essays on the proclivity of dominant discourses and institutions to marginalize and disempower the Third World subaltern. Its primary concern is to examine the ethics and politics involved in the process of othering, focusing specifically on the representation of the South. The key points can be summarized thus:

1   Even progressive western intellectuals, notably Foucault and
    Deleuze, who understand the relationship between discourse and
    power still engage in gross universalizations when speaking on
    behalf of the Third World 'masses' or referring to '*the* workers'
    struggle'. Even a radical critic like Foucault is prone to believing
    that oppressed subjects can speak for themselves, because he
    has no conception of the repressive power of colonialism and
    patriarchy. Such critics fail to recognize the significance of the
    international division of labour, which Spivak contends renders

universalizations such as 'the worker' meaningless, and the fact that they are favourably but unconsciously located within this.

2   Alternative representations (e.g. from within the South) are similarly problematic. Spivak uses the example of widow-sacrifice (*sati*) in colonial India. The British tried to abolish this practice as part of their 'civilizing mission' in a move that Spivak famously describes as 'White men saving brown women from brown men' (p. 297). The dominant, patriarchal Hindu position was to excuse the practice as acceptable because widows wanted to die – it was a courageous and pure act. Spivak's point is that one voice is completely ignored in these representations – that of the widow herself. She is caught between patriarchy and imperialism and disempowered and marginalized by both – she has no space in which she can speak.

3   Even when the subaltern attempts to speak, *she cannot be heard*. Spivak uses the case of Bhuvaneswari Bhaduri, whose suicide in 1926 is interpreted as *sati* resulting from illicit love. Bhaduri is represented as committing suicide because she was pregnant, even though she deliberately took her life whilst menstruating in anticipation of this representation. Spivak argues that the erasure of Bhaduri's 'voice' is an act of epistemic violence. It is also illustrative of the violence that denies the subaltern a voice, not because they are unable to speak, but because they are both denied space in which to speak and when they do speak they are not listened to.

The crux of Spivak's argument is an analysis of two related but different meanings of 'representation' – *speaking for* (as in the sense of political representation) and *speaking about* (as in describing, portraying or representing). The British attempt to abolish *sati*, for example, conflated the two types of representation in *speaking for* the Indian woman (in other words, representing her politically) and *speaking about* her (portraying the Indian woman as she 'really is'). Not only does this silence the Indian woman, it also masks the complicity of the British in this representational process, which is fundamentally a relationship of power. Indian elites are also complicit in this process, as are even those practitioners of the critique of subjectivity and 'the best prophets of heterogeneity and the Other' (western intellectuals such as Foucault and Deleuze), since both claim authority to speak for and speak about subaltern subjects (discussed further in Chapter 6). Spivak's intention is to challenge the easy assumption that the postcolonial scholar can recover the standpoint of the subaltern (Loomba 1998: 234). At the same time, she takes seriously the desire of postcolonial intellectuals to reveal injustice and oppression. She suggests that scholars adopt the Gramscian maxim – 'pessimism of the intellect, optimism of the will' – to combine scepticism of the possibilities of recovering subaltern agency with the political commitment to make visible the position of the marginalized. It might be impossible, but the only ethical position is to try.

In *A Critique of Postcolonial Reason* (1999), Spivak explores how European philosophers not only tend to exclude the subaltern, but actively prevent non-Europeans from occupying positions as fully human subjects. She explores the links between discourse and the inequities of global capitalism, in which women in the South are positioned as 'super-exploited' labour. She also coined the term 'strategic essentialism' to refer to a means of overcoming difference in temporary solidarity for the purpose of social action, a notion of particular significance for feminists and women activists.

## Critiques of postcolonial theory in relation to the question of development

Despite its significance in critiquing Eurocentrism and neocolonialism, postcolonialism has been subject to a great deal of criticism. Some critics (e.g. Kaiwar 2015) point to a periodization marking two distinct phases of postcolonial theory, or to two distinct generations of postcolonial theorists (e.g. Lazarus 2011). The first phase – inspired by Lazarus's first-generation writers such as Fanon and Achebe, liberation struggles around the world and radical theories, such as Marxism – occurred in the two decades or so after the end of formal colonialism in many African and Asian countries from the late 1950s when the 'postcolonial' was concerned with agendas of anti-colonialism, development and distributive justice. The second phase – inspired by a second generation of poststructuralist theorists from the late 1970s onwards – tended to eschew 'economism' and the 'metanarratives' of Marxist critiques of capital and focused instead on critiquing Eurocentrism and resisting universalism. The latter approach, critics argue, now dominates postcolonial theory and has been accused of becoming institutionalized, representing the interests of a western-based, metropolitan, intellectual elite (for example, Bhabha, Said and

Spivak were employed at universities in Europe and/or the USA when they made their most influential interventions) who speak the language of the contemporary western academy, perpetuating the exclusion of the colonized and oppressed (Ahmad, 1992; McClintock, 1992; Watts, 1995; Chibber 2013; Kaiwar 2015). This shift, it is argued, is apparent in the only limited reference to earlier traditions of criticism in what are now considered the key texts of postcolonial theory – of Fanon in Bhabha, of Fanon, Césaire and Middle Eastern scholars in Said's *Orientalism*, of Ngũgĩ and Achebe in Spivak's writings – which gives the impression that these key texts are almost exclusively located in western social theory (Moore-Gilbert 1997: 16).

There is some credence to this periodization in that it captures the changing theoretical influences on postcolonialism and some of the distinctions that we might draw between postcolonial theory located in the western academy and decolonial theory in, for example, Latin American contexts. However, there are several problems with recent critiques of postcolonialism. First, while much of the criticism from the Left shares a commitment with postcolonial theory to critique Eurocentric liberalism, racism and imperialism, it departs from it by insisting that postcolonial theory neglects 'subaltern' Europe – 'the struggles of poor and working people, of migrants and minorities, and of women' (Jani 2017: 112). In contrast, postcolonial theorists have tended to assert that the experiences of oppression in former colonies are different precisely because it was enacted and perpetuated through imperial power relations. While capitalism may have been the common structural force, exploitation of labour through slavery in the Congo was quite distinct from exploitation of labour under industrialization in Europe and had quite different outcomes; Europe's marginalized are thus not 'subaltern' in the same way as post-colonial subjects, as Spivak makes clear. Furthermore, the contemporary struggles of migrants and minorities in Europe cannot be understood through class relations alone, since those experiences are profoundly shaped by the fact that migrant and minority bodies are gendered, racialized, queered and so on – the very differences that postcolonial theory brings to the foreground.

A second problem with Marxist critiques of postcolonialism is that they tend to collapse quite different theorists and theories into a singular 'postcolonial theory'. Kaiwar's periodization, for example, places Said in the same postcolonial 'camp' as Spivak and Bhabha,

yet even on a theme considered core to postcolonial theory –
representation – all three differ in significant ways. In *Location
of* Culture, Bhabha (1994: 171, 173) positions his concern with
representation as post-Marxist:

> Postcolonial criticism bears witness to the unequal and uneven
> forces of cultural representation involved in the contest for political
> and social authority within the modern world order. Postcolonial
> perspectives emerge from the colonial testimony of Third World
> countries and the discourses of 'minorities' . . . [It] departs from the
> traditions of the sociology of underdevelopment of 'dependency'
> theory . . . The postcolonial perspective resists the attempt at holistic
> forms of social explanation.

Spivak is also post-Marxist, but unlike Bhabha, she draws more clearly
on Derrida to argue that the subaltern cannot speak for herself because
the subaltern is always a power/knowledge construct: always the
object of discourse, never the subject. Said's theory of representation
is quite different again and shifts over time. As Lazarus (2011: 197)
argues, in *Orientalism*, Said draws on Foucault, but vacillates between
theorizing colonial discourse as ideological misrepresentation (i.e.
that a real 'Orient' exists external to Orientalist discourse) and power/
knowledge construct (i.e. a more clearly poststructuralist position
that the Orient can only be understood as a discursive construction).
Lazarus (*ibid.*) also points out that in his later writings, Said sought to
distance himself from poststructuralism and 'wrote from premises and
on behalf of principles quite different from those generally prevailing
in postcolonialism studies'. Specifically, whereas Spivak sees all
acts of representation as potentially silencing, Said's commitment
to critical humanism also shaped his views on the responsibility of
the intellectual to use their status and privileges to speak on behalf
of those who need or request representation (see Said 1994). These
different views of representation have important implications for
decolonizing development studies. However, the point here is
that deploying periodization to critique postcolonial theory risks
generalization and caricaturing what is a complex and diverse body
of critical theory.

As a body of theory that might be applied to development, some
critics suggest that greater theoretical sophistication has created
greater obfuscation; postcolonialism is too theoretical and not rooted
enough in material concerns (Ahmad 1992). Emphasis on discourse

and representation is accused of detracting from an assessment of the material ways in which colonial power relations persist, not enough consideration is given to the relationship between postcolonialism and global capitalism, and economic relations and their effects elude representation in much of postcolonial studies. Some critics berate postcolonial theory for ignoring urgent life-or-death questions, including questions of inequality of power and control of resources, human rights, global exploitation of labour, child prostitution and genocide. They argue that postcolonial theory cannot easily be translated into action on the ground and its oppositional stance has not had much impact on the power imbalances between North and South. Critics accuse it of being preoccupied with the past, while new forms of Orientalism continue to disadvantage large parts of the world. However, many of these criticisms are overstated. As Lazarus (2011: 10) argues, like other radical critiques such as Marxism, postcolonialism has:

> stood as a firm opponent of 'mainstream' or politically institutionalized anti-liberationism, as expressed both in the frankly imperialist language of leading policy makers and intellectuals in the core capitalist states, and through the punitive policies enacted by such corporate agencies as the World Bank, the International Monetary Fund, and the World Trade Organization.

Criticisms are also overstated when considering the role that Marxism has played in much of postcolonial thinking and the fact that its origins were primarily in modes of literary and cultural analysis. In his later work, Said (2004: 142) was explicit in his concerns and the directions in which he felt postcolonial studies ought to go:

> In the various contests over justice and human rights that so many of us feel we have joined, there needs to be a component to our engagement that stresses the need for the redistribution of resources and that advocates the theoretical imperative against the huge accumulations of power and capital that so distort human life.

In addition, what has become known as postcolonial theory is located within a broader field than English literary studies. Dipesh Chakrabarty, for example, is a notable historian and original member of the Subaltern Studies group. Like Spivak, he is concerned with institutional power, but he differs by focusing attention back on the power of Eurocentrism. He writes (1992: 1–3):

> Insofar as the academic discourse of history . . . is concerned, 'Europe' remains the sovereign, theoretical subject of all histories, including

the one we call 'Indian', 'Chinese', 'Kenyan' and so on . . . Third-world historians feel a need to refer to works in European history, historians of Europe do not feel any need to reciprocate . . . The everyday paradox of third-world social science is that we find these theories, in spite of their inherent ignorance of 'us' eminently useful in understanding our societies. What allowed the modern European sages to develop such clairvoyance with regard to societies of which they were empirically ignorant? Why cannot we, once again, return the gaze?

Chakrabarty's aim is to 'provincialize Europe' – rather than to see Europe as occupying a central position in the generation of theory and knowledge, to reposition it as one centre among many in the production of knowledge forms. Like postcolonialism generally, this is not a rejection of western knowledge, its universalisms and its grand theories. Nor is it a call for cultural relativism or a notion that all forms of knowledge are equally valid. Rather, it requires the positioning of European knowledge within its historical context and resisting the temptation to universalize from particularities. However, there are difficulties in provincializing Europe because academic disciplines, including development studies, are inextricably bound to their European cradle.

Although Bhabha, Spivak, Said and Chakrabarty write from the point of view of literary/postcolonial critics, subsequent chapters demonstrate the significance and pertinence of their work for the field of development. Many who work in this field struggle with the same dilemmas raised by postcolonial theorists. These dilemmas include questions such as:

- What are the ethico-political implications of representations for the South, and especially for subaltern groups that form the focus of much of development-related work (Kapoor 2004: 628)?
- By what right and on whose authority does one claim to speak on behalf of others?
- On whose terms is space created in which these others are allowed to speak? Are we merely trying to incorporate and subsume non-western voices into our own canons?
- To what extent do our depictions and actions marginalize or silence these groups or mask our own complicities (perpetuating epistemic violence)?
- What social and institutional power relations do these representations, even those aimed at 'empowerment' set up or neglect (Kapoor 2004: 628)?

- To what extent can we attenuate these pitfalls?
- Postcolonialism attempts to overcome inequality by opening spaces for the voice of the non-western academic and what Spivak refers to as the 'native informant' to be heard. How do we solve the problem that poverty and lack of technology (e.g. books, email, fax machines) make this increasingly difficult?
- Is it possible within development studies to provincialize Europe?

Postcolonial theories have taught us that our discursive constructions are intimately shaped by our positioning, in terms of socio-economic factors, gender, culture, geography, history and institutional location. This positioning is unavoidable, but what Spivak demands that we have a heightened sense of self-reflexivity about how this positioning shapes our representations and with what effect for those we seek to represent (either politically by speaking for, or discursively by speaking about). As we shall see, these concerns are of utmost significance in development studies and practice:

- by focusing upon the conventions of writing and representation by which Northern disciplines and institutions 'make sense' of the world;
- by challenging the truth claims of modernism and demonstrating that the production of western knowledge is inseparable from the exercise of western power;
- through the growing struggle to loosen the power of western knowledge and reassert the value of alternative experiences and ways of knowing.

As we shall also see, these concerns open important spaces for consideration of the divergences and possibilities of dialogue with decolonial theory.

## Postcolonialism and decolonial theory

According to Achille Mbembe, contemporary calls to decolonize society are underpinned by two related but distinct approaches. The first, into which much of postcolonial theory can be located, seeks to critique the production and promotion of knowledge based on European traditions and experiences. It is the fight against 'epistemic coloniality': the endless production of theories that are based on

European traditions and produced by Europeans, 'a particular anthropological knowledge, which is a process of knowing about Others – but a process that never fully acknowledges these Others as thinking and knowledge-producing subjects' (Mbembe 2016: 36). The second, into which much of decolonial theory might be located, attempts to envisage alternatives to European traditions and experiences and demands that decolonizing efforts go beyond critique and towards the removal of enduring forms of colonial domination. Many decolonial theorists argue that these two approaches are incommensurable. This is because those scholars located in the global North face a double bind: decolonization as a force to dismantle the power structures of modernity can never be achieved from within its own theoretical orthodoxies and infrastructures. Or, as Audre Lorde (1984) famously put it, 'The master's tools will never dismantle the master's house'.

One problem lies in the way in which the language of decolonization has been superficially adopted in the social sciences and conflated with debates about social justice and civil rights. Decolonization is not simply a formal process of handing over the instruments of power, but is a 'long-term process involving the bureaucratic, cultural, linguistic and psychological divesting of colonial power' (Tuhiwai Smith 2012: 101). As Tuck and Yang (2012: 2, 7) argue: 'decolonization wants something different' and, in the settler colonial context, must involve the repatriation of land, '*all* of the land, and not just symbolically'. For this reason, decolonization implicates everyone and is profoundly unsettling (in the dual sense of reversing colonial settlement and being discomfiting). Tuck and Yang are critical of the way in which decolonization has become a metaphor, for example in debates about decolonizing knowledge, or decolonizing development: 'when a metaphor invades decolonisation, it kills the very possibility of decolonisation; it recentres whiteness, it resettles theory, it extends innocence to the settler, it entertains a settler future' (*ibid.*: 3). Decolonizing knowledge should be considered a means to the end of colonization, not an end in itself (Essen *et al.*: 2017). The aims of those who use decolonization as a metaphor without acknowledging that the terms of the debate about decoloniality and decolonization should be determined by those on the margins are thus incommensurable with the aims of those who seek decolonization in material terms. Decolonization as metaphor 'represent(s) settler fantasies of easier paths to reconciliation' (*ibid.*: 4) and ignores the reality that, for those whose lands and resources were stolen, reconciliation may not be the desired objective. This premature attempt at reconciliation (see Chapter 1) is a form of anxiety and 'is just as relentless as the desire to disappear the Native; it is a desire to not have to deal with' the problem of indigenous peoples who in material terms are still living under settler colonialism (*ibid.*: 9). It is also a move to innocence that attempts to 'relieve the settler of feelings of guilt or responsibility without giving up land or power or privilege, without having to change much at all' (*ibid.*: 10).

These critiques are an important reminder that attempts to decolonize discourses and practices from positions of relative privilege are not commensurable with the desire of those living under (neo) colonialism for material decolonization. Decolonizing knowledge and cultivating a critical consciousness about the relationship between power and knowledge – important as this is – does not equate with the more difficult politics of relinquishing stolen land and resources. As Povinelli (2011b; 2014) puts it, the politics of recognition does not equate with the politics of redistribution and may perversely be a strategic containment of potentially more radical futures. This fundamental difference *ought* to be discomfiting to those of us in positions of relative privilege concerned with decolonizing knowledge. We are being asked to consider how the pursuit of social justice through a critical enlightenment can also be a move to innocence by those who remain privileged by neo-colonial relationships and structures of power. We are being asked to reflect on whether 'our' concerns are diversions and distractions, which relieve us of 'feelings of guilt or responsibility and conceal the need to give up land or power or privilege' (*ibid.* 21). However, incommensurability does not preclude the possibilities of provisional solidarity and/or activism:

> We, at least in part, want others to join us in these efforts, so that settler colonial structuring and Indigenous critiques of the structuring are no longer rendered visible. Yet this joining cannot be too easy, too open, too settled. Solidarity is an uneasy, reserved, and unsettled matter that neither reconciles present grievances nor forecloses future conflict.
>
> (Tuck and Yang 2012: 4)

The contention in this book is that both postcolonial theory and decolonial theory can inform the critique of development whilst always being understood to be in tension. Decolonial theory is located in the realm of material decolonization and is articulated by theorists who are positioned politically and personally with decolonial agendas. These personal and political struggles are different to and perhaps incommensurate with the concerns of theorists who are positioned differently in relation to neocolonial power relations and who work from postcolonial perspectives (Noxolo 2017). However, as we shall see, postcolonial perspectives are not confined to the realm of the metaphorical. Rather, they provide important critiques of the material conditions that continue to entrench colonial and imperial power relations and give rise to profound inequalities and injustices.

Postcolonial theory is an important critique of Eurocentric modernity, which obscured 'the specificities of race and place, and invisibilized other epistemes to masquerade as universal and total' (Asher 2013: 832), but it is only one body of critique. Other critiques of Eurocentric modernity share similarities with postcolonial theory but are also differentiated in important ways. For example, as Asher (2013: 833) argues, Latin American decolonial theorists differentiate their work from postcolonial studies and the metropolitan knowledge of South Asian, African and Middle Eastern scholars in three main ways. First, they make a distinction by not starting with the eighteenth-century European Enlightenment, but with the beginning of the Conquest of the Americas in 1492 and its formative role in modernity. Second, they aim to go beyond critique and deconstruction to foster decolonial thinking and politics. Third, their liberation politics emerge from the world views of exploited and marginal groups rather than from privileged institutions of higher education. A further difference is highlighted by Mignolo (2007a, 2008) and Coronil (2008), who contend that postcolonial critiques of modernity emerge from *within* the modernist project and remain bound to modernity's Eurocentrism even as they critique it. Thus, while Said offers an important corrective to Marxist critiques by demonstrating how Europe discursively constructed the Orient, he is silent on the foundational role of Latin America in constituting Europe. Another important divergence is that although postcolonial theory addresses the material and socio-economic, it does so largely through the realm of the cultural; in contrast, the decolonial theory of sociologists Anibal Quijano and María Lugones and philosopher Walter Mignola is linked strongly to world systems theory and theories of development and underdevelopment (Bhambra 2014). These differences explain the lack of dialogue between the contemporaneous writings of postcolonial theorists such as Bhabha, Said, Spivak and decolonial theorists such as Dussel, Mignolo, Quijano, which is compounded by the fact that the latter were writing in Spanish and it is only relatively recently that their works have been translated into English. However, their political projects are also seen to differ.

According to Mignolo (2007b: 452), postcolonial theory is a 'project of scholarly transformation within the academy' and thus is grounded in Eurocentric poststructural theory (Foucault, Derrida and so on). In contrast, he argues that decolonial theory is a project of delinking from Eurocentrism and draws on non-European sources beyond the metropolitan academy. Examples include: Waman Puma de Ayala, the sixteenth-century Quechuan writer who documented and denounced the brutalities inflicted on the indigenous peoples of the Andes by the Spanish Conquistadors; the anti-colonial activism of Gandhi; the radical political and epistemological shifts enacted by Cabral, Césaire, Fanon, Anzaldúa. However, it is the contention here that postcolonial and decolonial theory need not be considered entirely conflictual and that in the field of development both bring insights that are capable of effecting radical change in theory and practice. As Asher (2013) points out, this is to some extent apparent in Escobar's writing, whose critique of development as colonialist in *Encountering Development* (1995) drew inspiration from Foucault, Said and Bhabha, among others, but whose more recent work (Escobar 2008, 2010) draws more on the 'ground-up' knowledge of social movements to critique development. However, while Escobar's decolonial politics are articulated around the urgent need to *make* the subaltern speak, he 'still ignores the problem of representation' (Asher 2013: 838; see also Chapter 6). Thus, Spivak's insistence on the need to 'grapple with the complexities of representing the subaltern and what circumscribes their speech and reception' (*ibid.*) remains critically important. In addition, there is a need for what Mollett (2017: 5) refers to as 'postcolonial intersectionality' to address the erasure of gender and race in accounts of social movements, and to create a fuller picture of the 'multiple and simultaneous struggles over livelihoods, access to natural resources and the right to be recognized as *human*'.

As Bhambra (2014: 119) argues, 'both postcolonialism and decoloniality are developments within the broader politics of knowledge production and both emerge out of political developments contesting the colonial world order established by European empires, albeit in relation to different time periods and different geographical orientations'. Both are concerned with contemporary global inequalities and the historical basis of their emergence. As Mignolo (2008) argues, understanding these global inequalities requires interrogating the colonial matrix of power that combines discourses of modernity (progress, development, growth) and the logic of coloniality (poverty, misery, inequality). The contemporary and intertwined global challenges of uneven development and environmental catastrophe requires closer dialogue between postcolonial and decolonial theory precisely because they approach modernity and coloniality differently. These differences may shape diverse but allied efforts to translate postcolonial and decolonial theory into politics that can effect progressive change. And, as discussed in subsequent chapters, there is already much common ground between the two in the emphasis they place on learning from below, planetarity and pluriversality and the intersection of race and gender in political economy.

## Conclusion

While it is impossible to claim an implicit coherence for postcolonialism, its theories and approaches coalesce around particular intellectual concerns rooted in anti-colonialism and anti-racism (e.g. Du Bois, Plaatje, Garvey), decolonization (e.g. Ngũgĩ, Cabral, Fanon) and postcolonial diaspora (e.g. Said, Bhabha, Spivak, Chakrabarty). These concerns are 'interconnected in a loose patchwork of themes and approaches' (Young 2001: 92) and through overlapping ethical and political concerns. Though much has been achieved in terms of freeing the world from colonial domination, postcolonial theories assert that discursive power (which also translates into material power) still lies with the West. As Young (2001: 428) argues:

> [I]njustice, inequality, landlessness, exploitation, poverty, disease and famine remain the daily experience of much of the world's population . . . [Postcolonialism] . . . operates out of a knowledge that was formed through the realities of such conditions: its politics of power-knowledge asserts the will to change them.

This politics provides fertile ground for exploring the potential for productive dialogue between postcolonialism and its concerns with power-knowledge, decolonial theory with its focus on transformative political praxis, and development studies, which retains a political and ethical interest in dealing with injustice and inequality on a global scale. It is the contention here that development provides fertile terrain for engaging with postcolonial and decolonial theory to move towards:

> Renovation and reconstruction of critical thought in ways that take into account the present-day relations between culture, politics, and economy, challenge the hegemony of Eurocentric perspectives, and promote dialogues and *thinking with* thought and knowledge 'others', including that of Afro and indigenous social movements and intellectuals.
>
> (Walsh 2007: 233)

Subsequent chapters speculate on what postcolonialism, through an engagement with other critical theories, might bring to the decolonization of development and development studies.

## Summary

- This chapter details a small number of the many histories, rebellions, political resistances, cultural movements and intellectual and theoretical debates that emerged during the nineteenth and twentieth centuries. These include pan-Africanism, the Harlem Renaissance, négritude and Black Consciousness, anti-colonial literature, women's anti-colonial activism, feminism, Subaltern Studies, postcolonial theory and decoloniality.
- These various movements produced both anti-colonial politics, culminating in the ending of colonial domination in a remarkably short period of time, and anti-colonial theory.
- Today we know this anti-colonial theory as postcolonial theory, which alongside sustained political practices emerging from decolonial theory seeks to build on the 'rich inheritance' of the radical legacy of these earlier movements and theories – their 'political determination', their 'refusal to accept the status quo', their 'transformation of epistemologies' (or systems of knowledge) and their desire to create 'new forms of discursive and political power' (Young 2001: 428).
- Postcolonial approaches have potential implications for development studies in terms of how they make sense of and intervene in the world and create space for alternative development knowledge.

## Discussion questions

1 How were various anti-colonial movements connected with each other throughout the twentieth century?
2 How do these connections disrupt the notion that ideas, knowledge and practices flowed from imperial metropoles into colonial peripheries?
3 What were the main differences between assimilationist and anti-imperial approaches of the political and cultural movements up to the 1970s?
4 What is the meaning and usage of the term 'discourse' in postcolonial theory?
5 How does postcolonial theory understand the relationship between power and knowledge?
6 What would you consider to be the key elements of contemporary postcolonial approaches that are of relevance to development studies?
7 What are the tensions between postcolonial and decolonial theories and how might these be navigated?

## Further reading

Haddour, A. (2006) 'Foreword: postcolonial Fanonism'. In *The Fanon Reader* London, Pluto Press, pp. vii–xxv. A succinct analysis of Fanon's changing philosophies and an excellent introduction to the key elements of his work (contained in the rest of the *Reader*).

Loomba, A. (1998) *Colonialism/Postcolonialism*, London, Routledge. Clearly written overview of the meanings of postcolonialism and the principal elements of postcolonial approaches.

Moore-Gilbert, B. (1997) *Postcolonial Theory. Contexts, Practices, Politics*, London, Verso. A comprehensive and accessible survey of postcolonial theory, with a careful analysis of the main arguments by and critiques of Spivak, Said and Bhabha.

Moraña, M.D. *et al.* (eds) (2008) *Coloniality at Large: Latin America and the Postcolonial Debate* Durham, Duke University Press. An excellent review of debates about how anti-colonial and postcolonial theories, including the work of the South Asian Subaltern Studies collective, relate to diverse Latin American colonial experiences.

Said, E. (1985), *Orientalism*, Harmondsworth, Penguin (first published 1978). Considered the foundational text in postcolonial theory and a must-read for anyone interested in postcolonialism.

Spivak, G. (1988) 'Can the subaltern speak?'. In C. Nelson and L. Grossberg (eds), *Marxism and Interpretation of Culture* Chicago, University of Illinois Press, pp. 271–313. Not an easy read, but I recommend attempting it, having read the various summaries in this chapter.

Tuck, E. and Yang, W. (2012) 'Decolonization is not a metaphor' *Decolonization: Indigeneity, Education, & Society* 1, 1, pp. 1–40. A powerful critique of the ease with which the language of decolonization can be appropriated in ways that re-centre white settler perspectives and a reminder that decolonization in settler societies requires something far more profound than other civil or human rights-based projects.

Young, R. (2001) *Postcolonialism: An Historical Introduction* Oxford, Blackwell. A comprehensive and accessible account of the diversity of anti-colonial political and cultural movements across the world and the variety of anti-colonial theories that they inspired. A much broader rendition of the origins of postcolonialism than one usually finds.

## Useful websites

www.lib.virginia.edu/area-studies/subaltern/ssmap.htm Bibliography of the writings of the Subaltern Studies group, 1988–2000.

http://frantzfanonfoundation-fondationfrantzfanon.com/ Website with links to the works of Frantz Fanon, analysis and debate inspired by his work, and links to other anti-colonialist thinking.

www.southernperspectives.net/ Website based in Australia/New Zealand promoting a south-south dialogue of ideas and profiling individuals and organizations that explore a southern perspective on a broad range of disciplines, including creative arts, humanities, professions, social and physical sciences.

https://globalsocialtheory.org/topics/decoloniality/ Introduction to the key ideas of the research collective on Decoloniality, organized by Walter Mignolo and Arturo Escobar, which brings together scholars of Latin American/European origin working in universities in the United States and Latin America and interested in ideas of dependency theory, colonialism, gender and critical theory.

# **3** A postcolonial history of development

## Introduction

As we saw in Chapter 1, until recently there has been little dialogue between postcolonialism and development, reflecting differences in disciplinary traditions, politics, wariness over motives and divergences in the languages and concepts used to articulate core issues. The main mutual criticisms are that development studies tends to impose ideas and practices without listening to those people in the South who become objects of development interventions, while postcolonial studies neglects material issues such as poverty. As Sylvester (1999: 703) puts it: 'development studies do not tend to listen to subalterns and postcolonial studies do not tend to concern itself with whether the subaltern is eating'. Similarly, Sharp and Briggs (2006: 6) argue:

> For many who have spent a professional lifetime working in 'development', either as theorists or practitioners, postcolonialism is seen to offer overly complex theories which are largely ignorant of the real problems characterizing everyday life in the global South. Alternatively, many postcolonial theorists consider development studies still to be mired in Eurocentric, modernist, and/or neocolonialist mindsets.

This chapter explores how the politics of postcolonialism diverge sharply from other critical theories of development. Although it shares similarities with dependency theories, its radicalism rejects established agendas and accustomed ways of seeing. This means that postcolonialism is a powerful critique of development (as an idea and as a practice), fuelling the mutual distrust between the two approaches. However, it is precisely because of these divergent traditions that a dialogue between development studies and postcolonialism offers significant potential for an alternative conceptualization of development.

## The origins of development

Development can be thought of as a 'continuous intellectual project as well as an ongoing material process' (Apter 1987: 7), rather than something that is measurable or quantifiable. This is not to suggest that defining and measuring development is insignificant, since this has been central to post-war development theory and policy. However, a postcolonial approach to the history of development theory places emphasis on the nature and consequences of development as a set of ideas, discourses and practices. As we saw in Chapter 1, prior to the nineteenth century development was seen as a natural, immanent, evolutionary process without intentionality. However, from the nineteenth century onwards development also began to be thought of as an intentional practice – 'a means to create order out of the social disorder of rapid urbanization, poverty and unemployment' (Cowen and Shenton 1995: 32) caused by capitalist development, or a 'will to improve' (Li 2007). The notion of development as intentional practice has shaped the development industry from the 1950s onwards, but development discourses have never been entirely free from ideas of development as an immanent process. The seemingly opposing ideas of immanent (evolutionary) progress and intentional development have often coexisted, rather than being two separate poles at which development theorists can be placed (Baaz 2005). There are two important points to note. First, as this chapter goes on to explore, development as a form of intervention in the global South does not begin when colonialism ends; nor does colonialism as a set of processes in the global South end when development begins. Second, development has never been solely a form of intervention in the global South. Rather, as Silvey and Rankin (2011) argue, development is a way of dealing with capitalism's 'surplus people' wherever they are found. The massive racialized incarceration and the prison-industrial complex is arguably a solution to uneven development in the USA (Gilmore 2007). In perhaps a more progressive vein, experiments in the provision of a universal basic income as a potential solution in the shift to a 'post-work' world are being tried in locations as diverse as India and countries in the European Union (see Chapter 5).

Crucial to understanding the interwoven nature of capitalism and development is recognition that the origins of development as an idea and a process have a much longer history than development as an intentional practice, dating back beyond the nineteenth century to the

European Enlightenment and the beginnings of European modernity. The seventeenth century witnessed the emergence of new ideas about the relationship between humanity, society and nature and sought to challenge traditional worldviews that were dominated by Christianity (especially the Catholic Church). Reason was seen to triumph over faith and superstition:

> In the century of the Enlightenment, educated Europeans . . .
> experienced an expansive sense of power over nature and themselves:
> the pitiless cycles of epidemics, famines, risky life and early death,
> devastating war and uneasy peace – the treadmill of human existence –
> seemed to be yielding at last to the application of critical intelligence.
>
> (Gay 1973: 3)

Enlightenment philosophers, particularly in France, but also in Italy, Scotland and the Netherlands, embraced the notion of critical inquiry and the application of reason. The world became knowable through the gathering of empirical knowledge and thus controllable through technological and medical progress. The idea of development began not with a singular set of ideas or projects, but with this tendency towards critical enquiry about the organization and structure of societies. The metaphor of the 'light of reason' shining into the dark recesses of ignorance and superstition in traditional societies gained pre-eminence. Reason exploded the notion of the Divine Right of Kings that had maintained feudalism in Europe and fuelled a new desire for legal and constitutional reform. New forms of political organization (democracy, institutions, governance) were called for that could establish civil liberties and freedoms in rapidly changing societies. Central to this philosophy was a belief in universality – the idea that all human beings were essentially the same and notions of progress and freedom should be applied universally.

Development emerged from these early debates, especially within the Scottish Enlightenment and the work of Adam Smith (Figure 3.1). Smith is widely regarded as the founder of classical economics and was one of the earliest philosophers to attempt to systematically study the historical development of industry and commerce in Europe. His theories were laid out most famously in *The Wealth of Nations* (1776), which contained a sustained attack on the doctrines of mercantilism and provided one of the best-known intellectual rationales for free trade, capitalism and libertarianism.

**Figure 3.1** *Adam Smith (1723–1790)*

Mercantilism was based on the idea that the prosperity of a nation depended on its supply of capital – usually in the form of gold or silver bullion – and that the global volume of trade was unchangeable. Thus, governments should aim to retain bullion through a positive balance of trade with other nations (exports exceeding imports) and should advance these goals by playing a protectionist role in the economy, especially through the use of tariffs to discourage imports.

In contrast, Smith advocated free trade, vigorously attacking government restrictions that he believed hindered industrial expansion and tariffs that maintained inefficiencies, suppressed competition and thus led to higher prices. The 'invisible hand' of the market should control and regulate economies instead of government intervention. Smith's ideas influenced the more laissez-faire responses of European governments in the nineteenth century, even though these governments still adhered largely to mercantilist ideas of trade. However, his theories of development and progress became enormously influential in the articulation of neoliberalism after World War Two, when liberal notions of free trade and the emancipating potential of free, self-regulating markets came to dominate institutions such as the World Bank (WB) and the International Monetary Fund (IMF). Smith's theories were forged at the time of early industrialization and what was seen as a transition from tradition to modernity and progress within Europe – they were essentially Eurocentric. However, after 1945 they were seen as universal and applicable in very different contexts beyond Europe.

Enlightenment philosophy, such as that expounded by Adam Smith, was not always particularly emancipatory. One key concept related to the question of human beings harnessing nature and natural resources for social change, particularly during the late eighteenth and early nineteenth centuries. However, mastery of nature was characterized by relationships of power. The antidote to the disorder of industrialization was the capacity to use land, labour and capital in the interests of society, based on a notion of 'trusteeship'. However, only certain individuals could be entrusted with this role, meaning that property needed to be placed in hands of the 'trustees' (bankers and land owners, who were exclusively men) in order that they could control the effective use and distribution of society's resources. Peasants were seen as a threat to the trustees, but could be appeased through the effective use of resources so that society could evolve and revolution be avoided.

Despite the problematic relationship between trusteeship and inequities of power, this notion has arguably underpinned much of post-1945 development. The global financial institutions, based in the USA, are the trustees of the modern age, controlling the distribution of resources to maintain order and harnessing the light of improvement in societies in the global South. As former WB President, James Wolfensohn, argued in 1996, 'Knowledge is like light. Weightless and intangible, it can easily travel the world, enlightening the lives of people everywhere. Yet billions of people still live in the darkness of poverty – unnecessarily' (cited in Power 2003: 75). As Escobar (1995a: 2–4) puts it, the post-war development project can be seen as 'the last and failed attempt to complete the Enlightenment in Asia, Africa and Latin America'.

One implication of the Enlightenment is that it firmly planted in the popular imagination the idea of the West as superior and more advanced along a singular path to progress and modernity. Europeans were placed at the pinnacle of human achievement: civilization. During the nineteenth century, evolutionary theory also became central to conceptualizing social change. Social development or development of societies was likened to Darwinian notions of natural evolution. Development ideas thus depicted non-western societies as inferior, but as having potential to progress in an evolutionary fashion. These ideas, which can be traced back to the seventeenth century, shaped European colonialism, were globalized by European colonialism, and continued to have a profound influence on development theories and policies after 1945.

## Development, imperialism and colonialism

From the sixteenth century onwards, the primary way in which European powers sought to increase their wealth was through imperialism – the political control of peoples and territories by foreign states – and colonialism – the exercise of this political control through the establishment of settlements in foreign lands (Box 3.1). In tandem with these economic imperatives was the moral imperative of spreading the light of European 'civilization' to 'benighted' areas of the world.

Modern European empires far surpassed in extent and effects previous empires, for example, the Roman, Mongol, Aztec, Inca, Ottoman

## Box 3.1 Imperialism and colonialism

'Imperialism' and 'colonialism' are often used interchangeably, but there are important differences between them (see Ashcroft *et al.* 1998):

> **Imperialism** – a system of domination over space, encompassing 'the practice, the theory, and the attitudes of a dominating metropolitan centre ruling a distant territory' (Said 1993: 9). For example, the British involvement in west Africa during the nineteenth century was largely based on imperialism. Although Britain seized vast territories in west Africa, unlike other parts of Africa it was never effectively colonized. British administrations were rarely established in situ (apart from those small-scale administrations established in coastal regions by trading companies and missionaries), and there was no large-scale settlement. Unlike southern and eastern Africa, therefore, there are no colonial descendants living in this part of the continent today. However, west Africa was a central part of Britain's eighteenth and nineteenth century empire, a focal point for slavery and a key area for resource extraction (cocoa for the manufacture of chocolate, palm oil for the manufacture of soap, and rubber extraction). Imperialism, then, is a general term, which can refer to economic, political and cultural inequalities and dependencies whereby a country, region or group of people are subject to the rule of a separate and more powerful force. Imperial power can be exercised by numerous agents. Most obviously, nation states can wield imperial power, for example, Britain, France, Belgium and Germany in nineteenth century Africa. Companies can also act as imperial powers. For example, the British East India Company was responsible for the annexation of most of the Indian subcontinent along with the conquest of Hong Kong, Singapore, Malaya and other surrounding Asian countries. Organizations are also powerful actors. As we shall see, the regulatory power of the WB and the IMF is considered by some as a form of neo-imperialism.
>
> **Colonialism** – a tangible manifestation of imperial power; refers more specifically to 'the implanting of settlements on a distant territory' (Said 1993: 9). Colonialism is almost always a consequence of imperialism, depending on conquest, territorial expansion and the process of colonization whereby people, goods and capital move from the metropolitan centre to a colony. Examples include Spanish and Portuguese colonialism in Latin America; French colonialism in what is now Algeria and Morocco; and British colonialism in India, North America, South and East Africa, Australia, New Zealand and the Caribbean. Colonialism took different forms in different places. The Spanish, Portuguese and British colonization of the Americas was largely through the complete destruction

and subjugation of indigenous communities; the French colonization of Africa and Caribbean islands was based on an administrative system that legally incorporated these territories as part of France; the British colonization of India was achieved less through military force, although this was always a threat, and more through creating a hierarchical administrative structure that incorporated and co-opted Indian elites. Colonialism, then, is the imposition of political control through conquest and territorial expansion over people and places located at a distance from the metropolitan power. Both imperialism and colonialism bind metropolitan centres and colonies together in an unequal power and dependence. However, colonialism represents the direct imposition of imperial rule through settlement and political control over a 'separate group of people, who are viewed as subordinate, and their territories, which are presumed to be available to exploitation' (Jacobs 1996: 16).

and Chinese empires. Between the sixteenth and twentieth centuries, Belgium, Britain, France, Germany, Italy, the Netherlands, Portugal and Spain were all colonial powers, with influence extending over vast parts of the world. By the 1930s, 84.6 per cent of the globe had been or was controlled by these states; 'Only parts of Arabia, Persia, Afghanistan, Mongolia, Tibet, China, Siam and Japan had never been under formal European government' (Loomba 1998: xiii).

Unlike earlier empires, the colonial power of European countries represented the territorial expansion of capitalism, drawing distant lands into a capitalist world system of production, distribution and exchange. In 1915, Lenin famously referred to imperialism as the highest stage of capitalism and also the final stage, since he believed that the rivalry between European imperial powers would lead to war and the revolutionary overthrow of the capitalist system. Whilst he was correct about war – the First World War was largely an internecine war over territory between Europe's competing imperial powers – Lenin's prophecy of the demise of capitalism proved incorrect. Modern European empires were capitalist empires; capitalism was expanded through the conquest and possession of other people's land and labour in the service of the imperial powers. They not only extracted wealth and goods from colonies, but also restructured their economies. The main stages of modern imperialism can be summarized as follows:

1   Sixteenth century: a crisis in feudalism and the emergence of new European states prompted a search for new forms of revenue,

especially silver and gold, and the need to develop trading links. The major powers were Spain and Portugal; the major consequence was the conquest and destruction of indigenous civilizations and the imposition of Roman Catholicism in South America. Economic, social and cultural power worked alongside political power, with mercantilism providing the impetus for the first phase of colonial expansion.

2    Seventeenth century: mercantile imperialism inspired new forms of settlement and trade. The major interventions were Britain in North America, and Britain, France and the Netherlands in the Caribbean. The Dutch and British began establishing trade routes to North America and South and East Asia, driven by a need to extract raw materials for their own manufacturing industries and, through greater levels of colonization than seen in the Spanish and Portuguese empires, to create markets for manufactured goods (especially textiles, tobacco and sugar). Mercantile imperialism came to rely on the slave trade. As we saw in Chapter 1, the trade itself was abolished in 1807. However, it was abolished much later in the British colonies (1833), the former Spanish colonies (mid-nineteenth century), Brazil (1888) and the United States (1865).

3    Nineteenth century: age of empire/high imperialism. This period witnessed a vast expansion of capitalist imperialism, based on the large-scale extraction of resources to support rapid industrial expansion in Europe and also to create new markets for European manufactured goods. The settlement of Europeans in overseas colonies was seen as a vital part of the creation of these markets. The major powers were Britain and France, which had developed global empires, with significant involvement by Belgium, Germany, Italy and Portugal, which sought to expand in Africa, in particular, after the loss of its South American colony (Brazil and the Spanish colonies fought for and won independence in the first quarter of the nineteenth century).

4    The age of empire ended with decolonization and independence in two phases. First, British settler colonies were granted dominion status: Canada (1867), Australia (1900), New Zealand (1907) and South Africa (1910). Second, nationalist and independence movements, coupled with the political and economic consequences of World War Two, led to large-scale dismantling of western European empires from the late 1940s. India gained independence and was partitioned in 1947; most African colonies were granted independence in the 1960s; Hong Kong was handed back to China

in 1997. However, as Table 1.2 illustrates, European colonies still exist.

Development was integral to European colonialism. Put very simply, the extraction of resources and labour from the colonies required new forms of social control of subject populations in both metropolitan and colonial regions. The notion of 'development' was used to justify the imposition of forms of social control because native peoples were cast as backward in comparison to Europeans. They thus needed to be developed and modernized through education and labour. Postcolonial theorists have highlighted the fundamental contradiction at the heart of the European modernizing project (see Chapter 2). Central to this was the idea of a universal humanism based in notions of democracy and human rights. However, while imperialism became the mechanism by which these normative 'goods' were to be made universal – brought to the 'uncivilized other' by the humanitarian western powers – they were simultaneously denied to the colonized on the grounds that they were 'not yet ready' (Chakrabarty 2000). In other words, the colonized were not only 'out there' – spatially distant – but 'back there' – behind in evolutionary terms. They were not fully human. Thus, linear notions of historical time and progress determined that imperialism could be justified as part of the modernizing project of exporting civilization and human rights (see Box 3.2) and, ironically, that the latter could be denied to colonized peoples because they were not yet modern. Moreover, because they were not yet modern, they needed to be subject to regimes of discipline.

---

## Box 3.2  The colonization of time

The imposition of European time – and cultures of time – around the world was critical to the globalization of ideas about development. This began with the regulation and standardization of time in Europe and the displacement of peasant time, culminating in the globalization of this standardized European time. The enormous scale and violence of colonialism becomes apparent when we reflect on how very particular, western European notions of measuring and accounting for time became globalized. There were two main understandings of time at work in imperialism that are deeply rooted in ideas about progress and development and still influence thinking today: i) time as linear and evolutionary; ii) time as immanent or everyday time, needing to be harnessed and used as a tool to control human lives.

European notions of time as linear are very different to cyclical notions of time in many other cultures (e.g. in African cultures, in Buddhist and Taoist philosophy, in Hinduism). Linear notions of time are rooted in the Christian belief in the path to either heaven or hell, and in the coming of the messiah at the end of times. But they are also a product of Enlightenment notions of human progress and development. A particularly powerful tool in imperialism was the European idea about the passage of time and its relationship to human development. Linear notions of time in relation to human progress emerged in the seventeenth century and the beginnings of modernity. Europeans developed a sense of their own progress, thought of themselves as more advanced along a linear path to modernity and at the pinnacle of human achievement – civilization – conveniently forgetting the civilizations that pre-dated modern Europe. A notion of universalism underpinned these ideas: that the European model of progress was applicable to all human beings and that all human beings had the potential to reach civilization, or to be civilized. As Ann McClintock suggests 'the world's multitudinous cultures are not marked positively by what distinguishes them, but by a subordinate, retrospective relation to linear European time' (1995: 255).

Evolutionary theory became bound up with the secularization of sacred time. Instead of describing the progress of human salvation, time now described the 'self-realization of man'. At the onset of European colonialism in the fifteenth century, notions of time became spatialized. Europeans believed they could experience the different 'ages of man' by travelling to 'primitive' parts of the world and, by observing the people living there, could chart all aspects of human development, completing a universal history of humanity. The early days of European exploration and encounters with other cultures convinced Europeans that their time was modern, moving and dynamic. In contrast, the time of the peoples they encountered was static, less developed and older (Fabian 1983). While western people were living in modern times, the people they encountered were depicted as primitives and barbarians, further back in evolutionary history. This politics of time (*ibid.*) was used to justify colonial domination of other cultures. Colonial empires were built out of supposedly helping 'backward' societies develop and progress to achieve the heights of civilization attained in Europe.

These notions of linear, evolutionary time were intertwined with the colonization of everyday, immanent time. This began with industrialization and technological advances in Europe, which required time to be standardized. Clock time emerged in Europe with the shift of production from agriculture and cottage industries to factories (Harvey 1990). Different societies had diverse ways of measuring time. For example, E.P. Thompson (1967: 58–9) argued that in Madagascar, time might be measured by 'a rice-cooking' (about half an hour) or a 'frying of a locust' (a moment); in Chile, the time to cook an egg was measured by the time it took to say an Ave Maria out loud. Clock time is also a social construct – 'Clocks, it is often forgotten, do not keep *the* time, but *a* time' (Nanni 2012: 1) – but in modern societies it is accepted as an objective fact. It organizes lives and allows for moral judgements about all manner of social behaviours: the ill-discipline of lateness or the disparaging of other cultures of time.

Globalization of time is often considered a product of capitalist development, but colonization of time was profoundly cultural as well as economic. As Nanni (2012) demonstrates, European missionaries played a primary role as the first colonizers who sought to 'civilize' through regimenting time. Time practices, and ideas about time in the colonial context, were linked to the moral economies of both capitalism and imperialism. From the fourteenth century onwards, modernizing European societies saw a marriage of convenience between religious views of time (particularly Puritan and Evangelical notions of the sinfulness of idleness and time-wasting, and redemption through time well-spent) and economic notions of time as commodity. In a civilized, capitalist society all time must be consumed, marketed, used. It was offensive for the labour force merely to 'pass the time': time must be spent. Consequently, disciplined time was gradually imposed in European factories and schools.

Time discipline was established in Europe through the division and supervision of labour, fines, bells and clocks, money incentives, preaching and schooling, and time propaganda aimed at working people. These tactics were also imposed on colonized people, but over a much shorter time and in substantially different cultural contexts. Colonial discourses and the moralizing discourses of leisured classes were mutually reinforcing. Widespread belief that the Europeans masses and non-European societies were somehow 'not attentive enough' to the passage of time functioned as a powerful legitimizing discourse for both puritanical time-disciplining at home, and civilizing and missionary projects in the colonies.

The supposed 'savagery' and 'primitiveness' of indigenous peoples were directly constructed around their perceived inability to harness, utilize and quantify time as an abstract notion based upon mathematics, rather than a practical tool based upon nature. Temporal moralities both shaped and reflected colonial discourses of difference. The supposed laziness of natives was used as a justification for their subjection to European regimes of time and work, most brutally under slavery. The Manacas Iznaga bell tower (Figure 1.1) on a former sugar plantation in Trinidad, Cuba, stands testament to the way in which the entire lives of slaves – working, sleeping, eating – were governed by time. Time was a way of establishing difference between the supposedly cultivated Europeans and supposedly benighted Others, and a tool of colonialism. Colonial peoples were represented as inferior and lacking in relation to European perceptions of temporal morality, and were then subjected to colonial and humanitarian attempts to 'reform' indigenous understandings of time. Missionaries were crucial to this process – as the bringers of 'bells, bibles and the civilizing mission' (Nanni 2012: 16), and as the gatekeepers of western civilization in colonial and frontier contexts. Their methods might have been preferable to slavery, but they were still colonizing minds:

> Time was both a tool and a channel for the incorporation of human subjects within the colonizers' master narrative; for conscripting human subjects within the matrix of the capitalist economy, and ushering 'savages' and superstitious 'heathens' into an age of modernity.
>
> Nanni (2012: 4)

The two understandings of time thus worked in tandem in imperialism: *everyday time* needed to be harnessed and used as a tool to control human beings (both morally and economically), which would then lead to the advancement of colonized people within *evolutionary time* towards civilization. We can only imagine the ruptures that were created by the sudden imposition of such alien notions of time. Colonial time sought to destroy indigenous time differently in different places. Nanni (2012) contrasts South Africa with Australia. In South Africa, settler-colonialism aimed to appropriate labour. The imposition of capitalist time, and especially the observance of the Sabbath to regiment not only a day of rest but six days of work, was the priority. In contrast, in Australia settler-colonialism was directed at the appropriation of land. Civilizing through imposing a new order of time was less concerned with putting Aboriginal peoples to work and extracting productivity from them (the settlers had convicts for this), but controlling them through regularity and predictability, and limiting their mobility by establishing cultural curfews. However, the rise of a colonial time regime did not completely obliterate indigenous knowledge and practice. Instead, the imposition of 'whitefella time' was always incomplete, constantly resisted, strategically appropriated and negotiated with by 'the colonized'.

The imposition of western time cultures and associated models of progress and development in the global South have always met with resistance and negotiation (discussed further in Chapter 6). However, by seeking to destroy indigenous time in the name of 'progress' and 'development', colonialism sought to destroy and de-legitimise indigenous culture, which has had dire social consequences. One of the greatest impacts has been on social reproduction. Time in relation to space shapes social reproduction in all societies, including gender roles, divisions of labour, the roles that people perform and where they perform them. This creates social order, or what Barbara Adam (2005) terms 'timescapes', which in the case of indigenous peoples had formed over millennia prior to the arrival of Europeans. The colonization of time by industrial clock time, beginning in the west and then globalized through conquest, imperial expansion and neo-colonial domination, has fundamentally altered timescapes everywhere. It imposed alien conceptions of time and space upon indigenous and pre-colonial peoples, altering forever the social framework within which their social reproduction could, if at all, take place.

New forms of discipline varied across time and space, but they included forced labour schemes, taxation, schooling and segregation in native quarters and were designed either to adapt and/or marginalize colonial subjects to the European presence (Patel and McMichael 2004: 235). In both the colonies and metropolitan societies new forms of social discipline were produced. Industrial capital produced new class inequalities between labouring populations and the middle-class citizen-subjects that were premised on a racist international inequality generated by colonialism. As Patel and McMichael (2004: 236) argue, 'it was this inequality, and its local "face" in the colonies, that fueled anti-colonial resistance' (see Chapter 2). At the heart of colonialism, therefore, was the idea of managerialism, which was instituted through a process of 'civilizing' people as a nation, a class, a race and a gender. This was done through control of individually coded bodies – where they work, how they reproduce, the language they speak – or what is termed biopolitics (Gilroy 2000). Biopolitics, or the regulation of the state of education, sexuality, criminality and gender, was central not only to colonialism, but has shaped development praxis in the postcolonial era (see Chapter 5).

Also central to European colonialism was the idea of race and racism. As Baaz (2005: 62–3) argues, there is much debate about whether this is a modern phenomenon that arose during the Enlightenment and the development of the sciences in Europe, or is located further back in European history with counterparts in 'non-western' societies. Significantly, however, the racism that was produced through colonial history was different from earlier notions, in and outside the West, both in scope and nature. During the eighteenth and nineteenth centuries, what was peculiar to European racism was the science of race and its coincidence with processes of European global imperialism (Pieterse 1992). According to Baaz (2005: 63):

> The characteristics attached to the human species in the science of race, and the evolutionary ladder on which these variations of the human race were situated, permeated and legitimized this process with truly *global* dimensions. This is the specificity of European colonialism [and its associated racism] (emphasis in original).

Most formal structures of colonialism have been overturned, but the legacies of colonial rule remain intact in many spheres of life in both metropolitan centres and former colonies. Political, administrative, legal, educational and religious systems reflect past European influence. As we shall see, economic influences have had profound and lasting effects, including the creation of whole economies focused on producing a single raw material, for example, sugar plantations in the Caribbean, tea in Sri Lanka, coffee in Brazil, rubber in Malaya, copper in Zambia. This caused enormous problems on independence since it meant these economies were vulnerable to crop failures, price fluctuations and international demand. In short, the former colonies were still dependent on the former imperial powers and especially vulnerable to crises in global capitalism. In addition, different conflicts around the world often have their roots in colonialism and its aftermath (e.g. Kashmir, Northern Ireland and almost every conflict in Africa).

While the political and economic basis and effects of colonialism cannot be under-estimated, its cultural basis and effects have also received a great deal of critical attention in recent years. As Dirks (1992: 3) argues:

> Although colonial conquest was predicated on the power of superior arms, military organisation, political power, and economic wealth, it was also based on a completely related variety of cultural technologies. Colonialism not only has had cultural effects that have too often been either ignored or displaced into the inexorable logics of modernisation and world capitalism, it was itself a cultural project of control. Colonial knowledge both enabled colonial conquest and was produced by it; in certain important ways, culture was what colonialism was all about.

Colonialism mobilized different webs of meaning, practices, objects of knowledge and ways of knowing that helped legitimate and perpetuate colonial power. A focus on the cultural basis of colonial power does not efface the violence of conquest and control. Rather, an examination of the cultural 'structures of meaning' – ideas, imaginings, images and representations – that underpinned colonial rule can reveal the ways in which the violence of conquest and control was exercised, justified and represented (Blunt and Wills 2000). Indeed, as Thomas (1994: 2) argues:

> colonialism is not best understood primarily as a political or economic relationship that is legitimized or justified through ideologies of racism or progress. Rather, colonialism has always, equally importantly and deeply, been a cultural process; its discoveries and trespasses are imagined and energized through signs, metaphors and narratives; even what would seem its purest moments of profit and violence have been mediated and enframed by structures of meaning. Colonial cultures are not simply ideologies that mask, mystify or rationalize forms of oppression that are external to them; they are also expressive and constitutive of colonial relationships in themselves.

All forms of colonialism mobilized a series of practical and discursive strategies through which space was claimed as colonial and characterized as possessing particular attributes. Non-European space was imagined as empty and uninhabited and thus available for exploration, exploitation and, ultimately, colonization. It was imagined as morally and culturally empty, but with the prospects of becoming 'civilized' through the processes of colonization (Figure 3.2). Colonial space was also rendered familiar and knowable, and thus controllable, through various processes of mapping, the creation of

**Figure 3.2**  *Singapore Cricket Clubhouse and Padang and City Hall, juxtaposed with modern skyscrapers*

Source: Author

administrative units irrespective of already existing tribal, ethnic and other affiliations, scientific exploration and categorization, and through the imposition of metropolitan names (Driver 2001; Godlewska and Smith 1994).

Edward Said (1993: 6) argues that colonial and imperial struggles were explicitly geographical struggles and geographical conflicts are not only military but cultural: 'it is not only about soldiers and cannons but also about ideas, about forms, about images and imaginings'. This is as relevant today as it was in the eighteenth and nineteenth centuries. The military might of soldiers and tanks cannot be isolated from the structures of meaning that have helped to legitimate both conflict and resistance to such conflict. By creating linkages with distant parts of the world, European imperialism and colonialism provided the basis for globalization in the twentieth century. It also established enduring power inequalities between North and South that have continued despite independence, limiting the autonomy of former colonies to determine their own futures and paths to development. Indeed, many critics would argue that the power of Europe and the US to intervene in the affairs of other countries in the name of development is a product of 'neocolonialism', or the continued dominance of the North.

## Post-war development

It is widely assumed that the concept of development 're-emerged' after World War Two, with concerted attempts to measure 'global poverty' and to eradicate inequalities between nations. However, as Kothari (2006) points out, the cultures that travelled over colonial space through their performance by colonial officers have been reworked in the post-colonial period. This challenges assumptions of epochal historical periodizations based on a clear disjuncture between colonial and development eras. In other words, colonialism and development should not be understood as two separate and distinct periods of history. There are temporal continuities between the colonial and post-colonial or 'development' periods in terms of discourses and practices.

Kothari traces the continuities between colonial and post-colonial/'development' periods through a case study of English colonial service officers who served in sub-Saharan Africa in the 1950s. Following independence, these men returned to England and pursued second careers in the emerging and burgeoning international development industry. Most secured posts in bilateral government agencies and multilateral UN and other international NGOs. Many were ambivalent about the colonial project; as colonial officers they were ostensibly representatives of colonial power and institutions, but they were also often nostalgically and romantically inclined towards the colonized and sometimes questioned colonial policies (Kothari 2006: 237). Despite this, they adopted various strategies and lifestyles to maintain distance from and authority over those 'others' that constituted the colonized/recipients of development. These strategies included their performance of 'Englishness' based on class and educational background (most colonial officers were recruited from Oxbridge). Officers were re-posted at regular intervals to maintain social distance from the local populace. These strategies, networks and forms of self-representation did not come to an end at the moment of decolonization; rather, they continued to be mobilized through ideologies, individuals and institutions. The personnel were often the same, as were the ideas. An assumption that development began after decolonization ignores the colonial genealogy of contemporary international relations.

The rise of 'Third Worldism' also contained some key continuities between colonialism and development. The 'Third World' was first identified at the Bandung Conference in Indonesia in 1955, where mainly Asian countries united to promote economic and political co-operation and to oppose colonialism. They used the term 'Third World' to generate unity and support in the face of the US and former colonial powers, and this gained popular currency as African and Asian countries acquired independence. However, Third Worldism embodied the contradictions of the time: the universal institutionalization of national sovereignty as the representation of independence of decolonized peoples, political confrontation with European racism, and a movement of quasi-nationalist elites whose legitimacy depended on negotiating their economic and political dependence with former colonizers. As Patel and McMichael (2004: 241) argue:

> Decolonisation was rooted in a liberatory upsurge, expressed in mass political movements of resistance – some dedicated to driving out the colonists, and others to forming an alternative colonial government to assume power as decolonisation occurred. In this context, development was used by retreating colonisers as a pragmatic effort to preserve the colonies by improving their material conditions – and there was no doubt that colonial subjects understood this and turned the ideology of development back on the colonisers, viewing development as an entitlement.

Colonialism disciplined native subjects through biopolitics, a type of government that regulates populations through the application and impact of political power on all aspects of human life. Development legitimized the disciplining of subject-citizens by postcolonial elites and rulers. In Africa, forms of discipline included 'tribalization'. This was a legacy of European colonialism that had combined forms of urban power excluding natives on racial grounds with forms of indirect native rule in rural areas through a reconstruction of tribal authority. Although independence had abolished racial discrimination and affirmed civil freedoms, power remained divided through artificial tribal constructs along ethnic, religious, gender and regional lines. Fanon (1967) explains how postcolonial elites in both Latin America and Africa, in the absence of coherent social relations following independence, chose the easiest option – the single party – which inevitably led to dictatorship and continued oppression of the formerly colonized, supposedly liberated ordinary citizens (see Chapter 5).

Although there are significant continuities between the colonial and post-colonial periods in terms of development discourses and practices, the post-war period witnessed a shift in emphasis (Crush 1995), from an overt mission to extend civilization and modernity through colonization to ideas of progress embedded within a development discourse that cultivates an 'apparently more humanitarian image' (Kothari 2006: 235). It saw the emergence of a different politics of naming, the labelling of whole areas of the world as 'poor' and the invention of global solutions to problems of impoverishment in the South.

This was partly a consequence of the end of European colonial rule in some parts of the world from the late 1940s into the 1960s, where development became a means of framing the challenges and opportunities for newly independent states. Many of these states sought to deepen their victory over colonial rule by embracing development as a national framework for building independence; many former colonial and other economically advanced countries saw underdevelopment as a threat not only to the stability of these new countries but also to themselves. Development was thus associated with progress and modernity, in contrast to the backwardness of underdevelopment, and was conceived of as an economic process (Willis 2005: 2–3), but with very clear links to global geopolitics.

The conflict between capitalism and communism during the Cold War provided an important context for post-war development theory and policy. It created a new world order based on a tripartite division into First (capitalist), Second (communist) and Third (non-aligned) Worlds and a perceived requirement for richer countries to resolve 'underdevelopment' and poverty in 'backward' areas to prevent them falling under the sway of communism. Development thus became the key principle upon which the USA would seek to build its own global empire after 1945, persisting in a different form into present-day US foreign policy, particularly in the Middle East, Iraq and Afghanistan, where development is seen as the antidote to 'terrorism'. As Power (2003) argues, development has to be considered as related to the geopolitics of race, where economically advanced countries take an interest in and consider the needs of poor countries in ways that often issue directly from their own preoccupations or strategic interests.

Even prior to the terrorist attacks on New York in 2001, the CIA (2000: 7) recognized that the global economy would continue to produce inequality and political instability or conflict: 'Regions, countries, and groups feeling left behind will face deepening economic stagnation, political instability, and cultural alienation. They will foster political, ethnic, ideological and religious extremism, along with the violence that accompanies it'. Implicit within this, however, is the common representation of countries categorized as poor as tradition-bound, unstable, unchanging, brittle, weak, frail and dangerous, perpetually in need of intervention from economically advanced countries. Chapter 4 problematizes these representations and explores the extent to which they are legacies of a past rooted in imperialism.

In addition to its embedding within global geopolitics, development was seen as definable and measurable, a perception that remains at the core of the major development organizations. For example, the UN Human Development Index (HDI) combines three variables to compare countries – life expectancy, education (mean and expected years of schooling) and income (based on purchasing power parity). The WB focuses on economic development and uses GNP per capita as the indicator of levels of development. A growing GNP became a kind of myth in development circles, 'a dogma, a shibboleth in economic reasoning and a golden calf and centre of economic worship' in the 1950s and 1960s (Weisskopf 1964). As Patel and McMichael (2004: 242) argue:

> As a bloc, the Third World was incorporated into a hegemonic project of ordering international power relations, where states adopted a universal standard of national accounting (GNP), and foreign aid disbursements subsidized state apparatuses and elite rule. In postcolonial India, 'Instead of the state being used as an instrument of development, development became an instrument of the state's legitimacy' (Bose 1997: 153).

However, reducing development to quantifiable and measurable indicators is fraught with problems. Monetary indicators such as GNP often mask more than they reveal about poverty and inequality. Average and aggregated measures are meaningless in terms of representing the real situation on the ground or in particular places. As Spivak (2014) contends, while including education in measures of human development is progressive, mean and expected years of schooling do not measure the *quality* of that schooling. The HDI also reveals nothing about the poverty that underpins most of its variables. Definitions of poverty are hotly contested and its conditions vary between different areas. As we shall see, of significance is the fact that these notions of development are based around the idea of linear progression, indicated as different degrees of departure from already established western ideals and experiences. As Reddy and Hogge (2002) argue, the measures adopted by many global development organizations remain powerful examples of how *not* to count the poor. Despite more than 60 years of 'development', development agencies still have no real idea how many impoverished people there are in the world and only recently have they begun to ask: 'How do poor people *themselves* view issues of poverty, development and well-being?' In addition, in many countries in the South, 'local' development officials have often been trained in Northern monetarist economies and have little empathy with poor people in remote areas.

## Development theories and ideologies since 1945

As we have seen, the origins of development theories and ideas about progress and modernity can be traced back to post-war ideas and discourses about 'underdeveloped' areas, which also have their origins in seventeenth- and eighteenth-century European Enlightenment rationality. Deep-rooted Eurocentric ideas have cast a dominant shadow over development thinking, and the idea of progress forged in the seventeenth century still remains the bedrock of development thinking. Modernity is still seen as an unfinished project in the South. However, this is not to deny the complexities of the debates about development since 1945 and the ways in which these dominant modes of thinking have been challenged. Table 3.1 provides a basic chronology of the main approaches to and understandings of development since 1945. As Katie Willis (2005: 26) argues, theories have not evolved in a linear fashion, with one emerging from the last without any contestation or conflict. Rather, numerous and competing notions about development have co-existed, with certain ideologies or theories becoming dominant because they are advocated by more powerful actors.

Development thinking involves three main elements (Hettne 1995):

1   development theories – logical propositions about how the world is structured, which explain past and future development;

*Table 3.1  Approaches to development since 1945*

| Decade | Main development approaches |
|--------|---------------------------|
| *1950s* | Modernization theories: based on European model and notions of 'trickle-down'. Structuralist theories: Southern countries need to limit interaction with the global economy to allow for domestic economic growth. |
| *1960s* | Modernization theories. |
| | Dependency theories: Southern countries are poor because of exploitation by Northern countries. |
| *1970s* | Dependency theories. Basic needs approaches: focus of government and aid policies should be on providing for the basic needs of the world's poorest people. Neo-Malthusian theories: need to control economic growth, resource use and population growth to avoid economic and ecological disaster. Women and development: development recognized as having differential effects on women and men. |
| *1980s* | Neoliberalism: focus on the market; governments should retreat from direct involvement in economic activities. Grassroots and community-based approaches: importance of considering local context and indigenous knowledge. Sustainable development: balancing the needs of current generation against environmental and other concerns of future populations. Gender and development: greater awareness of the ways in which gender is implicated in development. |
| *1990s* | Neoliberalism. Post-development: ideas about 'development' represent a form of colonialism and Eurocentrism and should be challenged from the grassroots. Sustainable development. Culture and development: increased awareness of how different social and cultural groups are affected by development processes. |
| *2000s* | Neoliberalism: increased engagement with concepts of globalization. Sustainable development. Post-development. Grassroots approaches. Decolonizing development. |

*Source*: adapted from Willis (2005: 27)

2    development strategies – adopted by actors and agents (from grassroots to states and international organizations);

3    development ideologies – different strategies and agendas reflect different ideologies.

Pieterse (2001b: 3) defines development as the notion of 'organized intervention in collective affairs according to a standard of improvement'. Each theoretical approach has its own ideas about 'collective affairs' and what kinds of interventions are required to bring about improvement locally, nationally or globally. Development theory is thus partly about what constitutes 'improvement', what 'appropriate intervention' means, and the power relations underpinning the will to improve (Li 2007). In thinking about the range of approaches since 1945, it is pertinent to question the framework from which the two terms 'more' and 'less' developed derive their authority. Each of these different approaches imagines very different scales and spaces in its conception of development. They envisage different political and economic mechanisms in their understandings of progress. They are based on very particular ways of imagining and understanding the world.

## Modernization theories

Modernization is not the same as modernity. Modernity refers to an orientation of being-in-the-world, to a concept of a person as self-conscious subject, to an ideal of humanity as species being, to a vision of history as progressive, with humans at the centre (Comaroff and Comaroff 2012: 118). In contrast, modernization posits a 'strong, normative teleology, a unilinear trajectory toward a particular vision of the future – capitalist, socialist, fascist . . . – to which all humanity should aspire' (*ibid.*: 118–9). The core perceptions underpinning modernization theories are aptly summarized by Abrahamsen (2001: 19):

> Before development, there is nothing but deficiencies.
> Underdeveloped areas have no history of their own, hardly any past worth recalling, and certainly none that's worth retaining. Everything before development can be abandoned, and third world countries emerge as empty vessels waiting to be filled with the development from the first world.

Post-war decolonization and the formation of the UN gave a stimulus to theorizations, dominated by the USA, about political and economic trajectories of non-western others. Many development approaches shared a belief that development is a process that has industrialization as its endpoint. Resource distribution was also seen as primarily linked to the state and to a notion that all people would benefit through 'trickle-down' and resource redistribution. The Third World was perceived to be without and modernization theory was the mean to rectify this, but modernization theory was always vague as to how 'trickle-down' was to happen. It was deeply ideological, attempting to provide an alternative set of values to communism and based on a dualistic world view, opposing 'traditional' to 'modern' lifestyles and 'indigenous' to 'westernized' cultures (no one could belong to both). It proposed linear, Eurocentric notions of development of capitalism in the world. It was based on the idea that once impediments and obstacles are removed, modernization, development and progress will proceed in a 'smooth, self-perpetuating process with desirable results' (Kaiwar 2015: 121–22). 'Trickle-down' would occur from the developed core to the underdeveloped periphery at national, regional and global levels. The key proponent of this view was Walt Rostow, a profound anti-communist whose model in *The Stages of Economic Growth* (1960), based on the British industrial revolution, depicted the level of development increasing over time, from 'traditional society', to 'preconditions for take-off', to 'take-off', to a 'drive to maturity' and culminating in an 'age of high mass consumption' (Willis 2005: 40–1). David McClelland in *The Achieving Society* (1967) approached modernization from a psychosocial perspective, arguing that modernization only occurs when a given society values innovation, success and free enterprise. Samuel Huntington's *Political Order in Changing Societies* (1968) challenged the conventional view of modernization theorists that economic and social progress would produce stable democracies in recently decolonized countries. Instead, he argued that political order was most important to maintaining state stability. Order is threatened when the level of political mobilization exceeds the level of administrative institutionalization within a society. Huntington believed that economic development would increase political mobilization quicker than the development of appropriate social, political and economic institutions. This in turn would lead to instability. His solution was a stronger emphasis on institution-building in a society's development, and the establishment of a stable government with only gradual political liberalization.

Modernization approaches became dominant because they had a clear practical bent that came from 'an unreflexive faith in the winning virtues of the West' (Sylvester 1999: 705). That the USA dominated in expounding modernization made sense given its geopolitical and economic ascent after 1945. Modernization was seen as an antidote to the failure of aid programmes to the newly independent countries in the 1940s and 1950s, particularly India. The economic model became dominant in the 1960s and 1970s as it fed into international relations and development institutions (including USAID), despite the fact that many were critical of this kind of modelling and its assumptions of universal applicability. Ultimately, apart from the rise of the Asian economies in Singapore, Hong Kong, Taiwan and South Korea, modernization theories failed:

> The Rostovian comet has flashed across our horizon, finally to vanish into the academic haze of some US university – and the poor are still with us, poorer and more numerous, while the economic systems which were to offer the magic escape from poverty have been crippled by the [Cold] war which Rostow himself helped to shape and their imperial nakedness is finally exposed for all to see.
>
> (Buchanan 1977: 364)

There were numerous reasons for this failure. Of critical importance was their undervaluing of 'traditional societies', which were much more complex and neither 'backward' nor as irrational as they were depicted. Instead, modernization theories depoliticized development by ignoring diverse histories and cultures. Remarkably, they took no account of the significance of histories of imperial and colonial domination. Instead, the history of the Third World began in 1945 and all that went before was deemed irrelevant to the planned future trajectory of 'developing' economies. They assumed that development could only be brokered by states and development institutions and could not emerge from the 'grassroots' or from people in 'developing' countries. 'Trickle-down' often never occurred because of unequal power relations between individuals, groups and states. Steeped as they were in assumptions of western superiority, they failed to recognize the problems of 'modern' societies that were fractured and divided by race, class and politics. Their faith in modernity and technological progress wrought ecological disaster in many areas. Arguably, therefore, it was profound Eurocentricity that meant modernization theories failed to have any impact (Box 3.3). Despite this, they dominated for decades after 1945, primarily because they were ideological. They went hand in hand with anti-communism and with geopolitical interventions to aid anti-socialist guerrilla movements in countries such as Angola, Afghanistan and Nicaragua, or to defend anti-socialist dictatorships in places like Iran (1953), Guatemala (1954), Brazil (1964) and Chile (1973). As discussed subsequently, many of the Eurocentric assumptions have persisted with mainstream development, especially the idea that the global South lacks modernity. However, critics have long challenged these assumptions: according to Comaroff and Comaroff (2012: 120), 'it is not that people in the Global South "lack modernity". It is that many of them are deprived of the promise of modernization by the inherent propensity of capital to create edges and undersides in order to feed off them'. These criticisms were first articulated in what came to be known as dependency theory.

---

## Box 3.3 The myth of modernization in Zambia

James Ferguson's *Expectations of Modernity* offers a profound critique of the 'modernization myth' and what happens when it is 'turned upside down, shaken and shattered' (1999: 13). During the 1960s and 1970s, Zambia was lauded as the bright light of Africa's future. It was a 'middle-income country' with prospects for 'full' industrialization and even admission to the ranks of the 'developed' world. Processes of industrialization, urbanization and modernization appeared to be well established. However, Zambia's economic boom during this period was fuelled almost entirely by the export of a single raw material – copper. This left the economy vulnerable to fluctuations in the global market and when the price of copper fell drastically in the 1970s the consequences were devastating. Rather than continuing to develop, Zambia's economy rapidly de-industrialized, leaving workers struggling to get by. When ex-miners and other unemployed workers attempted to 'go back' to their rural communities they found themselves unable to re-integrate into societies they no longer understood. The effects on ordinary Zambians of this fleeting glimpse of modernity and associated promises of prosperity were profound. They were ultimately trapped in an era in which the modernist project had demonstrably failed and yet had been changed by it in such a way that returning to their previous cultural practices was impossible.

The idea of Zambia's progress towards a modern, industrial nation, embodied in the copper mining industry, was merely an imposition of European styles of life on Zambia which benefited only the European powers in need of cheap copper. Discourses and models of modernization, based on linear notions of progress, were unable to accommodate the realities of stagnation and decline in countries such as Zambia. Here, the collapsed hopes and continuing poverty of a country once thought of as the vanguard of Africa's industrial development belies the simplistic faith in the benign effects of globalization. As Ferguson (1999: 13) puts it:

> The mythology of modernization weighs heavily here. Since the story of urban Africa has for so long been narrated in terms of linear progressions and optimistic teleologies, it is hard to see the last twenty years of the Copperbelt as anything other than slipping backward: history, as it were, running in reverse. How else to account for life expectancies and incomes shrinking instead of growing, people becoming less educated instead of more, and migrants moving from urban centres to remote villages instead of vice-versa?

One problem with the myth of modernization in Zambia was its failure to recognize that models of industrialization based on the British industrial revolution could not simply be exported to other contexts. Ignoring the specificities of the Zambian economy and its particular position in the global economy ultimately spawned a process of de-development. Copper is still central to Zambia's exports, but the country now receives about US$1 billion a year in foreign aid, much of which was conditional on the state accepting privatization (Power 2003: 88). The privatized copper industry still faces severe problems, with continual threat of mine closures and unemployment. In 2000, a debt-relief package of US$3.8 billion was agreed, making Zambia one of the highest recipients of official aid in Africa (*The Economist*, 1 June 2002, cited in Power 2003).

Source: adapted from Ferguson 1999

## Dependency theory

Dependency theory originated in the 1950s, primarily in USA, Brazil, Chile and Columbia, later becoming popular in Africa, the Caribbean and the Middle East. Iran's economic programme in the 1980s, for example, reflected ideas on 'de-linking' from capitalism and self-sufficiency propagated by dependency theorists such as Samir Amin and Andre Gunder Frank. It emerged as a reaction to modernization theories, viewing the poverty of countries in the South as a product of their integration into the 'world system'. This system was seen as perpetuating the wealth of the 'core' (global North) at the expense of periphery (global South). It highlighted the growing dependence of poorer countries on international capitalism and often drew on Marx's writings about uneven development and exploitation. Much of the attention of dependency theory was focused on Latin American countries. These had been independent for nearly a century, but still remained poor despite the promises of modernization. Dependency theorists raised questions about how colonies had been stripped of resources, their land systems reorganized and labour relations reconfigured by colonialism and the parasitic and exploitative relations that had been put in place between core and periphery. Dependency on the metropolitan core was viewed as increasing underdevelopment of the periphery; 'take off' was impossible because of unequal relations and histories of colonialism.

Buoyed by the apparent success of Chinese socialism, many newly-independent countries in Africa, as well as in Latin America and Asia, adopted a more socialist agenda for change (e.g. Tanzania,

Mozambique, Vietnam, Cuba). The Economic Commission for Latin America (ECLA) was set up to advance the development strategy of the dependency school, particularly through the work of Raúl Prebisch, and even the ILO was forced to consider redistribution as part of its growth strategies. Dependency theories thus held some sway in the early 1970s, but they did not inspire many development policies, with the exception of Chile and Cuba. They offered only a variation of the tradition/modernity model; like modernization theory, they relied on an oversimplified dichotomous view of the world and relations between core and periphery (Rist 1997). Rather than seeking to challenge this world view, dependency theory reinforced a sense of 'us' and 'them'. Moreover, it was still a form of economic determinism, overlooking social and cultural variations, and placing overwhelming emphasis on the state in preventing inequalities and redistributing wealth. Despite saying much about underdevelopment, it also lacked a clear statement on what development might be, and its call for de-linking from the world capitalist economy was anachronistic at a time when the world economy was undergoing further globalization. However, while the simplistic and deterministic constructions of dependency theory have been discredited, as discussed below, its intellectual legacy can still be seen in post- and anti-development, and even some strands of postmodern and postcolonial writings (Simon 1997).

## Impasse, neoliberalism and the Washington Consensus

Modernization and dependency seemed to check-mate each other (Schuurman 2001), producing an 'impasse' in development theory in the 1980s and 1990s rather than alternative ways forward. Whatever theoretical positions were being advocated and translated in policy and practice, from modernization to socialist alternatives, the lives of the overwhelming majority of people in the South were not improving. Indeed, in some countries conditions of daily existence and levels of development and indebtedness actually worsened. Some writers sought to go beyond the impasse, but it has been more common to see a return and retreat to liberal orthodoxies (Hettne 1995; Munck 1999), such that neoliberalism came to dominate thinking and continues to do so today.

Neoliberalism became hegemonic precisely because the impasse in development theory coincided with a crisis in the political economy. The oil-indebtedness of the former colonies was spiralling, threatening the international trade and banking systems. The US economy was seriously weakened by inflationary spending on the Vietnam War, which precipitated a decision to decouple the US dollar from the gold standard. The global economy was in recession between 1978 and 1982, during which time the showcase of modernization in the Middle East – Iran – underwent an anti-modernization revolution. In 1979, Communist China began to open its economy to market forces and throughout the world new social movements began to burgeon, organized around development, environmental and feminist political issues. These global instabilities and shifts brought about 'an altered globalism' (Smith 1997: 173) in which the old locations and concerns of development were no longer the means of control.

They also gave rise to the ultra-rationalist advocates of public sector reform, economic structural adjustment and good governance (personified in the USA and UK by Reagan and Thatcher), through which neoliberalism came to dominate, particularly through US-based international financial organizations. Neoliberals advocated that in order for markets to work effectively, continued loans to the South were required but under strict conditions designed to align these economies to the North. Thus, during the 1980s and 1990s, the Washington consensus dominated development theory and policy, based on fiscal and monetary austerity, elimination of government subsidies, lower taxation, privatization, free trade, foreign direct investment (FDI) and so on. Countries in the South scrabbled to create a favourable investment climate – pandering to the needs of wealthy corporations and taking expenditure away from more critical social concerns, whilst being told that this was necessary short-term austerity for long-term prosperity.

Free trade, based on unrestricted movement of capital and goods, remains the basis for this globalized neoliberal approach to markets and development. The rationale is that if a country is economically competitive and efficient, it should be able to compete for a share of the global economic pie. However, free trade was always a myth. Northern economies are protected by trade barriers (in the form of subsidies for its own goods and tariffs on imported goods) that are supposed to be the very anathema of free trade (see Chapter 5). The International Finance Institutions (IFIs) are committed to working more closely with each other to reduce trade barriers, but many see their ideological coming together, with trade liberalization as the mechanism for global economic growth and stability, as the 'death of development'.

## Post-Washington consensus and resistance from the South

Critics of neoliberalism note that these ideologies have a strong US flavour and that most FDI is registered in US dollars. Because of the dominance of the IFIs, Washington is often seen as 'the undisputed political, economic and ideological centre of the world' (Fine *et al*. 2001: x); the WB is often considered to development theology what the Papacy is to Catholicism. Neoliberalism is almost like a world religion, with its dogmas, doctrines, priesthoods, law-giving institutions and its 'hell for heathens and sinners who dare to contest the revealed truth' (George 1999: 9). For example, when Joseph Stiglitz (WB economist and Nobel Laureate), described the WB's approach to privatization as akin to 'bribarization' he was fired. However, the 1990s witnessed growing discontent with the policies of institutions disseminating neoliberalism. The purported long-term prosperity never materialized and instead poverty reduction has made limited headway. Neoliberalism has also failed in parts of the former Soviet-bloc countries of East and Central Europe. 1997 witnessed the catastrophic meltdown of Asian economies, with crises in Brazil, other parts of Latin America and Russia.

As a consequence, an emergent 'post-Washington consensus' arose based on the belated recognition that market 'failures' and institutions are important conditioning factors in the outcomes of development. The global economy is still dominated by the IFIs, but the post-Washington consensus is seen to contrast with earlier neoliberal doctrine in recognizing 'development' as a complex social process, that institutions work differently in each society and that social relations with institutions and markets are important questions. Whereas neoliberalism ignores inequalities of power, the post-Washington consensus is founded on the notion of 'social capital' – the idea that capital is not simply embodied in land, factories and buildings but also in human knowledge and skills – and power. Therefore, relations between social groups are relevant in the making of development.

The impasse in development theory, coinciding as it did with political-economic crisis and the complete restructuring of the management

of the world economy, created the conditions for the hegemony of
neoliberal orthodoxy within mainstream development thinking.
However, it did produce greater recognition of the diversity of
approaches to and conceptualizations of development, with greater
awareness of environmental concerns, gender equity and grassroots
approaches (Willis 2005). These alternatives have challenged
neoliberal orthodoxies. Development theorists have been particularly
critical of the WB, arguing that it ignores the fact that 'development'
brings about enormous changes in social relations because its notions
of social differentiation and social change are still rather primitive.
It has been consistently criticized for its refusal to acknowledge the
significance of social class and that women are often central to the
forms of social capital that development agencies are keen to mobilize
in poverty reduction programmes. Think of the role of women in
agriculture, community development, natural resource use and so on.
Moreover, while the WB persisted with 'top-down', 'growth first'
policies, networks, community ties, experiences of participation and
new kinship relations were being forged across the South, especially in
Latin America, to deal with the economic crises precipitated by global
capitalism and structural adjustment.

The failures of neoliberalism also brought dominant, Northern
theorizations about development under sustained attack. Some strands
of post-structuralist and postmodern theory advocate disengaging from
development altogether. They call grand theories like modernization
and world-systems theories into question, rejecting universal claims
to knowledge in favour of particularities and difference. Many
scholars and activists concerned with development issues in the South,
especially poverty and economic development, dismiss postmodernism
as a 'First World' preoccupation, if not indulgence, with little practical
application for the South's development problems. However, others
have engaged with postmodernism as a way of moving beyond the
impasse in development thinking. The likes of Arturo Escobar and
some feminist scholars (see Marchand and Parpart 1995) have been
notable in this regard. A renewed emphasis on diversity and difference,
the role of culture in development, as well locally appropriate and
bottom-up approaches to development owe their origins to some
extent to postmodern critiques (Simon 1997). Postmodernism has also
continued to inform radical theories emerging from anti-development,
post-development and postcolonialism, which have accused
development of being nothing more than a neocolonial project that

works in conjunction with political and economic power, sustaining the continued dominance of the global North.

## Post-development

Post-development approaches emerged as a radical reaction to development and a rejection of the idea that attaining the affluent lifestyle of high mass consumption for the majority of the world's population is possible. These approaches reject the idea that the North has the right to administer development and democracy to the South. As discussed subsequently, they thus have much in common with postcolonial approaches, but are aligned particularly with decolonial theory and its insistence on delinking from the West. Postdevelopment theorists criticize the fact that the UN is dominated by countries in the North, G8 states still have over 40 per cent voting power in the WB, compared with the 4 per cent of the whole of sub-Saharan Africa. Individual politicians still have the power to determine development policy at the IMF and WB. In short, post-development theorists lean towards anti-development, are deeply suspicious about the motivations of development organizations and government development policy in the North, and see development as fatally flawed (Rahnema and Bawtree 1997).

Criticisms from within radical development NGOs add fuel to these arguments. For example, War on Want (2004) criticized the UK government for championing privatization in developing countries, arguing that privatization of public services has led to increased poverty. Despite this, developing country governments continue to come under intense pressure to commit their public services to privatization – often as a condition of receiving development assistance, loans or debt relief from international financial institutions and donor governments. The UK's Department for International Development (DfID) has invested heavily in the international privatization programme, creating new bodies and financing mechanisms to advance the cause of privatization across the developing world. It channels large sums of the UK aid budget every year to multinational corporations such as PricewaterhouseCoopers and KPMG in order to drive forward the privatization of public services in developing countries. The consultancy arm of the Adam Smith Institute (ASI) – the right-wing consultancy firm behind the Conservative government's privatization of Britain's public services in the 1980s – received over £34 million from the UK aid budget between 1998 and 2004. Between 2011 and 2017 this rose to £450 million. The War on Want report suggested that the benefits of such business to private sector consultants are clear, yet it is far from clear that these companies are a suitable choice to provide pro-poor reform solutions in the developing world. The report considered DfID's commitment to the privatization of public services to be incompatible with its then stated commitment to poverty reduction. This has proved to be correct: for example, a report by the Global Justice Now (2016) found that electricity consumers in Nigeria faced price increases of up to 45 per cent because of 'a controversial energy privatization programme supported by UK aid through a multimillion-pound project implemented by ASI'. In February 2017, DfID cut off funding to ASI after it emerged that it had tried to profiteer by exploiting leaked department documents and to 'unduly influence' a parliamentary inquiry by engineering 'letters of appreciation' from beneficiaries of its projects. However, despite criticisms of private firms using DfID (i.e. tax-payers') money to pursue a free-market agenda in developing countries, UK aid policy has become more explicitly oriented towards business interests of little value to beneficiaries.

In 2013, a statement by the then Secretary of State for International Development, Justine Greening, about DfID's Global Growth Fund captured the shift away from tackling poverty towards more explicit concern with UK economic interests. Greening announced: 'international development is 100 per cent in our national interest . . . We're helping developing economies grow faster but can we be smarter about the UK locking into the business opportunities those emerging economies present? Yes . . .' (Press Release 29/09/13; see also DfID 2015). And in August 2016, Greening's successor, Priti Patel, announced a new 'partnership' with India, which promises 'support for India to boost economic growth, jobs and trade, which will also benefit Britain' (Press Release 15/08/16). As one commentator puts it, 'Cash earmarked to help people in poor countries will instead be offered to leaders of middle-income giants such as India, China and South Africa, to get them to buy British exports' (Chakrabortty, *The Guardian* 23/08/16). As Noxolo (2017: 2) argues, the mindset behind the recent shifts in policy is 'disturbingly colonialist' (see Chapter 6).

Post-development theorists are critical of economic inequalities and dominance masquerading as aid. They are also critical of the positivist and empiricist methods of the IFIs, particularly the ways in which they define and problematize poverty whilst ignoring the fact that it is culturally and historically variable. One of the most influential theorists, Arturo Escobar, uses the example of Colombia to argue that prior to colonialism countries of the South were not impoverished and thus had no need of development. Issues such as low life expectancy, inadequate sanitation and housing were only seen as problems by the North, which wilfully misrepresents these issues as problems through distorted images in global poverty alleviation campaigns. Escobar's *Encountering Development* (1995a) contains a powerful and detailed critique of US-aided state-led developmentalism that remains one of the most sophisticated accounts of the links between ideology, discourse and practice.

Post-development represents a fundamental questioning of the knowledge and 'regimes of truths' that the North has accumulated about the rest of the world, criticizing assumptions of universality and attempting to re-centre attention on people, places, subjectivities and identities rather than assuming homogeneity. Development is perceived as a range of knowledge, interventions and worldviews, or as discourses and techniques related to the power to intervene, transform and rule over 'others' (see also Cowen and Shenton 1995; Crush 1995; Esteva and Prakash 1998; Li 2007). Post-development re-politicizes development to 'unveil the shroud of humanitarianism that obscures the way the Third World has been produced and controlled by the discourses and practices of development' (Van Ausdal 2001: 578). It proposes alternatives to the 'triumph of the western economy' based on the role of minority cultures and social movements.

A key feature of post-development is its rejection of the concept of development, focusing on its perceived failures and fallacies. The condemnation of development has been harsh: according to Esteva (1987: 135) 'development stinks'; similarly, Sardar (1996: 37) states that 'there is something rotten at the core of the very concept of development'; for Verhelst (1990: 1) it amounts to 'a rape' wherein 'whether by force or seduction' westernization penetrates and dominates the South; Apffel-Margelin (1996: 2) refers to it as 'colonization of the mind'. As with postcolonial critiques, this condemnation centres on the perceived Eurocentrism and cultural imperialism of development, and the apparent seamlessness between colonialism and development.

Post-development has been criticized for putting too much emphasis on westernization and failing to consider other centres that shape development discourse (e.g. East Asia, Japan, the Middle East and more recently China, Brazil and India) – the 'worship of progress' is not just a western preserve. Critics have also problematized its focus on local and grassroots initiatives (Pieterse 2001a), which are often romanticized. Like dependency theories, post-development simplifies 'development' and underestimates the complexities of motives involved (Sidaway 2002). It implies a conspiratorial, intentionalist reading of development, almost as if there is an evil genie controlling a global development project. It is often seen as weak empirically and shallow in its grasp of development institutions (Watts 2003). It ignores some of the critical development literature that, rather than rejecting development, has called for the reform of development institutions. It also tends to neglect the ways in which the critique of development has shaped the development industry. As Pieterse (2001b: 79) argues:

> At a time when there is widespread admission that several development decades have brought many failures, while the development industry continues unabated, there is continuous and heightened self-criticism in development circles, a constant search for alternatives, a tendency towards self-correction and a persistent pattern of cooptation of whatever attractive or fashionable alternatives present themselves. Accordingly, the turnover of alternatives becoming mainstream has speeded up.

Indeed, critiques from within have arguably had greater impact on development. For example, Streeten (1995) contributed to a volte-face by the WB (1997) in recognizing the deleterious impact of its extreme neoliberal policies. By comparison, the impact of post-development on mainstream development has arguably been insignificant and, despite persistent critiques, the notion of development envisaged by modernization has been reinvented in contemporary neoliberal discourses. The IFIs continue to advocate the idea that an economy must be free from the social and political 'impediments' and 'fetters' placed upon it by states that seek to regulate markets. The connections to the Enlightenment and to the writings of Adam Smith are thus enduring and continue to shape development thinking and practice today (Power 2003: 94).

While neoliberalism still dominates, there is some cause for optimism. Post-development approaches attempt to give voice to the excluded and seek to break with the dominance of western rationality and some of the legacies of Enlightenment ideals that have shaped post-war development. They break with the Eurocentrism of modernization theories and attempt to disrupt and dislodge the overwhelming theorization of development from the North. In these respects, they have much in common with dependency theory (Blaney 1996) and postcolonial approaches. And the tendency to portray development as a monolithic 'project' controlled by an evil genie in anti- and post-development is politically strategic and instrumental, since it makes the search for alternatives in other realms essential. As Simon (2006: 13) argues, 'this is the origin of postcolonial perspectives on development and those strands of post-development that have sought to move beyond *critique* of conventional development into more active "re-learning to see and reassess the reality" (Escobar 2001: 153) of the global South'. As discussed in Chapter 6, post-development, postcolonial and decolonial ideas have forged alternative notions of development that have also begun to influence development policy in some countries, particularly in Latin America.

## Postcolonialism and development: convergences

In many ways, as subsequent chapters demonstrate, postcolonialism is potentially a more constructive engagement with development than post-development. Whereas the latter proposes outright rejection of development, postcolonial approaches suggest that it is impossible to stand outside of dominant discourses such as development and instead there is a need to change the discourses from within. With this in mind, it is possible to identify areas for possible convergence between the two bodies of thought and practice.

### Postcolonial challenges to dominant understandings of North–South relations

Postcolonialism effectively extends dependency and post-development theory by reworking globalization as mutual dependency, by criticizing processes of marginalization, and by provincializing the North. The latter depends on decentering modernity as distinctively and universally western to acknowledge multiple modernities, privileging peripheral voices (Robinson 2003a; 2003b) and exploring how European thought might be reinvigorated from and for the margins (Chakrabarty 2000; Comaroff and Comaroff 2012). Chakrabarty's critique of historicism, for example, is particularly pertinent for development studies. Historicism has shaped much of western understandings of modernity, which see history progressing in linear stages based on a model of 'first the West, then the rest'. Thus, there is no place for the traditional in the modern, since by being modern those elements that render a society traditional have been superseded. As discussed, this notion has shaped relations between global North and South since the nineteenth century, wherein the North is perceived to be developed and modern and the South to be 'lagging behind'. However, as Chakrabarty (2000) illustrates, if we theorize from contexts outside of the North then these ideas begin to break down.

The story of modernity in India, for example, is less a story of a linear progression from a traditional society to a modern one, but rather a complex entanglement of modernity and tradition. In European Enlightenment thinking peasants could not become citizens without first being educated. Thus, basic rights to education in Britain preceded universal suffrage by almost fifty years. In contrast, in India, peasants became citizens on independence despite being largely illiterate. This countered imperial notions of colonized subjects being 'not yet ready' for universal human rights on the grounds that they were illiterate. Therefore, the postcolonial challenge to provincialize Europe is not to negate the value of universalisms – human rights, citizenship and democracy – since these have been fundamentally important to the formation of independent, postcolonial nations. Rather, it is to recognize that European models of development and modernity cannot easily be universalized, and that alternative ideas about the nature of human rights and citizenship may, in fact, emerge from places like India. The story of capitalism is similarly prone to such critique. Notions of 'late capitalism' (predicated on notions of progress that still position the North as most advanced and the South as lagging behind) fail to acknowledge that 'late capitalism' might, in fact, be driven from places like India, China, Malaysia or Singapore (Ong 2006).

Comaroff and Comaroff (2012: 121) take this argument a step further by arguing that rather than the global South 'catching up' with the global North, the global North is 'evolving' towards the global South: 'Africa, South Asia, and Latin America seem to be running ahead of the Euromodern world, harbingers of *its* history-in-the-making'. They argue that the former peripheries of the global economy are now the new frontiers in which mobile capital meets with little regulation, low labour costs, highly flexible and informal economies. These spaces now perform outsourced services for the North (call centres in India, for example), but are also developing cutting edge information technology empires of their own. And the new idioms of work, time and value emerging in the spaces of these new frontiers are now taking route in the global North: the working conditions in the so-called 'gig economy' increasingly resemble those of the global South in their informality, precariousness, low wages and lack of regulation. It makes little sense, therefore, to cleave to the notion that European economic models can and should be universalized, but this idea remains hegemonic within international development.

The challenge to provincialize development emerges from the core strategies of postcolonialism (summarized in Box 1.5), namely:

● Destabilizing the dominant development discourses of the West on the grounds that these are unconsciously ethnocentric, rooted in western cultures, reflective of a dominant western world view, and profoundly insensitive to the meanings, values and practices of other cultures.
● Problematizing processes of speaking and writing by which dominant development discourses come into being (e.g. terms such as 'the Third World') and contesting the power of the North to control, produce and deploy development knowledge.
● Challenging binary divisions of 'First World'/'Third World', 'West'/'rest', 'developed'/'developing' by emphasizing interconnectedness, that the South is integral to Northern 'modernity' and 'progress', and that modalities and aesthetics of countries in the South have partially constituted languages and cultures in the North.
● Attempting to recover the voices of marginalized and oppressed peoples in the South, particularly through an understanding of their potential and actual agency and resistance in development theory and practice. This involves paying heed to 'the ineluctable reality that many disadvantaged people across the world desire much of what *they* understand by the modern' (Comaroff and Comaroff 2012: 120).

# Convergences around social and cultural dimensions of development

The emergence of postcolonial approaches in development is part of a broader and concerted effort by theorists in recent years to shift culture from the margins of development studies and to demonstrate that it is, and always has been, central to development processes and their impacts on peoples and societies (Allen 1992; Porter *et al.* 1991; Radcliffe 2006; Schech and Haggis 2000). For example, colonial governmentality worked through local culture to attempt to construct a new sort of colonial subject; paradoxically, traditional cultures were also seen as an impediment to progress, innovation and 'development'. Modernization theories of the 1950s and 1960s were unable to see beyond culture; theorists 'read culture out of their own theory of the modern' (Watts 2003: 434). Cultural difference has consequences for growth, from the cultural particularities of the South East Asian 'tiger' economies (Rigg 1997) to the perceived deficiencies of African moral economies (Escobar 1995b; Ferguson 1994). Development theories no longer describe a transition from tradition to modernity where cultures converge, but recognize modernity and development as embedded within specific cultural contexts and are thus sensitive to the different starting points of transition and its outcomes. In other words, development is always culture and site specific – 'it is irreducibly *cultural geographic*' (Watts 2003: 435). It is culturally and ethnographically grounded – in projects, development institutions and so on (see Goldman 2005). In addition, theorists and activists seeking to posit alternatives to dominant notions of 'development' are sensitive to how cultural diversity makes a difference to imagining and re-imagining development, and in conceiving alternatives.

Despite this 'cultural turn' within development, it might be argued that there is still a large gap between how culture in the development process has been conceptualized and implemented. Rao and Walton (2004) acknowledge that cultural notions are now routinely incorporated into practice. However, they also point out that academics (they refer to anthropologists specifically), on the one hand, seem focused on critiquing development rather than engaging with it constructively while, on the other hand, policy economists either treat culture as emblematic of a tradition-bound constraint on development or ignore it completely. While the criticism that academics rarely engage constructively with development practice might be overstated, it would seem that there are some gaps between theory and practice that require closing (McEwan 2006a). In recent years, postcolonial approaches to development have begun to close this gap (Box 3.4).

---

## Box 3.4 Postcolonial development studies in Zambia and Barbados

Garth Myers (2006) uses a case study of Lusaka, Zambia, to develop a postcolonial approach to urban development. He focuses on the idea of exclusionary democracy and the concept of the domestication of difference, exploring how both can be traced from the colonial period into the contemporary context. The political and planning dynamics of the last years of colonialism shaped state–society relationships in contemporary Lusaka. This ensured that the processes creating exclusionary democracy and solidifying difference between groups continue in albeit ambivalent and incomplete ways in the present. Myers thus develops a postcolonial approach that remains connected to material issues – because so many Lusakans lack basic services and human rights, a search for the roots of exclusionary politics and differentiation that drive inequalities remains pertinent. He proposes a direct engagement with 'what is happening politically' as a result of policy, which is itself a set of discursive tactics rather than simply apolitical policy analysis draped in the language of 'good governance'. When ordinary people have their homes demolished in the twenty-first century and their governments justify their actions by recourse to late colonial ideas of 'sanitization' in the city (as Robert Mugabe did in Zimbabwe in 2007), such policies must be directly confronted for the colonialist exclusionary differentiation they encapsulate. As Myers (2006: 306) argues, in such ways, postcolonialism has 'great potential for contributing to an understanding of the ongoing transformations of the politics of urban development in the world and may be more policy relevant than many more explicitly policy-driven works'.

Similarly, work on the significance of transnationalism in development theory and policy has also begun to draw on postcolonial approaches. For example, Phillips and Potter (2006) expand on the opportunities afforded by the adoption of postcolonial discourse in development studies, drawing specifically on issues of transnationalism, hybridity and inbetweenness in their research on the migration of young, second-generation and foreign-born 'Bajan-Brits' to Barbados (the homeland of their parents). They focus on issues of 'race' and gender to examine the identities and experiences of return migration among this cohort from an interpretative perspective framed within postcolonial discourse. In particular, they draw on Fanon's *Black Skin, White Masks* to explore the complexities of the returnees' identities – shaped by both the considerable socio-cultural problems of adjustment they encounter and their positions of relative economic privilege, and by having been born and/or raised in the UK and being black. Accordingly, such migrants are both advantaged and disadvantaged, both transnational and national, and black but, in some senses, symbolically white.

Postcolonial theory presents significant challenges to the universalist assumptions that have underpinned modernization theories of development. As Simon (2006) argues, rather than acknowledging cultural diversity – the ways in which local, indigenous, cultures in the South differ from those of Northern industrial capitalist cultures – development studies viewed this diversity as a significant obstacle to development to be overcome by large-scale, technologically driven aid and investment programmes. 'Homogenization of the western model was the implicit and explicit objective' (Simon 2006: 14). As we have seen, Said (1985) and the Subaltern Studies collective in India (Chapter 2) were amongst the first to provide an explicitly postcolonial perspective on Eurocentric notions of modernity and progress. In relation to Africa, for example, development imports a 'reductive repetition', wherein the diversity of African historical experiences and trajectories, socio-cultural contexts and political situations are reduced to a 'set of core deficiencies for which externally generated "solutions" must be devised' (Andreasson 2005: 971). Thus, explanations of Africa's economic marginalization fail to interrogate the complex historical and contemporary processes that perpetuate this. The need to insist on Africa's fundamental inadequacies, rather than the inadequacies of Northern models, becomes urgent. Thus, explanations range from technocratic ones, such as the failure to erect structures of 'good governance', to root out corruption, to attract foreign direct investment, to vulgar travelogues depicting the inherent chaos and terror of the 'Dark Continent'. The blame for the failure of the western modernist project to successfully conquer the African continent is placed squarely on 'African culture'. The solutions, therefore, must continue to be generated externally.

Universalizing neoliberal approaches continue to advocate global cultural convergence and homogenization through the assimilation of western capitalist consumer cultures. As Pieterse (2004: 7–21) argues, these neoliberal meta-narratives of development constitute the contemporary versions of modernization theory. However, since the mid-1990s, postcolonialism has helped focus attention on the need to understand the complexities and differences at local levels that highlight human agency, multiple perceptions and the existence of multiple and alternative types of knowledge.

## Convergence around concern with marginalized peoples and progressive politics

The problem of development as a neocolonial project is not necessarily one of intention or aims. The intentions and aims of many individuals involved in development since the 1950s have been honorable and well meaning. Institutions like the WB do not set out to impose neoliberalism or exacerbate problems. However, the claim to expertise to diagnose problems and devise solutions is a claim to power and continuity from the colonial period into the present (Li 2007). Development praxis 'may perpetuate colonialist and western-centred discourse and power relations, even as it seeks to focus attention on the marginalized' (Sharp and Briggs 2006: 7). While the development specialist may be materially trying to improve the situation of less privileged groups in the South, the effect of their discourse (see Chapter 4) and practice is often to reinforce racist, imperialist conceptions, to erase complexity in the search for technical solutions, and, in some cases, to prevent less privileged people speaking and being heard (Alcoff 1991: 26) (see Chapter 6). As Chapter 5 illustrates, power relations cannot be ignored since these determine and inflect relationships with 'distant others'. These are profoundly unequal in terms of economic exchange and exploitation, political influence and the geographies of knowledge and culture, which have roots deep in a history whose legacies cannot simply be transcended by good intentions. This raises the question of whether development is inescapably problematic and doomed to failure: 'If we see development as a potentially neocolonial enterprise, is it ever possible to recover the sense of the development project as an altruistic endeavour with vital emancipatory potential?' (McKinnon 2006: 22).

Postcolonialism has been accused of offering no way out of this conundrum and, like many 'post-ist' critiques, of being disabling and nihilist. If we are always already constructed through the discourses of power and knowledge that have historically created North/ South, is it ever possible to step outside this to effect change? Anti-development theorists conclude that there is nothing redeemable about development, that it will always work in the interests of the North and perpetuate inequality, and thus should be resisted (Escobar 1995a: 98). The proposed alternative is localism based around community and indigenous responses, but this too fails to acknowledge that power relations work at local levels, restricting the abilities of particular groups and individuals in speaking and acting. It also fails to acknowledge the connections between the local and other scales. In contrast, postcolonialism need not reject development outright. Indeed, the accusation that postcolonialism is nihilist and disabling is somewhat unfair. For Spivak (1987), the historicizing of power relations between individuals and institutions can actually inspire critical curiosity into the processes that have allowed some voices to be heard, and this can be transformative. Many argue that development, while problematic, still contains elements of optimism that are vital to progressive politics (Simon 2006). It is also important to remember that development has always been informed by a greater diversity of ideology, discourse, policy and practice than the caricature of a monolithic modernization project suggests. What is needed, therefore, is for development studies and postcolonial approaches to be mutually informative. Development studies requires a more expansive notion of power and agency, while postcolonialism requires a better understanding of the structuring role of resources and institutions in the creation and maintenance of networks of power (McFarlane 2006).

## Challenging critiques of postcolonial theory in the context of development

From the perspective of development, some of the criticisms levelled at postcolonial theory (see Chapter 2) are pertinent, but they also require some rebuttal. Realists accuse postcolonialism of ignoring issues of the human rights and freedoms of marginalized people. Concerns with representation, text and imagery are perceived as too far removed from the exigencies of the lived experiences of millions of impoverished people (Jackson 1997). In dismissing the universalist assumptions of political economy, postcolonial approaches have also been accused of ignoring the material ways in which colonial power relations persist: 'it is remarkable . . . that a consideration of the relationship between postcolonialism and global capitalism should be absent from the writings of postcolonial intellectuals' (Dirlik 1994: 353; see also Ahmad 1992). Even scholars sympathetic to the ethics and politics of postcolonialism have pointed out that debates about postcolonialism and globalization have largely proceeded in relative isolation from one another, and to their mutual cost (Hall 1996: 257). Economic relations and their effects elude representation in much of postcolonial studies (Eagleton 1994). As Jacobs (1996: 158) notes, the theoretical abstractions of postcolonial theory do not always adequately connect to the specific, concrete and local conditions of everyday life and are not easily translated into direct politics. This apparent neglect of material concerns and political strategies has generated the fiercest criticism of postcolonialism, with critics accusing it of ignoring urgent life-or-death questions (San Juan 1998) and of solidifying the fundamental schism between western theorizing and the practical needs of impoverished people globally.

The argument that postcolonialism is too rooted in discourse might have some credence given the origins of postcolonial theory in the humanities. However, of course, this also ignores the fact that discourse itself is intensely material. Indeed, this has been demonstrated with examples ranging from the ordering of imperial and postcolonial urban spaces, to the materialities of travel and emigration, to concerns with embodiment, identities, cultural politics and reconciliation (see Blunt and McEwan 2002). Similarly, intersections between postcolonialism and feminism have had some influence in development and have demonstrated how discourse informs lived experience in ways that are relevant to women everywhere, whether they are striving for economic empowerment whilst having simultaneously to renounce 'normality', or facing the conundrum of attaining citizenship whilst becoming alienated subjects (Quayson, 2000). This is compounded for those women most marginalized by global inequalities. For example, Spivak (1985a, 1999) draws out the connections between the silencing of 'Other' women, who are often spoken for, about and against, and their marginalized position within global economies. Postcolonial feminisms have made important contributions in exploring the links between the discursive and the material in creating possibilities for effecting change (Rajan 1993; Rose 1987).

Clearly, it is not sufficient to confine analysis to texts alone, but there are connections between the relations of power that order the world and the words and images that represent the world. The challenge for postcolonial development studies is to respond to the potential of 'mixing up conceptual elaboration with substantive detail' (Philo 2000: 27) and of dealing simultaneously with the material and immaterial, the cultural and the political. To have greater immediacy in critical development studies, postcolonial approaches need to engage more with questions of inequality of the power and control of resources, human rights, global exploitation of labour, genocide and so on. Of course, this raises questions about where postcolonial studies (which has its origins in an engagement with the representational) ends and critical development studies (alert to global inequalities and so on) begins. This is worthy of further debate, but at this juncture it is sufficient to recognize that concern for material practices and spaces need not be disconnected from discourses, texts, imaginings and counter-imaginings, since there are fundamental entanglements between the two.

Postcolonial theory necessarily positions itself in critical opposition to global inequalities. It is perhaps ironic, therefore, that one of the most persistent criticisms is that it has failed to consider the relationships and tensions between postcolonialism and global capitalism (Dirlik 1994; Eagleton 1994; Hall 1996). According to Parry (2002: 78) 'the sanctioned occlusions in postcolonial criticism are a debilitating loss to thinking about colonialism and late imperialism. This dismissal of politics and economics which these omissions reflect is a scandal'. With respect to the more cultural issues of the politics of recognition, postcolonial approaches might appear radical and progressive, but from the perspective of political economy and the politics of distribution they look less progressive, 'for they offer no means for challenging the economic system' (Sayer 2001: 688). The relative neglect of class in favour of identity politics in postcolonial analyses has also been criticized as a potentially serious omission, both in terms of the conflicting class interests within post-independence political formations and the international alliances forged by the new indigenous ruling classes. However, one response to these criticisms is that without Marxism some of the best ideas that postcolonialism has produced (e.g. in the work of Fanon, Bhabha and Spivak) would be much poorer (see Chapter 1). In addition, postcolonial theorists such as Said and Chakrabarty have consistently argued for a postcolonial criticism that is worldly and attuned to both discursive and material concerns. The two should not necessarily be seen as antagonistic, as some critics of postcolonialism would suggest.

A further criticism is that apart from the adoption by some post-development theorists of discursive approaches (e.g. Escobar 1992, 1995b; Esteva 1987), postcolonialism has not easily been translated into action on the ground and its oppositional stance has not had much impact on global power imbalances or inequities. However, as Rattansi

(1997: 497) argues, 'it is simply untrue to say that global capitalism has been ignored in postcolonial research, although . . . what postcolonial studies has been about is finding non-reductionist ways of relating global capitalism to the cultural politics of colonialism'. Postcolonial research has explored the constitutive relation between imperialism, colonialism and global capitalism (e.g. Chatterjee 1996; Miyoshi 1997; Said 1993; Spivak 1988). While class relations are not often explicit, in many studies they are implied. As discussed in Chapter 2, much of Spivak's work, for example, has been concerned with exploring connections between the micro-spaces of academe and the macro-spaces of the global economy and international division of labour, and between the discursive construction of gender and the doubly subaltern position of women in the former colonies.

Similarly, the failure of discursive approaches to engage with critiques of capitalism is also to some extent overplayed. As Ashcroft (2001) argues, despite the centrality of representation, the significance of postcolonial analysis is its insistence on the importance of the material realities or lived experiences of postcolonial life that are directly related to economic issues. He uses the example of the consequences of the rise to prominence of tropical sugar for Caribbean societies and cultures to demonstrate the link between the material and the discursive in processes of postcolonial transformation (Box 3.5), and how political economy and cultural approaches might work in tandem to critique the dominant order (Perrons 1999). Similarly, Young (2001: 428) makes a powerful case that while postcolonial critique challenges established, Eurocentric knowledge in the cultural sphere, it also continues to work in the spirit of anti-colonial movements by further developing its radical political edge to reinforce global social justice. Its 'politics of power–knowledge asserts the will to change' the injustice, inequality, landlessness, exploitation, poverty, disease and famine that remain the daily experience of much of the world's population. As Yeoh (1991: 462–3) suggests, the task of interrogating the relationship between postcolonialism and global capitalism is a crucial and relatively neglected one that 'requires a more critical and simultaneous engagement with both registers' (see also Hall 1996; Slater 1998). Increasing dialogue between postcolonialism and development studies promises this kind of engagement.

Postcolonialism is significant for development studies because it foregrounds the workings of power in the global capitalist political economy and exposes its cultural and ideological underpinnings.

## Box 3.5  Discourse and material economies: sugar and colonialism

The interrelation between the material economies of colonialism and the transformative dynamic of colonial discourse has been profoundly important. Although the transformation of representation is crucial, such practices are situated in a material world with, in most cases, urgent material implications. The story of the extraordinary rise of sugar, both in the economy and in the diet of Britain, illuminates what post-colonial transformation actually means in the lives of colonized peoples. The sugar industry had a catastrophic effect on the Caribbean environment, culture and people. Yet out of the ruins caused by European obsession with sugar, an obsession which had extraordinarily damaging effects on tropical plantation colonies, arose a culture so dynamic that it has acquired a peculiar place in global culture.

The picture is not entirely bleak, however. The history of the Caribbean might appear to be one of social and political decay, exploitation, disruption and ruin. However, a postcolonial approach allows us to ask 'how did these displaced, traumatized and diasporic cultures break down the brutal binaries of the colonial plantation society and produce one of the most vibrant cultures in the world today?' These societies transformed themselves by utilizing the heterogeneous range of cultural backgrounds and influences that constituted them, employing many of the social strategies of resistance built up through centuries of plantation slavery. Sugar workers, under conditions of extreme exploitation, poverty, hardship and dislocation, developed forms of cultural resistance that came to characterize the vitality of the Caribbean. Ultimately, the transformative effects of these cultures circulated into imperial and global centres.

In sugar, therefore, we find an extreme display of the transformative development of post-colonial futures. As an overwhelmingly colonial product, a product consumed everywhere, a product whose production devastated environments, displaced huge populations of diasporic peoples, revolutionized patterns of consumption, it is an unparalleled linchpin of economic history. Sugar focuses the dynamic reality of post-colonial strategies as no other single product has ever done.

Source: Adapted from Ashcroft (2001: 67–80)

It pays particular attention to cultural politics, particularly the imbrication of race, class and gender with power (Chowdhry and Nair 2002). This allows an alternative critique of global power hierarchies and relations. It also allows development studies to engage with the cultural politics of the colonial past and postcolonial present.

## Summary

- There have been historical divergences between postcolonialism and development; the lack of dialogue between the two can be explained by the oppositional stance adopted by postcolonial scholars to development on the grounds of its western-centricity.

- Postcolonialism can be placed within a tradition of critiques of post-war development as an idea and as a practice, which includes dependency and post-development theories.

- Applying postcolonial critiques to development does not imply anti-development; rather it aims to understand the power of development ideas, knowledge and institutions and their consequences in particular places at particular times.

- While development has been criticized for not listening to subaltern voices, postcolonialism has been subject to criticisms that it is becoming institutionalized, representing the interests of a Northern-based intellectual elite rather than people in former colonies, and that it is too theoretical and not rooted enough in material concerns.

- Emphasis on discourse is accused of detracting from an assessment of the material ways in which colonial power relations persist, but this underplays the consideration given to the relationship between postcolonialism and global capitalism.

- Despite mutual criticisms, the tensions between postcolonialism and development studies are starting to prove productive and increasing convergences are emerging around social and cultural dimensions of development.

- Postcolonialism poses significant questions for development studies, particularly around issues concerning the material effects of discourse and notions of agency and power.

- Postcolonialism addresses important concerns such as the impact of colonial practices on the production and reproduction of identities, the relevance of cultural politics in understanding domination and difference, and the relations between global capitalism and power.

## Discussion questions

1  In what ways have Enlightenment ideas influenced post-war development thinking?
2  Explain the continuities between imperial and colonial ideologies and notions of development after decolonization.
3  What are the similarities and differences between post-development theories and postcolonial approaches?

4   How and why are development studies and postcolonialism beginning to converge?
5   What are the main criticisms of postcolonialism in the context of development and how pertinent are these?

## Further reading

Cowen, M. and Shenton, R. (2006) *Doctrines of Development* London, Routledge. An informative examination of the history of the idea of development and of the doctrines that governments have used to practice development policy.

Crush, J. (ed.) (1995) *Power of Development* London, Routledge. A collection of essays that conceptualize development as a discourse and explore the language and rhetoric of development texts.

Pieterse, J.N. (2001b) Development Theory: Deconstructions/Reconstructions London, Sage. Covering issues of Eurocentrism, critical globalism, intercultural transaction, de-linking and post-development theory, and connecting issues of development with critical theory.

Power, M. (2003) *Rethinking Development Geographies* London, Routledge. A critical examination of the spatiality of development thought and practice, including a chapter on postcolonialism and development.

Radcliffe, S. (2015) Dilemmas of Difference. Indigenous Women and the Limits of Postcolonial Development Policy Durham, NC, Duke University Press. An exploration of the relationship of rural indigenous women in Ecuador to the development policies and actors that aim to help ameliorate social and economic inequality, but because of their inability to recognize and reckon with the legacies of colonialism, reproduce the very poverty and disempowerment they are there to solve.

Rahnema, M. and Bawtree, V. (eds) (1997) *The Post-Development Reader* London, Zed Books. Brings together key writings on post-development by scholars and activists from both North and South.

## Useful websites

www.adamsmith.org The Adam Smith Institute, promoter of free market ideology. Aims to increase awareness of the work of Adam Smith and the role of market-led economic development.

www.brettonwoods.org Bretton Woods Committee, provides information about international financial matters and the role of the IMF, WB and WTO in global economic policy.

www.dfid.gov.uk/ UK Government's Department for International Development.

www.globaljustice.org.uk/sites/default/files/files/resources/the_privatisation_of_ uk_aid.pdf Report by Global Justice Now (2016) on the privatization of UK aid.

www.imf.org International Monetary Fund, provides information on the approaches and activities of the IMF, with useful information on structural adjustment programmes.

www.socialistinternational.org The Socialist International, a worldwide organization of socialist, social democratic and labour parties. Promotes quite different ideologies to those above.

www.usaid.gov United States Agency for International Development.

www.waronwant.org/ War on Want, charity and activist organization that fights poverty in developing countries in partnership and solidarity with people affected by globalization.

www.worldbank.org World Bank information and research on development practice and policy.

# Discourses of development and the power of representation

## Introduction

Postcolonialism was until relatively recently neglected within development studies. As Kothari (1996: 13) argued, 'most development academics and practitioners have never heard of Said, Spivak, Bhabha, Fanon and other post-colonial thinkers' and texts on Eurocentrism are 'rarely found on reading lists on Development Studies courses in Development Studies departments and institutions'. It is only relatively recently that this has begun to change and postcolonialism has brought development discourses into critical view as unconsciously ethnocentric, rooted in European cultures and reflective of a dominant western world view. This chapter examines how postcolonial approaches challenge the very meaning of development as rooted in colonial discourse, depicting the North as advanced and progressive and the South as backward, degenerate and primitive. It explores further the notion that postcolonialism is a much-needed corrective to the Eurocentrism of a great deal of writing on development. It also examines the ways in which postcolonialism inspires counter-discourses that seek to disrupt the cultural hegemony of the West and its knowledge forms, including development knowledge.

Postcolonial critiques cast light on the ways in which development studies have assumed the power to name and represent other cultures, societies and peoples today. They problematize the ways in which the world is known, particularly the homogenizing of the South into the 'Third World', and challenge the unacknowledged and unexamined assumptions at the heart of western disciplines that are profoundly insensitive to the meanings, values and practices of other cultures. Language is fundamental to the way in which interventions, such as those concerned with development, are ordered, understood and justified. By foregrounding the power of language, postcolonialism offers ways of understanding what development is and does, and why it is so difficult to think beyond it. As discussed in Chapter 3, postcolonial and post-development theories are often inspired by similar streams of social theory. Both offer critiques of development informed by postmodernism and post-structuralism and both share common interests in diversity and difference. Some critics see these theories as irrelevant to interpretations of cultures and societies in the South because they are 'infused with French social theory' (Paolini 1997: 84). However, the recognition of difference and the significance of context is important in countering the universalizing tendencies of both colonial and contemporary development discourses. The significance of postcolonial theory lies in its critique of the Eurocentrism of certain kinds of scholarship about the South and its material consequences.

## The language of development

Development texts are written in a representational language, using metaphors, images, allusion, fantasy and rhetoric to create imagined worlds that arguably bear little resemblance to the real world (Crush 1995). Consequently, development writing often produces and reproduces misrepresentation. Postcolonialism seeks to remove negative stereotypes about people and places from such discourses, requiring that categories such as 'Third World', 'developing world' and 'Third World women' are problematized and rethought, and an understanding of how location, economic role, social dimensions of identity and the global political economy differentiate between groups and their opportunities for development. An understanding of the discursive power of development is central to this.

### Discourse and 'othering'

Discourse refers to the ensemble of social practices through which the world is made meaningful. It is not confined to words and pictures, but encompasses their material effects. Discourses shape the contours of the 'taken-for-granted world'; they are produced and reproduced through representations to naturalize and universalize a particular view of the world and to position subjects in it. Discourses always provide partial, situated knowledge. They are embedded in power relations, but are always open to contestation. Moreover, discourse determines what it is possible to say, the criteria of 'truth', who can speak with authority and where such speech can be spoken. Discourse is important to development, not least because our knowledge of the world is constructed in a variety of ways, through experience, learning, memory and imagination. This raises important questions: what 'experiences' and whose 'learning' has been brought to bear in understanding the world? What are the material consequences of this?

As discussed in Chapter 1, the idea of the subaltern has been central to postcolonial theory, particularly the exploration of the processes that have rendered the peoples of colonized countries (and continue to render them in contemporary contexts) as subordinate, inferior and without agency or voice. These processes are overwhelmingly discursive. Said's *Orientalism* (1978) referred to this process as 'othering', and critiqued the constellation of false assumptions underlying western attitudes towards other parts of the world (see Chapter 2). The term 'Subaltern' contains within it the notion of alterity, meaning 'otherness', which refers to the systematized and hierarchical construction of difference between groups of people based on such factors as 'race', ethnicity or culture. Said exposed the subtle and persistent Eurocentric prejudices against non-European peoples and their cultures, which he argued had served as an implicit justification for Europe's colonial and imperial ambitions. The denigration of oppressed, colonized peoples and their rendering as subaltern subjects was crucial to the self-definition and assumed superiority of the colonizers.

Central to western philosophy has been the notion that the 'self' is defined by constructing the 'other'. Binary oppositions have shaped western knowledge forms as far back as the Enlightenment and even ancient Greek philosophy. These binaries are not innocent, but are bound up in the logic of domination; they have material consequences. As Haraway (1991: 177) argues:

> (Binaries like) self/other, mind/body, culture/nature, male/female, civilised/primitive, reality/appearance, whole/part, agent/resource, maker/made, active/passive, right/wrong, truth/illusion, total/partial, God/man . . . have all been systematic to the logics of domination of women, people of colour, nature, workers, animals – in short, domination of all constituted as others, whose task is to mirror the self.

Binaries function by establishing the normal, normative (good), self which is mirrored by the abnormal, deviant (bad), other:

| | | |
|---|---|---|
| Self | / | Other |
| Man (male/masculine) | / | Woman (female/feminine) |
| Developed | / | Undeveloped/Underdeveloped/ Developing |
| Culture | / | Nature |
| Active | / | Passive |
| Civilized | / | Savage |
| Progress | / | Backwardness |
| White | / | Black |
| Middle class | / | Working class |
| Heterosexual | / | Homosexual |
| Able bodied | / | Disabled |
| Rational | / | Hysterical |
| Sane | / | Lunatic |
| Christian | / | Non-Christian |
| West (Occident) | / | East (Orient) |
| Democratic | / | Barbaric |
| Good | / | Evil |
| Normal | / | Abnormal |

These relationships are both horizontal and vertical. Being identified with the qualities on the left often implies greater advantages; being identified with those on the right means deviating from the norm, being more likely to inspire fear and thus being subject to logics of domination. For example, in nineteenth century Europe, women were seen as passive, closer to nature and less developed than men in evolutionary terms: 'All psychologists who have studied the intelligence of women . . . recognize today that they represent the most inferior forms of human evolution and that they are closer to children and savages than to an adult, civilized man' (Gustave Le Bon 1879, cited in McClintock 1995: 54). Women were considered prone to hysteria (from the Greek word *hystera* (uterus)), and to madness (lunacy from the Latin *luna/lunaticus* linking madness with the phases of the moon; women's menstrual cycle coincides closely with the lunar cycle, hence they were perceived to be governed not by reason but by nature). These traits are in opposition to the norm, which renders them threatening and requiring control. Pseudo-scientific arguments based around such thinking had material effects. For example, women in Britain had the legal status of minors, without rights to property or enfranchisement, and were denied access to formal education on grounds that they were biologically and psychologically ill equipped to enter public spheres of politics and education. The pseudo-science of 'race' in the nineteenth century was equally pernicious and its material effects even more enduring.

Binaries have been used to differentiate whole swathes of the world from each other and were deeply entrenched in imperial and colonial discourses:

> The identification of the Northern temperate regions as the normal, and the [colonial] tropics as altogether other – climatically, geographically and morally – became part of an enduring imaginative geography, which continues to shape the production and consumption of knowledge in the twenty-first century world.
>
> (Driver and Yeoh 2000: 1)

Postcolonial theory examines the centrality of alterity to imperialism and colonialism, global geopolitics and development theory and practice, with the construction of First World/western/Northern/ advanced/developed/modern 'self' and Third World/Southern/ backward/developing/traditional 'other' framing interventions in both colonial and post-colonial contexts.

Western colonial projects were based partly on an imagination of the world that legitimized and supported the power of the West to dominate 'others'. By representing other societies as 'backward' and 'irrational', the West emerges as 'mature', rational and objective. The notion of 'the Tropics' was invented alongside a range of other labels (e.g. the 'Orient') to view certain societies as 'other' and different from the European or 'western' self. As discussed in Chapter 2, this 'othering' has been theorized as 'Orientalism' (Said 1985) – an imperial vision of particular places and peoples that had enormous material consequences (Box 4.1). One of Said's most important insights was that techniques of othering were a political vision of reality whose structure promoted the difference between the West and the rest. Parts of the world were discursively produced as the West's inferior Other, which in turn constructed and reinforced the self-image of the West as superior. The two were discursively locked together as each other's opposite – the Other was irrational, sensual, backward, feminized and despotic; the West was rational, progressive, dynamic, masculine and democratic (Baaz 2005). Colonial discourses extended and cemented the divisions between Europe and its colonies.

## Box 4.1 **Imperial discourse: the fiction of *terra nullius***

Modern applications of the term *terra nullius* (translated from Latin as 'no one's lands') stem from sixteenth- and seventeenth-century discourses describing land that was unclaimed by a sovereign state recognized by European powers. During the eighteenth century, it gave legal force to the settlement of lands occupied by 'primitive' people, where no system of laws or ownership of property was held to exist. Thus colonization was soon justified as 'bringing faith and civilization' to the 'savages' and 'barbarians', even if that meant the subjugation or even destruction of indigenous polities, religions and cultures that had existed over millennia. During the nineteenth century, and well into the twentieth, this doctrine was bound up with imperial racisms, some of which were inspired by social Darwinism. Consequently, many European imperialists believed that what they called the 'lower races' were doomed to extinction, particularly those indigenous inhabitants of the continents into which European countries were expanding – Siberia, Central Asia, North America, South America, North Africa, South Africa, Australia and the islands of the Pacific. *Terra nullius* was a legal fiction used to justify this expansion and was especially common in defence of the British invasion of Australia. Aboriginal lands were declared *terra nullius* since they were inhabited by people who would soon die out. These discourses created a self-fulfilling prophecy, legitimizing acts of dispossession, mass violence and genocide in the name of British imperialism:

> In 1836 Major Thomas Mitchell passed through the land of the Jardwadjali people [in Victoria, Australia] . . . .. During this search for exploitable land, Mitchell claimed he was exploring a *terra nullius* – an empty land – despite having contact with local indigenous people, some of whom his party murdered. Mitchell wrote: 'It was evident that the reign of solitude in these beautiful vales was near a close; a reflection which, in my mind, often sweetened the toils and inconveniences of travelling through such houseless regions'. He described the country he saw as 'resembling a nobleman's park', and as an 'Eden' awaiting 'the immediate reception of civilized man'. Its 'primitive' inhabitants would be swept aside in order to add to the wealth and power of the British empire . . . .. Although the land had been occupied for thousands of years, Mitchell was able to map a 'socially empty space' (Birch 1996: 180).

Colonized Others were, in many ways, defined in terms of their difference from the West. The same dichotomies – nature/culture, emotion/reason, irrationality/rationality, body/mind, passivity/ innovation, feminine/masculine, barbarism/civilization, unrestrained sexuality/control – informed representations of the colonized Other, whether in Asia, Africa or the Americas:

> Despite the enormous differences between the colonial enterprises of various European nations, they seem to generate fairly similar

stereotypes of 'outsiders' – both those outsiders who roamed far away on the edges of the world, and those who . . . lurked uncomfortably nearer home. Thus laziness, aggression, violence, greed, sexual promiscuity, bestiality, primitivism, innocence and irrationality are attributed (often contradictorily and inconsistently) by the English, French, Dutch, Spanish and Portuguese colonists to Turks, Africans, Native Americans, Jews, Indians, the Irish and others.

(Loomba 1998: 107)

The savage Other – to be feared, controlled, disciplined and civilized – was inside as well as outside of European cultures. In both the sciences and the arts, the profiles of the colonized (especially Africans), came to match those of the Irish, Jews, women, children, criminals, prostitutes, the mentally ill and the working classes in Europe (Gilman 1985; McClintock 1995). The colonized were feminized and rendered childlike. Othering affected different colonized peoples in different ways and differentiated among colonized peoples, primarily through the *invention* of ethnic groups. Rwanda provides one of the clearest examples of where discourses of race, gender and evolution were used to differentiate between colonized peoples in a policy of 'divide-and-rule' (Box 4.2). Here, because ethnicity was constructed to differentiate people, the colonial presence was experienced very differently depending on how a person was defined. While it is now anathema to racialize people based on physical characteristics, the legacies of such differentiation continue into the present, in the case of Rwanda to catastrophic effect.

One example of this continuing discursive othering is the reporting by western media of the ongoing conflict in Mali. This started in January 2012 with insurgent groups in the north fighting against the Malian government for greater autonomy for the Tuareg people. Islamist groups took advantage of the situation to take control of areas of the north of the country. The government of Mali asked for foreign military help to re-take the north and in January 2013, the French military intervened, followed by the deployment of armed forces from African Union countries. By February, the Islamist-held territory had been re-taken, but Tuareg separatists have continued to fight the Islamists. Having signed a peace agreement with the government in 2013, the Tuareg rebels pulled out, claiming that the government had reneged on its commitments to the truce. A ceasefire agreement was signed in 2015, but sporadic violence still occurs. The reporting of this conflict in the West was heavily criticized on social media. As one commentator writes:

> The media narrative about Mali today is that of just the 'typical' African conflict – heavily armed dark-skinned men in dusty Jeeps, warring tribes destroying ancient cities, terroristic violence in the desert. Yet another over-simplistic version of Black-on-Black violence with an international twist and a dash of al-Qaeda. While this narrative provides the [western] media with easy sound bites and conveniently reinforces stereotypes, it is also patently untrue. Instead, what's going on today in Mali is a complex tapestry of local issues, minority grievances, Islam in Africa, and the far-reaching effects of colonialism.
>
> (France François February 4 2013, www.ebony.com/news-views/
> mali-in-conflict-384#axzz2K7W2iEvE)

Discourses of otherness were central in legitimating colonial domination and expansion. However, colonialism rested on a core contradiction between these discourses of otherness, which needed to fix the Other as always, irrevocably different, and a civilizing mission that rested on the possibility that the Other could be redeemed and become 'just like us' (see Chapter 2). Colonial discourses were characterized by an 'ironic compromise of mimicry' based on a desire for a reformed, recognizable Other (Bhabha 1994: 86). The colonized should become like the colonizer but, simultaneously, remain different – 'almost the same, but not quite'. This apparent contradiction was a critical element in legitimating colonial expansion and in constituting 'the white man's burden' to civilize the Other. It also provides a justification for continued economic and geopolitical interventions in the name of development, and still frames the way in which the world is often understood through western eyes.

---

## Box 4.2  Colonial ethnicization of Rwanda and post-colonial consequences

The assassination of Rwandan President Habyarimana on 8 April 1994 triggered a three-month campaign of violence. An estimated 800,000 Rwandans were killed in 100 days. Both the scale and brutality of the genocide and the refusal of outside powers to intervene to stop it shocked the world. Most of those killed were Tutsis; most of the perpetrators of the violence were Hutus. Colonial history played a critical role in the atrocity. The German colony of Ruanda-Urundi used racial science to differentiate between Hutu and Tutsi, who were discursively constructed in very different terms (Prunier 1995: 6–8):

*Hutu*: 'very typical Bantu features . . .. They are generally short and thick-set with a big head, a jovial expression, a wide nose and enormous lips. They are extroverts who like to laugh and lead a simple life'.

*Tutsi*: 'has nothing of the Negro apart from his colour . . .. Gifted with a vivacious intelligence, the Tutsi displays a refinement of feeling which is rare among primitive people. He is a natural-born leader, capable of extreme self-control and of calculated goodwill . . .. They differ absolutely by the beauty of their features and their light colour from the Bantu agriculturalists of an inferior type'.

In reality, the two groups were very similar – they spoke the same language, inhabited the same areas and followed the same traditions. However, the colonial construction of racial divisions left a legacy of ethnic tension that Rwanda has struggled with ever since.

Racism often predominated in representations of the Rwandan genocide. The role of European colonialism was ignored and media discourses characterized this as a spontaneous uprising rather than an orchestrated political campaign (Power 2003: 170). This was the 'dark continent' at its most savage. Yet the social inequities that existed between Tutsi and Hutu were not far removed from those that existed between rich and poor in early nineteenth century Britain. British society was allowed to evolve. The working classes gradually gained more rights and freedoms, primarily through parliamentary reform, and the threat of revolution was abated. In contrast, Rwandan social divisions were frozen in time, concretized as ethnic divisions by colonial discourses and encouraged by colonial rule. At independence, social inequalities in Rwanda had been frozen and exacerbated for at least 70 years. Although the story of Rwanda is complex, discourse is significant in explaining the continual ferment since independence, the atrocities that occurred in 1994, and stereotypical western media representations.

## 'Worlding'

The material dimensions of the historical legacy of imperialism, especially the way in which the colonies were incorporated into the international division of labour and the profound global inequality and socio-economic impoverishment that resulted, are clearly of significance in understanding the condition of postcolonialism today. Postcolonial theorists like Said and Spivak add the importance of the cultural production of European imperialism to this concern with political economy, emphasizing its importance in shaping domination of colonial societies. Spivak refers to this process as 'worlding'. The worlding of 'the Third World' is a continuation of imperial power relations into the present: 'For Spivak, the epistemic violence of imperialism has meant the transformation of the "Third World" into a sign whose production has been obfuscated to the point that Western superiority and dominance are naturalized' (Kapoor 2004: 629). In other words, that the Third World exists and is inferior to the First is taken for granted and the dominance of the West is seen as natural rather than as problematic.

This pattern of obfuscation has permeated international development. For example, modernization theories, which dominated development thinking from 1945 into the 1980s, almost totally ignored colonialism (see Chapter 3). The Third World was assumed to be without history. The legacy of the erasure of local cultures and social systems, the devastation wrought by slavery and the imposition of mono-economies based on resource extraction, and the incorporation of the colonies into a profoundly unequal international division of labour were deemed of little or no relevance. First World patterns of economic development could thus serve as both guide and goal. The pervasiveness of this thinking is visible in the structural adjustment and 'free-trade' policies of the Bretton Woods institutions (WB and IMF) and the World Trade Organization (WTO), under which countries must liberalize socio-economic and trade regimes. These policies ignore the legacies of the history of imperialism and the unequal footing on which this history has placed countries of the South within the global capitalist system. They also promote what Spivak (1988: 291) refers to as a 'self-contained version of the West', which ignores its complicity in imperialism and the fact that its own development was bound up with the political economy of empire.

According to Spivak, the 'worlding' of the Third World allows governments, producers and consumers in the North to overlook the interrelationships between advanced economies, on the one hand, and imperialism, globalization, unequal development and exploitative working conditions in the South on the other. It disconnects the South from the North and either 'ignores colonialism, or situates it securely in the past to make us think it is now over and done with' (Kapoor 2004: 630). Moreover:

> [T]he student [in the North] is encouraged to think that he or she lives in the capital of the world. The student is encouraged to think that he or she is there to help the rest of the world. And he or she is also

encouraged to think that to be from other parts of the world is not to be fully global.

(Spivak 2003a: 622)

We produce the Third World and the subaltern through discourse to suit our own image and desire. When we act in accordance with personal, professional, institutional or organizational interests, representations say more about those doing the representing than those represented; 'they construct the Other *only in so far as* we want to know it and control it' (Kapoor 2004: 636).

This is important when considering how the South has been constructed as particular spaces for investigating progress and development. Many systems of ordering the world are Eurocentric. They are partial and frequently negative truths which deny cultural difference and stereotype the South as an overpopulated world of problems:

> What is the geography of the Third World? Certain common features come to mind: poverty, famine, environmental disaster and degradation, political instability, regional inequalities and so on. A powerful and negative image is created that has coherence, resolution and definition. But behind this tragic stereotype there is an alternative geography, one which demonstrates that the introduction of development into the countries of the Third World has been a protracted, painstaking and fiercely contested process.
>
> (Bell 1994: 175)

Juxtaposed with this is the idea of 'the West', a set of images that provides a standard or model for comparison (e.g. development and progress), and criteria for evaluation against which other societies are ranked and around which powerful positive or negative images cluster (Hall 2002). The idea of the West is a means for organizing global power relations and enabling certain kinds of knowledge, such as the theory and practice of international development. It was through contact with other places with very different histories, ecologies, patterns of development and cultures that the idea of the West took shape. 'The rest' became the Other or the alter-ego because it was outside global modernity. As European empires grew, non-Europeans began to be represented as 'uncivilized' or 'backward', thus justifying European control. An undeveloped, savage or backward Other gave Europeans a sense of being modern and fully developed.

It is worth bearing in mind that it is not only the West that constructs a self and other. For example, Ayatollah Khomeini's Iranian revolution in the 1980s was a reaction to collective discontent about 'Westoxification' and a desire to produce alternative political systems to those developed in the West based on Islamic traditions rather than western canons and doctrines (Dabashi 1993). A repulsive rejection of the West animated the Shi'ite revolution in Iran and more recently has given rise to extremism within Sunni Islam in the form of the so-called Islamic State in parts of Iraq and Syria. The imagined West became the epitome of moral corruption, illegitimate global domination and plundering the wealth and sovereignty of other states. Northern discourses, however, operate at a different scale and have greater power to script and order the world, both discursively and geopolitically.

The power of development discourses also lies in the fact that they were accepted and appropriated by many newly independent countries. Thus development and modernization became objects of desire for formerly colonized nations. Discourses of development were often wrapped up in the formation and construction of post-colonial nation states. Akhil Gupta (1998) argues that in India, the dominant discourse that enshrined national development as the key to progress after independence in 1947 (particularly under Nehru's government) is a central condition of the post-colonial state of India today. On independence, the Indian government appropriated dominant, western discourses about development and 'underdevelopment'. Its state-guided development project to address 'underdevelopment' was not simply about addressing India's structural location in global economic relations, but was an important aspect of India's post-colonial national identity. Thus underdevelopment figured in these state-guided discourses not simply as a material relation of social and economic position, but also as a state of mind, a form of identity and a notion of citizenship. Gupta argues that some discourses in India held that national development involved 'mimicking' the historical trajectories of the British imperial rulers. Consequently, these hegemonic discourses have perpetuated the uneven materiality of the lives of Indian people and other decolonized peoples in the South.

## Development discourses and representation

The power to represent other places (to name, describe, publish, claim and construct knowledge) was instrumental in reinforcing a sense of difference between the West and non-West, North and

South, which also translated into a sense of superiority and justified various political interventions that underpinned imperialism, and, later development. Through discourse, Europeans created a 'western-style for dominating', inventing the authority and legitimacy to subjugate and oppress nonwestern peoples through colonialism. These discourses contained what became acceptable ways of referring to other parts of the world and a set of assumptions and generalizations about the peoples and cultures in these regions. Aspects of colonial representations prior to 1945 pervaded development discourses from the 1950s – the colonies became the Tropics, then the 'developing', 'Third' and 'post-colonial' world. Rather than an objective, detached intellectual endeavour, scholarly discourse on North–South relations in international relations and development thus 'becomes imbued through and through with the imperial representations that have preceded it' (Doty 1996: 161).

Consequently, development was dominated by particular discourses that governed foreign policy relations between North and South. Thus, between 1945 and 1970, modernization theory articulated a discourse of 'lack' – the idea that the former colonies were deficient, should develop by modernizing and should follow the same trajectories as western economies in order to 'catch up', both in economic terms and in terms of 'civilization'. There is nothing new in this: as Grosfoguel (2007) argues, hierarchies of superiority and inferiority began with colonial expansion and the sixteenth-century characterization of 'people without writing' to eighteenth- and nineteenth-century characterizations of 'people without history', to twentieth-century characterizations of 'people without development'. More recently, the discourses have shifted to depictions of 'people without democracy'. All of these characterizations of non-European 'others' have been used as justifications in the name of development for intervention by western powers in countries of the global South. And this was not a benign process: as Grosfoguel (2012: 97) contends, these interventions went from 'convert to Christianity or I'll kill you' in the sixteenth century, to 'civilize or I'll kill you' in the eighteenth and nineteenth centuries, to 'develop or I'll kill you' in the twentieth century, to 'democratize or I'll kill you' at the beginning of the twenty-first century. As Power (2003: 66–7) suggests, we might consider how different development policy and practice might have been if countries in the South were 'seen as *rich* in economies, connections, cultures and lives whose contributions, diversity, wealth and worth are not captured adequately by being imagined as more or less *developed*'.

How would global development look if the discourse were different: for example, if 'end extreme poverty' was replaced with 'end extreme wealth' in driving global development interventions by the UN, WB and IMF? Instead, development created a space in which only certain things could be said or even imagined about 'Third World others'. This notion of the Third World as unique, different, other, haunts development studies today. Modernization theories of development envisaged rational bureaucracies, legal systems, economies, military institutions and electoral democracies, but calling this 'development', imposing this on other places and not allowing for the evolution of indigenous alternatives, simply concealed the violence being done to these cultures.

The notion of the Third World is also bound up with discourses of crisis, which legitimate Northern intervention. Aid is seen popularly in many western states as a response to crisis. Driving aid is what some analysts have referred to as the 'CNN effect' in which sensationalist reporting by news media shapes the responses and opinions of individuals in the North. The effect of this is that many people in rich countries are not interested in giving development aid unless it is packaged as humanitarian assistance (Dogra 2014; Ignatieff 1998). There are multiple images of 'Third World poverty' and humanitarian crises, which tend to paint a picture of negativity, failure and lack. For example, according to a VSO survey (in Power 2003: 171), 80 per cent of British people associate the developing world with 'doom-laden images of famine, disaster and Western aid'. At the heart of many development discourses is the unproblematized notion of 'the poor'. What precisely it means to be poor is rarely explained, but simply reflects Northern assumptions. Development agencies have rarely considered how the people they label as poor *themselves* view issues of poverty, development and well-being. Nor have they reflected on the effects of such labelling.

Humanitarian appeals and development discourses often use images of children. These are persuasive and emotive, implying that all children share the same attributes and needs, appealing to universal aspects of the condition of childhood. As Erica Burman (1995: 22) argues, while children's rights in much of the world are in need of promotion and protection, such images are problematic and contradictory: 'children are typically abstracted from culture and nationality to connote such qualities as innocence, and the quintessential goodness of humankind untainted by the cruel, harsh contaminating world'. The binary between innocence/goodness and experience/contamination is a product of a specific western philosophical legacy. The representation of lone black children in aid appeals works to pathologize their families and cultures, blaming them for failing to fulfil their duties to protect and care for these children (Holland 1992). Thus, colonial legacies blend into humanitarian concern, where to qualify for 'help' parents are either invisible or infantilized as incapable (Burman 1995).

Development discourses also tend to globalize middle-class Northern agendas. Child labour is a much-cited example of an issue driven by Northern concerns rather than by Southern realities. Ethical and fair-trade codes that prevent children from formal employment can increase the financial insecurity experienced by some households. This is especially problematic where there is a high proportion of child-headed households, for example in parts of sub-Saharan Africa where the adult population has been decimated by HIV-AIDS. To survive, children are often forced to work in less visible and unregulated areas of the informal sector. The exploitation of child labour requires attention, but also sophisticated locally specific responses rather than blanket bans to appease Northern consumers. Moreover, liberal human rights discourses that problematize child labour also obfuscate the workings of global and national capital regimes (Chowdhry 2002). These discourses depend on the creation of an 'illiberal', uncivilized, 'Third World' in which human rights abuses are perpetuated. However, in focusing attention on such abuses in particular 'Third World' sites they fail to situate these within the broader machinations of the neoliberal global economy and ignore western complicity in their perpetuation.

Similarly, in the 1990s population control became integral to aid and development programmes, with international loans linked to family planning programmes (Duden 1992). Although this was presented as a basic human right that empowers women, in practice this was often far from the case. Poor women seeking employment in poverty alleviation schemes were forced to accept dangerous birth control drugs and devices (Burman 1995). This further colluded in pathologizing the South, blaming cultures for 'overpopulation' and associated environmental degradation. As the case of India illustrates, the problem is not overpopulation, but rather the distribution of resources:

> Massive sums have been provided by the USAID to push controversial contraceptives like Norplant/Net-O-En which are banned in most of the Western countries. Advocacy for 'population control' has been a crucial concern . . . Poor women are also blamed

for causing environmental crisis by breeding like 'cats' and 'rats'. It is
time to ask our policy makers, to what extent can top-down population
control programmes that violate basic human rights of Indian women
be justified? By victimiing the victims of the patriarchal class society
aren't we ignoring the major causes of environmental crisis such as
industrial toxic wastes, chemical fertilizers, nuclear armaments, over-
consumption of the affluent in both the first as well as the third world?
Moreover, what about the first world where population growth has
declined yet environmental conditions have deteriorated?

(Patel 1992, cited in Burman 1995: 27–8)

Inherent within such discourses is a normative model of the northern
nuclear family, even though this is nowhere near universal in many
northern countries and becoming less so. For example, the WB in the
1990s aimed to reduce population growth by 'stimulating a demand
for smaller families' (cited in Williams 1995: 172).

Discourses of hazard and vulnerability are common in development
and are particularly effective in representations of humanitarian
crises. As Bankoff (2001: 19) argues, 'Tropicality, development
and vulnerability form part of one and the same essentialising and
generalising cultural discourse that denigrates large regions of
the world as disease-ridden, poverty-stricken and disaster-prone'.
Western societies are 'unable to escape from the cultural constraints
that continue to depict large parts of the world as dangerous places
for *us* and *ours*, and that provide further justification for Western
interference and interventions in others' affairs for *our* and *their* sakes'
(Bankoff 2001: 20). Media reporting of the Indian Ocean tsunami on
26 December 2004 is illustrative of the power of these discourses.
Initial reports focused on technological lack – the absence of an early
warning system, in contrast to those in the USA. This was claimed
to have contributed to the catastrophic loss of life. Blame was placed
squarely with the underdeveloped Other, where governments had
'ignored calls for an early warning system', and thus abrogated
their responsibilities to their own vulnerable people (*Guardian* 27
January 2004). Initial television news reports were conveyed through
recordings and images captured by tourists, ensuring that the focus
remained on the impacts on westerners rather than on Sri Lankans,
Indians or Indonesians. The subsequent focus on the massive
outpouring of aid from benevolent westerners to help the stricken
peoples of South East Asia ensured that the massive relief efforts
within the affected countries were ignored. By 29 December, the

*Guardian* newspaper was already publishing maps of who was giving what to where, emphasizing the passivity of the affected countries in the South and the dynamism and generosity of the enlightened individuals, government and aid organizations in the North.

Even ethical and fair trade rely on discourses of 'self' and 'other' – the 'haves' (Northern consumers) and the 'have nots' (Southern producers). On the one hand, by buying ethical products affluent Northern consumers can 'immerse themselves in a foreign world of fantasy, where workers are content and the earth is clean' (Dolan 2005: 369). On the other hand, the South is imagined as steeped in backwardness, corruption and economic chaos and in dire need of salvation by Northern benevolence. Desire for ethical action in the North is triggered not by a sense of commonality, affinity or solidarity but by images of downtrodden Southern workers, reinforcing a sense of cultural difference. Thus, progressive initiatives that can make a tangible difference in the South still fortify centuries-old self/other distinctions by reminding consumers in the North of their ethical obligations to the poverty-ridden South. They are also shaped by Northern definitions of what is 'fair' and 'ethical', while excluding 'local' understandings of social organization.

These are just a few examples of discursive representations that have shaped development interventions in the South by Northern agencies. Aid and development are not the same thing, but interactions between the two say a great deal about how people perceive the 'problems' of other places and worlds of difference. The rationale for aid in the public mind remains emergency relief, but it works in the strategic interests of donors in projecting economic, cultural and political power beyond the boundaries of other countries. Representations of Africa are perhaps most stark in illustrating this.

## From imperial to development discourses of Africa

There are striking similarities between representations of Africa in the nineteenth century and popular representations today. Several familiar tropes, or common themes, persist in these discourses. For example, it was generally believed that all Africans resembled one another physically (Bolt 1971). The individual was lost in the 'tribe', and the 'tribe' within the 'race'; all were equally inferior. To erase the existence of difference was to create a homogeneous mass of people to which the same stereotypes could be applied. It also erased the existence of distinct African cultures and traditions. Consequently, 'the African' was constructed and analyzed, imperial discourse created knowledge about 'the African', and thus subjected African peoples to regimes of power (Butchart 1998). The erasure of African histories further emphasized the supposed primitive nature of African societies. European ignorance of the rich histories of Africa was thus recast as a *de facto* absence of history. Despite the presence of these histories within oral traditions and material cultures, the lack of a written history attested to the barbarity of the African. Since the histories of African peoples were not written down, imperialists could erase their existence, or claim that any history was insignificant. Therefore, Europe could justifiably colonize Africa and write its own history of the continent.

A particularly prevalent negative image of Africans during the nineteenth century was their representation as indolent children (Russett 1989). If Africans were perceived as permanently childlike, then it could be argued that it was natural and just that the more 'advanced' nations should become permanent guardians. If Africans were incapable of exploiting their own resources, European powers were justified in governing and developing Africa themselves. Through colonialism non-Europeans could be brought to adulthood, to rationality, to modernity (Blaut 1993). Images of infantilism and general incompetence were thus used to legitimate colonialism and imperialism under the guise of paternalism:

> The child analogy was useful to whites for it denied to Africans the privileges reserved for adults. It both reflected and strengthened the idea that African cultures did not represent worthwhile achievements and were too loosely formed and inchoate to offer any significant resistance to an inrush of westernisation. Most important, the analogy acted as a sanction and preparation for white control, for its main implication was paternalism which denied the African the right of deciding on his own future.
>
> (Cairns 1965: 95)

Scientific racism, which insinuated that Africans had advanced in evolutionary terms no further than a European child, brought a new dimension and scientific authority to racial distinctions and to racial stereotyping (Broks 1990).

These tropes of representation persist in some development discourses. The connection between Africans and childhood still operates in contemporary aid discourses. In her study of Scandinavian aid workers in Tanzania, for example, Baaz (2005) suggests that they continue to represent Tanzanians as irresponsible, carefree and not to be trusted. The motif of disease also has a powerful hold on western imaginations, with contemporary depictions of Africa often echoing nineteenth century images of the 'Dark Continent'. For example, economist Jeffrey Sachs, in a chapter entitled 'The voiceless dying: Africa and disease' (2005: 194, 197), writes:

> . . . beyond anything I had experienced or could imagine, disease and death became the constant motif of my visits to Africa. Never, not even in highlands of Bolivia, where illness is rife, had I confronted so much illness and death. India had never evoked the same sense of death in the air . . . I began to suspect that the omnipresence of disease and death had played a deep role in Africa's prolonged inability to develop economically . . . Africa's growth rate has been among the lowest of any world region during each major subperiod since 1820. That includes a long stretch before Africa fell to European colonial rule in the 1880s and the period since independence. Could the exceptional burden of illness be a significant reason? . . . Malaria [HIV/AIDS, tuberculosis] also causes poverty . . .
>
> In *The End of Poverty* London, Penguin.

For nineteenth century explorers, Africa was disease-ridden, barbaric, threatening, uncontrollable and a moral wilderness (McEwan 2000). Sachs takes this further. Not only is Africa a land of disease and death, but this is the reason for its backwardness. The key questions to consider are: what are the underlying assumptions that can attribute poverty to disease, rather than vice versa? What are the consequences of these kinds of representations for development interventions in poorer countries?

Ambivalent images of Africa – on the one hand as passive, childlike and obedient, but on the other savage and dangerous – persist in contemporary media representations where images of starving children are juxtaposed with images of brutish, primitive others, especially in reports of African wars. These wars are taken out of their geopolitical and economic contexts, and their regional and international significance is underplayed in characterizing them as a product of African history and culture. Whereas wars perpetrated by Northern powers (e.g. in Iraq and Afghanistan) are rationalized in anodyne descriptions of 'targeted' bombing and collateral damage,

African wars are portrayed as incomprehensible. Reports of recent conflicts in DRC, Sierra Leone, Liberia and Darfur, for example, have focused on sensationalist accounts of cannibalism, amputations and killing with machetes. Africa is still seen as inherently a place of savagery and danger.

Images of danger are not confined to the media, but are also recycled in the preparatory courses for development workers, where experienced workers advise new entrants: 'The relating of experiences often seem to focus on danger and fear, thereby constructing the image of a heroic, daring and capable Self' (Baaz 2005: 144). Images of unpredictability and danger are promoted both by embassies through their guidelines and warnings, and by country-specific advisory risk assessment sites. Such warnings contribute to the 'culture of fear', which as Baaz argues is related to inequalities of wealth – westerners are more 'at risk' because they are relatively wealthy while working in a context of poverty. Moreover, with very few exceptions, stories about crimes committed against white people are circulated, while those committed against blacks are not (Baaz 2005: 146). This is not to claim that development workers do not challenge these images, as many of them undoubtedly do, but denying that the South is a place of danger still asserts a rational, realistic, courageous Self, not unlike those 'heroic' nineteenth century explorers.

Discourses of indolence and passivity have shaped development in the post-colonial period. These scripted the 'white man's burden' to awaken colonized peoples from their passive and lazy disposition and to infuse them with a work ethic and energy. Similarly, in the early modernization version of development, people in 'traditional' societies were represented as unreceptive to new ideas and uninterested in new information. According to Rostow (1960), development required not only 'attitudinal changes towards science' and 'propensities to calculate risk', but also a 'willingness to work' (in Baaz 2005: 122). Thus, a major task of development was to fill this lack and to instill a work ethic. Media representations of Africa, in particular, continue to perpetuate such stereotypes by focusing on starvation, disaster and war and promoting images of 'danger, ill-fate and apathy' (Pieterse 1992: 209). Popular discourses promote the westerner as the potential saviour – the energetic, efficient actor. Many development agencies today are critical of such representations, but the international aid industry continues to reproduce them.

In efforts to appeal to people's generosity, the Southern recipient – often symbolized as a starving child – has been represented as a passive, helpless object in need of assistance from affluent Northerners. The representation of the Ethiopian famine in 1984–1985 was perhaps the most (in)famous example of this and politicized the issue of representation within development. Media representations juxtaposed now iconic images of starving children (Figure 4.1) and passive, helpless Africans with frenzied activity in Europe and the USA, especially around Bob Geldof and the Live Aid concerts in July 1985. As Lidchi (1999: 92, 95) argues:

> Northern audiences, whose popular belief was that Africa was poor and undeveloped, now had proof. African nations were visibly passive nations constantly threatened by the possibility of a 'natural disaster' and always in the need of assistance, and dependent on Western goodwill . . . African *subjects* were represented as the passive

**Figure 4.1** *Iconic image of the Ethiopian famine (Birhan Woldu)*

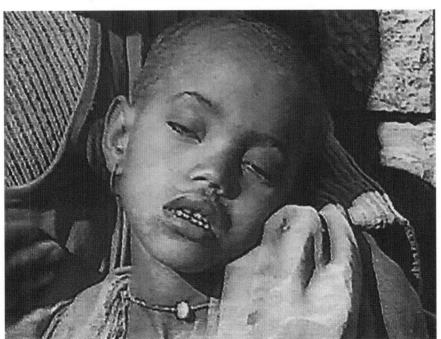

Courtesy of bbc.co.uk

recipients of aid – *objects* of development – who had no voice, no identity and no contribution to make during the crisis. The West, in contrast, was constituted as being full of active subjects, development workers, fundraisers, journalists or world citizens.

Despite intense criticism, both the 2005 Live 8 concerts – organized to mark the twentieth anniversary of Live Aid and to put pressure on the G8 Summit to address Africa's continuing problems – and the 2014 Band Aid 30 single released to raise money to tackle the Ebola crisis in West Africa remained deeply problematic in terms of how Africa and Africans were represented (Boxes 4.3 and 4.4). Attempts to foreground Africa in contemporary debates still struggle to avoid stereotypical representations.

---

### Box 4.3 Live 8 2005 and Africa (extract from BBC News, 'Live 8 logic attracts criticism')

(British band) Blur frontman Damon Albarn has become the latest critic of next month's Live 8 concerts, saying they will perpetuate the idea of Africa as a 'failing, ill' place. He has followed several fellow musicians, teachers, black rights campaigners, police and an international aid charity in questioning the logic behind Bob Geldof's event. Live 8 concerts will be staged in London, Edinburgh, Philadelphia, Paris and Rome in an effort to pressure G8 leaders to tackle debt in Africa. Reviving memories of Geldof's 1985 Live Aid concerts, the new event will include performances by Coldplay, Madonna, U2, 50 Cent, Stevie Wonder and Sir Paul McCartney . . .. Campaign group Black Information Link branded the London concert's line-up 'hideously white' for having only one ethnic minority artist – Mariah Carey – among its 22 performers at that stage. 'It seems like the great white man has come to rescue us while the freedom fighters never get a mention,' said black musician Patrick Augustus . . .. [Youssou] N'Dour is the only major African artist due to perform at any of the five concerts. Fellow Senegalese star Baaba Maal wrote in The Independent: 'I do feel it's very patronising as an African artist that more of us aren't involved. If African artists aren't given a chance, how are they going to sell records and take the message back to Africa?' . . .. Geldof's method of putting pressure on leaders to improve aid, drop debts and ease trade restrictions in Africa was also queried. John O'Shea, chief executive of international aid charity Goal . . . told the Guardian newspaper . . . 'There is a fire raging – we need someone to put out the fire, not hand out chocolate'.

http://news.bbc.co.uk/go/pr/fr/-/1/hi/entertainment/music/4079958.stm (10/06/2005)

## Box 4.4 Band Aid 2014 and Ebola

In 2014, Bob Geldof reformed Band Aid to raise funds to prevent the Ebola crisis in West Africa spreading throughout the world. As in previous incarnations, the group covered the track 'Do They Know It's Christmas?' altering the lyrics slightly:

| Original lyrics | New lyrics |
| --- | --- |
| But say a prayer | But say a prayer |
| Pray for the other ones | Pray for the other ones |
| At Christmas time it's hard | At Christmas time it's hard |
| But while you're having fun | But while you're having fun |
| There's a world outside your window | There's a world outside your window |
| And it's a world of dread and fear | And it's a world of dread and fear |
| **Where the only water flowing is** | **Where a kiss of love can kill you and** |
| **The bitter sting of tears** | **there's death in every tear** |
| And the Christmas bells that are ringing | And the Christmas bells that are ringing |
| Are clanging chimes of doom | Are clanging chimes of doom |
| **Well, tonight thank God it's them** | **Well tonight we're reaching out and** |
| **instead of you** | **touching you** |

Band Aid 2014 also attracted familiar criticisms, with the Liberian academic Robtel Neajai Pailey (based at SOAS in London) arguing on BBC Radio 4's Today programme that the lyrics were meaningless since most victims were Muslim, reinforced stereotypes, and painted the continent 'as unchanging and frozen in time'. Pailey argued that the song is 'incredibly patronizing' and 'reeks of the "white saviour complex" because it negates local efforts.'

More pointedly, Paul Richards (2016) argues that the interventions by the West worsened the spread of the disease. He argues that western commentators continued to stereotype Africans as mired in tradition and superstition (e.g. beliefs concerning washing the dead), which produced doom-laden predictions about the spread of disease by western media and scientists alike. The latter's obsession with biosafety meant that they ignored the sacred dimensions of customary funerals and generated friction between affected communities and central governments in West Africa. Richards argues that money and technologies mattered less than indigenous knowledge in ending the epidemic. He explains that the calm, rational response by local chiefs who understood risks and adapted burial practices led to the containment of the disease, and he argues that most remote areas where western interventions were slowest to arrive were in fact the quickest to bring the disease under control.

Questions can be asked about whether Band Aid, despite its good intentions, is a form of 'poverty porn' and whether it is ethical. In answering this, a key issue is that it and similar charitable efforts do nothing to address the structural causes of poverty and do not equate with activism. Rather, they conflate humanitarian aid with foreign aid, which is why a substantial proportion of the UK public and right-wing newspapers like the *Daily Mail* are opposed to foreign aid, despite it always having the interests of UK businesses at its heart (Mawdsley 2017). They also misrepresent poverty, making a complex human experience understandable, consumable and easily treatable and perpetuating a saviour/victim binary. Both the 'saviours' and the 'victims' are deceived and the wrong people are empowered: the 'saviours' believe they have agency to solve the problems of others, while 'the saved' are robbed of their agency. Such campaigns also misrepresent the poor, who are defined by physical suffering or lack of material resources, when in fact people living in poverty are more likely to define their condition psychologically and emotionally. However, campaigns such as Band Aid are successful in focusing attention on crises that might otherwise be ignored and in raising large sums of money. The moral dilemma is whether their profitability is worth the perpetuation of false ideologies and stereotypes.

Development agencies acknowledge the problems of such representations and are involved in attempts to present more positive images, emphasizing the strength, dignity and self-determination of people in the South and seeking to foreground rather than hide the complexities of poverty. However, the racism that underpinned colonial discourses of difference has not been completely eradicated, but has shifted towards cultural racism in aspects of contemporary development. Baaz (2005: 47), for example, quotes a Danish aid worker in Tanzania:

> There was someone here who tried to make me say that black people are a bit . . . [*makes a sign with his hand signifying 'stupid'*]. But I told her that for me it doesn't matter if they are red, yellow or black or whatever colour; in any case, the blood has the same colour . . . I can't see that there is a difference between black or white . . . as individuals or as people. The difference could be that . . . and this is not the fault of the individual – that the culture is not the same, or that they have not reached the same level that we have.

Thus, while it is no longer seen as acceptable to refer to 'race' or skin colour in attributing intellectual inferiority to the Other, the assertion of sameness (we all have the same colour blood) is brushed aside by an account of difference located in culture. Although taking issue with the notion that Tanzanians are 'stupid', the aid worker contradicts this by asserting that 'they have not reached the same level that we have'. Cultural traits, rather than racial ones, are thus seen to differentiate 'us' from 'them' and former racial borders now tend to coincide with contemporary cultural borders. Despite efforts to challenge such stereotypes, they persist within popular and political imaginations.

The heterogeneity of African cultures, histories, experiences and practices is staggering (as it is in Latin America and Asia), yet this receives marginal recognition in development. Instead, modern development scholarship returns to the essential notion of Africa's inadequate characteristics: 'Reductive repetition becomes an effective tool with which to conflate the many heterogeneous characteristics of African societies into a core set of deficiencies' (Andreasson 2005: 972). These deficiencies are seen to be internal and intrinsic and, therefore, the solutions must originate externally. The fact that modernism has failed spectacularly to impose itself on the continent testifies not to the problems inherent in Northern models and their imposition upon Africa, but to Africa's fundamental deficiencies. 'Africa (in the singular) fails because it is unable to incorporate Western thought and practice into its chaotic societal structures' (Andreasson 2005: 973).

It is not surprising, therefore, that those teaching university courses on Africa face a constant struggle to encourage students in the North to think of Africa as a diverse, rich mosaic instead of referring to 'a country like Africa'. This reduction is part of a long history of misrepresentation of the continent. Importantly, the authority to construct Africa in various discourses is still located in the former imperial powers (see Box 4.5). Historical images of Africa inform contemporary representations of the continent; 'the metaphor of Africa as the Dark Continent continually (re)makes and represents the continent as Other' (Jarosz 1992: 105). Images of famine and a Malthusian population crisis, of Africa as a land of environmental catastrophe, disease and death continue to resonate, particularly in academic and popular mass media accounts of HIV-AIDS (Jarosz 1992: 111–13). The process of othering Africa continues to a large extent in contemporary academic, popular and some development discourses that represent it as a repository of disease, war and pestilence, as the 'hopeless continent' (*The Economist* May 2000) or as a 'basket case' (*African Business* January 1999). However, commentators in African countries have increasingly resisted this, not least because such representations continue to have material effects (Box 4.6).

## Box 4.5 'How to write about Africa' (extracts from a parody by Binyavinga Wainaina, 2005)

Always use the word 'Africa' or 'Darkness' or 'Safari' in your title. Subtitles may include the words 'Zanzibar', 'Masai', 'Zulu', 'Zambezi', 'Congo', 'Nile', 'Big', 'Sky', 'Shadow', 'Drum', 'Sun' or 'Bygone'. Also useful are words such as 'Guerrillas', 'Timeless', 'Primordial' and 'Tribal' . . ... Never have a picture of a well-adjusted African on the cover of your book, or in it, unless that African has won the Nobel Prize. An AK-47, prominent ribs, naked breasts: use these. If you must include an African, make sure you get one in Masai or Zulu or Dogon dress.

In your text, treat Africa as if it were one country. It is hot and dusty with rolling grasslands and huge herds of animals and tall, thin people who are starving. Or it is hot and steamy with very short people who eat primates. Don't get bogged down with precise descriptions. Africa is big: 54 countries, 900 million people who are too busy starving and dying and warring and emigrating to read your book. The continent is full of deserts, jungles, highlands, savannahs and many other things, but your reader doesn't care about all that, so keep your descriptions romantic and evocative and unparticular . . ... Taboo subjects: ordinary domestic scenes, love between Africans (unless a death is involved), references to African writers or intellectuals, mention of school-going children who are not suffering from yaws or Ebola fever or female genital mutilation.

Throughout the book, adopt a *sotto* voice, in conspiracy with the reader, and a sad *I-expected-so-much* tone. Establish early on that your liberalism is impeccable, and mention near the beginning how much you love Africa, how you fell in love with the place and can't live without her. Africa is the only continent you can love – take advantage of this. If you are a man, thrust yourself into her warm virgin forests. If you are a woman, treat Africa as a man who wears a bush jacket and disappears off into the sunset. Africa is to be pitied, worshipped or dominated. Whichever angle you take, be sure to leave the strong impression that without your intervention and your important book, Africa is doomed.

Among your characters you must always include The Starving African, who wanders the refugee camp nearly naked, and waits for the benevolence of the West. Her children have flies on their eyelids and pot bellies, and her breasts are flat and empty. She must look utterly helpless. She can have no past, no history; such diversions ruin the dramatic moment. Moans are good. She must never say anything about herself in the dialogue except to speak of her (unspeakable) suffering . . ... These characters should buzz around your main hero, making him look good. Your hero can teach them, bathe them, feed them; he carries lots of babies and has seen Death. Your hero is you (if reportage), or a beautiful, tragic international celebrity/aristocrat who now cares for animals (if fiction) . . ... Avoid having the African characters laugh, or struggle to educate their kids, or just make do in mundane circumstances. Have them illuminate something about Europe or America in Africa. African characters should be colourful, exotic, larger than life – but empty inside, with no dialogue, no conflicts or resolutions in their stories, no depth or quirks to confuse the cause . . ... Remember, any work you submit in which people look filthy and miserable will be referred to as the 'real Africa', and you want that on your dust jacket. Do not feel queasy about this: you are trying to help them to get aid from the West. The biggest taboo in writing about Africa is to describe or show dead or suffering white people . . .

Readers will be put off if you don't mention the light in Africa. And sunsets, the African sunset is a must. It is always big and red. There is always a big sky. Wide empty spaces and game are critical – Africa is the Land of Wide Empty Spaces. When writing about the plight of flora and fauna, make sure you mention that Africa is overpopulated. When your main character is in a desert or jungle living with indigenous peoples (anybody short) it is okay to mention that Africa has been severely depopulated by Aids and War (use caps) . . .. Always end your book with Nelson Mandela saying something about rainbows or renaissances. Because you care.

## Box 4.6  Africa as contemporary dark continent?

*1.  Extracts from 'Hopeless Africa',* The Economist, *11 May 2000*

At the start of the twenty-first century, Freetown symbolizes failure and despair. The capital of Sierra Leone may be less brutalized than some other parts of the country, but its people are nonetheless physically and psychologically scarred by years of warfare, and this week they had to watch as foreign aid workers were pulled out. The United Nations' peacekeeping mission had degenerated into a shambles, calling into question the outside world's readiness to help end the fighting, not just in Sierra Leone, but in any of Africa's many dreadful wars. Indeed, since the difficulties of helping Sierra Leone seemed so intractable, and since Sierra Leone seemed to epitomize so much of the rest of Africa, it began to look as though the world might just give up on the entire continent . . .. Since January, Mozambique and Madagascar have been deluged by floods, famine has started to reappear in Ethiopia, Zimbabwe has succumbed to government-sponsored thuggery, and poverty and pestilence continue unabated. Most seriously, wars still rage from north to south and east to west. No one can blame Africans for the weather, but most of the continent's shortcomings owe less to acts of God than to acts of man. These acts are not exclusively African – brutality, despotism and corruption exist everywhere – but African societies, for reasons buried in their cultures, seem especially susceptible to them. Sierra Leone manifests all the continent's worst characteristics. It is an extreme, but not untypical, example of a state with all the epiphenomena and none of the institutions of government. It has poverty and disease in abundance, and riches too: its diamonds sustain the rebels who terrorize the place. It is unusual only in its brutality: rape, cannibalism and amputation have been common, with children often among the victims . . .. In itself, Sierra Leone is of no great importance. If it makes any demands on the world's attention, beyond the simple one of sympathy for its people, it is as a symbol for Africa.

Source: http://www.economist.com/opinion/displaystory.cfm?story_id=E1_PPPQNJ

*2. A voice from Africa: extracts from 'Praise Africa, don't bury it',* Anver Versi, African Business, *July 2000*

Africans are justifiably angry over the blanket slandering of their continent. Over the last few months, Africa has been pilloried by the western press over events in Zimbabwe and Sierra Leone, and the whole of Africa has been termed a 'basket case', a 'hopeless continent'. This loaded reporting is not simply an unacceptable blow to an African's pride and self-esteem, it threatens to wreck African economies. Who would want to invest in a hopeless continent? Who will spend holidays in a basket case Africa?

There are 53 states in Africa which is the second greatest land mass on earth. Sudan alone is larger than the whole of Western Europe; the DRC is larger than all the European member states combined. If you could place all European, North American and Latin American states into Africa, you would still have room to fit the Gulf states. If Europe has problems dealing with its minuscule states, imagine the challenge that governing countries the size of Nigeria, Egypt, Sudan, the DRC, South Africa poses. If Europe has still to come to terms with its ethnic diversity – despite two world wars in the last century – imagine the challenges facing leaders of a continent which has a far greater ethnic diversity in one country, Nigeria, than all of Europe, North America, South America, the Middle East and some Asian countries put together. If Europe still has economic and employment problems after three centuries of the colonial era, industrialization, and trade monopolies, imagine the task facing Africa which at independence had practically no industries, few schools, the worst of possible terms in overseas trade, and a population that had no opportunity to accumulate meaningful capital. Add to this a burden of debt for loans which many countries were forced to take out, the structural adjustment programmes which the IMF now admits led to increased poverty, and the destructive impact of the Cold War whose legacy is still being felt in the on-going conflicts in Angola, Sudan and the DRC. Then look at all the problems that confronted African states at independence in the 1960s and compare them to the problems Western and even East European states faced at the same time, you would be forgiven for thinking that Africa had no chance . . .

Yet, incredibly, a surprising number of African states have and are succeeding against the odds . . . Botswana and Tunisia have recorded the fastest growth rates in the world this year. Africa, according to the African Development Bank, will grow by between 4% and 5% this year. This figure is greater than for any other region of the world. The vast majority of African countries now have democratically elected governments. African stock markets, despite their modest size, were the most profitable last year. The return on investment is higher in Africa than anywhere else in the world . . .. [N]ever has so much formal education been spread so fast to so many people as in Africa, despite structural adjustment programmes that have slashed spending on education . . .. Since independence, millions of new classrooms, housing units, offices and clinics have been built. Air and seaports have been developed and hundreds of thousands of kilometres of new roads have been laid. Starting from point zero, several African countries, Mauritius and Tunisia to name just two, have become among the most competitive industrial centres in the world . . .

Despite its problems, including endemic diseases, wars and famine in some parts, outright looting of national resources in others, Africa's performance from less than a standing start 40 years ago has been exemplary. Many nations, including the US, are now prepared to invest substantially in Africa. Some voices . . . are calling for a Marshall Plan for Africa. They want to see an end to the sticking-plaster approach to Africa's problems. They want to see Africa given a fair opportunity to stand as an equal in the community of nations. Other voices, unfortunately equally powerful, can see nothing good emerging from Africa. It is therefore our duty, and that of others with a voice that can be heard, to make sure that Africa's many achievements are trumpeted at least as loudly as its shortcomings.

Source: http://www.africasia.com/archive/ab/00_0708/praise_africa.htm

While popular stereotypes of Africa and Africans persist, some representations have begun to shift in recent years, mainly in response to a recognition that some African countries have burgeoning economies that present considerable market opportunities for western businesses. Following the global financial crisis of 2008, which affected economies the global North most acutely, Africa began to represent a continent of opportunity. For example, in 2011, *The Economist* issued something of an apology to its depiction of Africa as 'the hopeless continent': 'Since *The Economist* regrettably labelled Africa 'the hopeless continent' a decade ago, a profound change has taken hold. . . [Today] the sun shines bright. . . the continent's impressive growth looks likely to continue' (December 3 2011, www. economist.com/node/21541008). Underpinning this new positive spin are statistics showing that over the past decade Africa's trade with the rest of the world increased by over 200 per cent, the continent's foreign debt decreased by 25 per cent, foreign direct investment grew by 27 per cent in 2011; despite the ongoing political crises in North and parts of Central Africa, African economies are projected to grow by between 4.2 per cent (UN 2012) and 5 per cent (IMF 2012); there are currently more than 600 million mobile-phone users and increasing literacy and improving skills have resulted in a 3 per cent growth in productivity. According to the McKinsey Global Institute, 'The rate of return on foreign investment is higher in Africa than in any other developing region' (www.twnside.org.sg/twnf/2012/3856. htm). Global consultancies such as Goldman Sachs, Deloitte and McKinsey have been keen to promote the idea of Africa as no longer a continent populated by impoverished and often starving people, but by burgeoning 'middle classes' with money to spend. Who are these middle-class people? According to the WB, any person living on above $2 per day – just above its own poverty line. Here we have the discursive construction of the 'new African consumer'. As one African commentator wryly observes:

> In a time when growth rates of industrialised countries stutter and when the Chinese and Indian engines of the global economy are somewhat slowing down, financial analysts and investment consultants can't get enough of the one thing that they have dismissed for so long: Africa. . . Close your eyes and let your imagination do the rest: hundreds of millions of purses loosening their strings.
>
> (Enaudeau 2013)

This discursive shift is clearly a strategic move and it is not coincidental that it comes at a time when western economies are being left behind by the likes of China, India and Brazil in terms of investment and development in African countries.

Yet, in their efforts to now present a more positive image of African development to encourage investment, these newer discourses remain partial representations. While overall volumes of trade have increased, Africa's share of global trade continues to decline. As Achille Mbembe (2013) argues, economic growth in many African countries is fuelled by a mineral boom, which instead of creating jobs is deepening social inequalities. The continent still faces enormous challenges in terms of investment in basic infrastructure and political instability remains a threat. Economists are fond of saying that one in three Africans are now middle class, but questions remain: Who are they talking about?; Is this how they would choose to represent themselves?; and, Is this yet another example of western observers speaking for and about people in the global South? Enaudeau (2013) contests this persistent will to represent by arguing that Africa's 'middle classes' are perfectly able to speak for themselves:

> . . . At the bottom of the pyramid are those on whom narratives are imposed and who have limited means to resist; at the top are those who have decided on their narrative and are writing their memoirs already; and in between is where the action is, where narratives overlap, clash or fuse because Africans are playing the field unencumbered by the nay-sayers or the yay-sayers. There is much to be learned about that life; and who better to tell these stories of in-betweenness than members of the middle classes themselves, African journalists, artists, bloggers and academics?

Significantly, the new discourses also fail to interrogate the relationship between the kind of development taking place in African countries and the gross inequalities they are not only perpetuating, but exacerbating. For example, thirty years after Live Aid and the Ethiopian Famine, a report by global consultancy organization Deloitte (2014) celebrates Ethiopia's economic 'growth miracle' and attempts to reassure investors:

> eager to avail themselves of the rich opportunities that Ethiopia now offers. . . that past threats to the country's economic and political stability belong firmly in the past, and that solid measures have been put in place to mitigate these risks in the future'. The report praises improvements in poverty reduction, health and education, and infrastructure and governance improvements.
>
> (Deloitte 2014: 11)

Ethiopia has indeed experienced an economic boom, which has generated millionaires faster than any other African country, a real estate boom (especially in Addis Ababa), and GDP growth rates of over 100 per cent. Yet this is a partial story that fails to address both the repressive governance and the profound inequalities that this economic growth is exacerbating:

> A generation after the famine that pierced the conscience of the world, Ethiopia is both a darling of the international development community and a scourge of the human rights lobby. Even as investment conferences praise it as a trailblazer the entire continent should emulate, organisations such as Human Rights Watch describe it as 'one of the most repressive media environments in the world'.
>
> (*The Guardian* 22/10/14)

Whilst being complicit with the discourses generating by the likes of Deloitte, the Ethiopian government is also keen to hide the fact that it is once again one of the hungriest countries in the world. More than 50 per cent of the population survives on less than $1 per day. In 2010, 2.8 million people required food aid. At the same time, the Ethiopian government sold off 3 million hectares of the most fertile land to rich countries and wealthy individuals. In February 2016, The World Food Programme announced that over 10 million Ethiopians require urgent humanitarian assistance. This rose to over 11 million by October 2016, at which point the government declared a state of emergency in response to increasing food insecurity and escalating violent protests across the country. This does not appear to have improved the situation: higher temperatures linked to climate change and successful failures of seasonal rains mean that Ethiopia still teeters on the verge of an unprecedented humanitarian disaster. In December 2017, according to Oxfam (www.oxfam.org/en/emergencies/ethiopia-food-crisis), 8.5 million people are facing severe hunger, particularly in the Southern Somali region. 700,000 are on the verge of starvation and this number is likely to increase as the latest forecasts have predicted below average rains for 2018. 9.2 million are expected not to have regular access to safe drinking water, which will lead to disease and increases in infant mortality. Water and food shortages in neighbouring countries put even more pressure on the receiving areas and with over 780,000 refugees (as of February 2017), Ethiopia is currently one of the largest refugee-hosting countries in Africa.

While drought is clearly a significant factor, at the heart of Ethiopia's problems is the nature of development that has been widely celebrated and encouraged. International organizations have encouraged Ethiopia to produce cash crops – like the highly successful partnership between Ethiopian producers and Dutch investors that has created the world's biggest flower export industry, which exports roses to the Dutch markets (Deloitte 2014). This has created employment for Ethiopian workers, and international investors and a small number of the Ethiopian elite have profited enormously, but the wider costs in Ethiopia have been extreme. To produce the cash crops, state authorities have displaced hundreds of thousands of indigenous peoples, abused human rights, destroyed traditions and degraded environments. In short, Ethiopia has been subject to a land grab by both the government and overseas investors and, as Li (2014) points out, this was part of a wider land grab across Africa that was actively supported by the WB. Since 2008, the Ethiopian government has leased or sold enormous expanses of fertile agricultural land to what it refers to as 'foreign investors' – international corporations, domestic agents and fund managers. Among these 'foreign investors' are nations, such as China and the Gulf States, who are securing their own food security at the expense of Ethiopia's rural population (Hall 2011). The available land for domestic crops has been substantially reduced and this has created greater dependence on food aid than ever before. Food is being exported while people starve. Discourses of Afro-optimism should therefore be subject to similar interrogation as those of Afro-pessimism. And, as I discuss in Chapter 8, engaging with African scholars and their ideas about African futures is far more likely than western economic discourses to produce a deeper understanding of both development in African countries and solutions to global challenges more broadly.

## Critiquing development discourse

While there are ever-pressing material issues such as poverty in the world, concerns with the *language* of development might seem esoteric. However, as we have seen, language is fundamental to the way we order, understand and intervene in the world, and to how we justify those interventions (Crush 1995; see also Escobar 1984, 1988, 1995a; Pieterse 1991; Slater 1992a, 1992b). In addition, dominant discourses have had enduring material effects. Postcolonialism allows us to 'deconstruct the mythical material futures of development that were predicted for many countries' (Power 2003: 121), echoing earlier radical critiques:

> a decade or so ago, as the 'underdeveloped' were administered larger and larger doses of the magic medicine of 'development' [we] began to notice how this therapy always – and strangely – enriched the developers and impoverished the supposed beneficiaries of the process.
>
> (Buchanan 1977: 363)

For Buchanan, writing in the 1970s, development had become a 'dirty word'. In response, development theorists sought increasingly to critique what he called 'the generation of smooth-talking, formula fixated and model-making technocrats' based in the 'white North'. Postcolonial critiques have continued to build on these critiques, drawing particular attention to relationships of power in discursive representations of the South and their material effects.

Yet caricatures of the South persist, particularly in representations by the IFIs. Their contemporary appeal lies in their simplicity and generality. As Williams (1995: 173) argues, 'Development strategies rest on a common diagnosis of the problems of quite different countries, and the transfer of standardized policies and second-hand technologies from one country to another'. Complexity and variation cannot be managed within either the practice or the discourse of development. Postcolonial critiques destabilize these dominant discourses of development, which are ethnocentric and often originate in imperial Europe. They provide a corrective to development discourses that prioritize western histories and universalize western paths to 'industrialization' or mass consumption (McEwan 2000). This tradition of critique continues, largely because Development Studies is still western-centric, dominated by hegemonic (white, male, Northern) voices and discourses. The texts of development are strategic and tactical, promoting and justifying certain interventions and delegitimizing and excluding others (Crush 1995). Power relations are clearly implied in this process; certain forms of knowledge are dominant and others are excluded. A postcolonial approach to development discourses, therefore, can say a great deal about apparatuses of power and domination within which development texts are produced, circulated and consumed. Development discourse promotes and justifies very real interventions with very real consequences – Ethiopia is a case in point. It is, therefore, imperative to explore the links between the words, practices and institutional expressions of development, and between the relations of power that order the world and the words and images that represent the world.

Postcolonial critics, Said and Spivak in particular, have argued that literary writing (especially about the South) struggles to avoid reproducing various forms of western hegemonic power. The same may be said of the field of development. Post-development theorists (e.g. Escobar 1984, 1995b; see also Crush 1995) have long demonstrated that working in development inevitably positions us within a 'development discourse', where the North's superiority over the South is taken for granted and western-style development is the norm. An 'us and them' dichotomy shapes our representations: 'we' aid/develop/civilize/empower 'them'. This can be seen in Northern representations of 'others' in immigration and security discourses, in the sexualization and racialization of female migrant labour, and in discourses about child labour and human rights. Changing this relationship is not brought about simply by good intentions or by semantics – changing the terms of reference to 'beneficiaries', 'target groups', 'partners' or 'clients', instead of 'poor', 'underdeveloped' or 'disadvantaged' does not by itself alter the discourse or dismantle the us/them power relationship (Kapoor 2004: 629). Rather, development theorists would do well to engage with ideas coming from those other places that have been marginalized by dominant Northern modes of theorizing. These other ways of knowing the world 'are not irrelevant to us because they emanate from other societies and cultures' (Potter 2001: 425–6) (see Chapter 5).

## Development discourse and the 'new imperialism'

The need to decolonize development discourses is, if anything, even more urgent in the context of contemporary global geopolitics and the 'new imperialism'. While there are continuities between colonial and postcolonial discourses as they are manifested within the development literature, the twenty-first century has witnessed a significant shift in discourses of development. The terrorist attack on the World Trade Center on 11 September 2001 precipitated a 'new' era of threat, crisis and insecurity. The confidence in the dream of liberal democracy and the promise of modernity appears to have been shattered. Safety, freedom, democracy and prosperity are now perceived to be as much under threat from global terrorism as they were from Nazism seventy years ago. Accompanying this geopolitical shift has been a shift in discourses of development and modernity (Biccum 2005; Duffield 2001, 2007). In international political circles, development has been incorporated into emerging metropolitan discourses and policy frameworks of security. An emerging spate of academic literature, translated into public and popular discourse, narrates these shifts 'as a "new" era of imperialism, for which it is largely apologetic' (Biccum 2005: 1005).

This literature argues that following the collapse of communism and the old world order in the late 1980s and early 1990s the world has descended into a state of chaos. Robert Kaplan's thesis (1994) was one of the most influential right-wing treatises that posited the world generally, and Africa in particular, as on the verge of anarchy, conflict and chaos. West Africa, specifically, was a hopeless place of overpopulation, squalor, environmental degradation and violence, overpopulated by barbarians and starving children (see Hartmann's critique (1999)). Kaplan's apocalyptic and Malthusian vision depicted Africa as mired in tradition – the main obstacle to and the reason why Africa has failed to cope with modernization. Such partial, stereotypical and uncontextualized accounts 'convey a dominant image of Africa – that is a place of "misery", "chaos" and "brutality", the recurrence of which is almost predictably systematic' (Beattie et al. 1999: 233). Echoing nineteenth century imperial discourses, African governments and individuals are seen as incapable of harnessing

Northern notions of law and order, markets, good governance, transparency and democracy that are considered the prerequisites to development. Kaplan ignores the fact that development trajectories worsened *despite* the general trend towards democracy and liberalization in many African countries since the 1990s (UNDP 2003). Instead, he asserts that a new world order is required to restore stability. A 'new' imperialism can be justified as a means of restoring order, without which the threat to democracy, freedom and prosperity will remain.

Despite fervent criticisms of such ideas, the vocabulary of mainstream development discourse has shifted to normalize this 'new' imperialism, justify the (re)colonization of the Middle East, and position neoliberal capitalism as the only option for global governance. The 'new' imperialism is justified as a means through which to deliver 'development' to places like Iraq and Afghanistan – ironic given the destruction precipitated by the 'war on terror' (Grosfoguel's (2012) 'democratize or I'll kill you'). The more 'developed' these places are – in other words, with western style democracies and plugged into global capitalism – the less prone to terrorism their people will be. The menace of poverty and the threats this poses need to be nullified by a new development agenda. The 'new' imperialism is thus couched in terms of a new civilizing mission, and western governments use the discourse of development to narrate themselves as civilized, developed and modern (vis-à-vis a barbaric Other that is now named radical Islam). They are, by dint of these virtues, equipped to lead metropolitan efforts to restore global order and security. The lessons of postcolonialism, as developed through the likes of Bhabha and Spivak, thus have particular political significance in revealing, understanding and challenging the 'new' imperialism. They reveal a number of paradoxes; firstly, in the logic underpinning 'new' imperialism (Biccum 2005: 1008–9):

● that freedom, liberation and civilization can be brought to a people by conquering them;
● that autonomy can be garnered through oppression;
● that a free, liberal and democratic society at 'home' is made possible through despotic rule abroad;
● that there is a need to mask the connections between domestic civil liberties and external military and political repression

through the creation of a false self/other, colonized/colonizer, developed/un(der)developed and, in former USA President Bush's terms, good/evil binary;

- that the universal ideal of one developed world, or a globalized world under neoliberal capitalism, is known not to be an answer to poverty and will never arrive;
- that it is possible that policies advocated by the 'new imperialism' might be the cause of the menace of poverty and not its solution.

The second paradox is the reinvention of poverty in discourses of development. The aim of lifting 'millions out of poverty' (DfID cited in Biccum 2005: 1014) echoes the apologetic literature on the 'new' imperialism. Any responsibility for the creation of poverty within global capitalism is effaced by positing imperialism as a solution and positioning the West as the benevolent hand that will lift the impoverished out of their natural state of degeneracy. Poverty exists with no history and no causal factors; it is simply a threat and a new global challenge that legitimates the 'new' development policies of the West. The third paradox is the emergence of a 'new' Dual Mandate within development discourses. In the nineteenth century this underpinned colonial policy and today it justifies the 'new' imperialism on the basis of moral obligation and economic necessity. Poverty is a threat to 'us' – it is 'our' moral duty to lift 'them' out of their primitive state. The final paradox is the new 'worlding of the West as world' (Spivak 1987). A threat to 'our' security is a threat to global security; 'western interests' are constructed as universal; global interests are domesticated as 'our interests' and, therefore, 'western interests'. And securitizing 'our' futures again structurally exploits other countries.

As Biccum (2005) argues, this message is now marketed through government departments (e.g. DfID), whose promotional literature asserts that a 'new' development agenda will protect 'us all' (i.e. the North) from the menace of poverty. The solutions are thus both capital-centric and western-centric. Some within the development field might counter that western-centrism is overplayed given the emergence of Southern initiatives such as the New Economic Partnership for Africa's Development (NEPAD) (see Chapter 6). In reality, however, this is dominated by South Africa and Nigeria, who have appropriated Northern agendas. NEPAD's focus on human rights, democracy and free trade offers little that is new, nothing that emerges from distinct African contexts, or challenges dominant, Northern development discourses that are increasingly couched in the language of global security.

While terrorism and poverty is currently constructed discursively as a threat to 'western interests', China is also invoked as a new economic threat with potential to challenge the North's historical relationship with other developing countries, particularly in Africa. As Mawdsley (2008) argues, contemporary representations of China in British newspaper reports evoke classic framings of the 'Yellow Peril', in which the Chinese are represented as unscrupulous, inhumanely cruel, despotic, devious and inscrutable. Popular books heighten the sense of alarm, with titles such as *Red Dragon Rising: China's Military Threat to America* (Timperlake and Triplett 2002), *China: The Gathering Threat* (Menges 2005), *The China Threat: How the People's Republic Targets America* (Gertz 2002) and *Hegemon: China's Plan to Dominate Asia and the World* (Mosher 2000). That China is fast becoming a major player in global development is cause for further anxiety. China's economic boom is driving demand for resources, especially in Africa, and this is likely to place China and the West (especially the USA and a post-Brexit UK eager to revive Commonwealth trade relations) in competition. It also underpins a new Orientalism that draws on familiar themes.

China's approach to development in Africa differs from that in the North. While the latter portrays Africa as a problem and/or threat, Chinese discourses convey respect and mutual opportunity. Unlike the North, China does not have a blanket policy towards the continent, but pursues bilateral arrangements with individual states. Core principles underpin these arrangements: (i) the global order is unjust and inequitable and China must act together with other developing countries; (ii) all countries have a right to self-determination and no right to meddle in the internal affairs of other states; (iii) China endorses peaceful multilateralism. While the North focuses on negative representations of China's impact in Africa (undermining local manufacturing, lack of concern with labour rights and human rights, transparency and environmental protection, exploitation, corruption and degradation, support for corrupt and despotic regimes in Sudan and Zimbabwe), Africans are more likely to have a positive assessment of China's present and future role (Shelton 2006, cited in Mawdsley 2008).

What emerges from Northern discourses about China's role in Africa are patterns in which certain tropes are reiterated: African weakness and vulnerability, Northern trusteeship, Chinese ruthlessness. Clear discursive themes are apparent (Mawdsley 2008): a tendency to homogenize China and Chinese actors; a tendency to focus excessively on China's interest in oil; a preference for focusing on China's negative impacts in Africa; a tendency to portray Africans as victims or villains; a complacent account of the roles and interests of Northern actors in Africa. In short, European and US interests in Africa are presented as a strategic need to save Africa from China, based in an ethical concern for post-colonial Africa. The North may have supported corrupt regimes in the past, but it has learned its lesson. While colonialism was morally wrong and economically exploitative, it at least had developmental/paternalistic dimensions and well-intentioned elements, unlike China's 'new scramble for Africa' and its 'aggressive', colonial ambitions. As Mawdsley (2008) argues:

> There is room for dialogue and negotiation on all sides. But the current journalistic vogue for demonizing China and largely exculpating the West (a most unsatisfactory arbiter of what 'responsible power' should look like) does not pressure politicians, policy-makers or corporations to reflect on structural inequalities perpetuated by the West, on what China has to offer, or the need to respect African sovereignty.

Also missing from dominant discourses of development is any real sense of self-reflection. The EU's role in the South has increasingly been to denounce corruption, 'bad governance' and the oppression of minorities. However, significant erasures persist in such discourses: Greece, Italy and Belgium rank highly on the list of the world's most corrupt countries; it is only a few decades since the last military putsch in Europe; minorities across Europe still suffer from political and legal forms of discrimination (Karagiannis 2004: 135). Emphasizing these traits in the South, while understating them in the North, serves a distinct purpose. It constitutes the South as ripe for intervention, as an economic reservoir of raw materials, and as an object of study.

## Challenging representations: gender and development

Some of the most strident criticisms of Northern representations and discourses of development have emerged within women's movements in the South. We saw in Chapter 3 that various women's movements around the world have sought to challenge the hegemony of western feminism, and this extends to challenging the hegemony of western development discourses. Gender and development has been an important area where feminist politics have increasingly overlapped with postcolonial imperatives. The ways in which Northern women represent their Southern counterparts, and the power relationships inherent in this, have increasingly been brought under scrutiny. If the 'Third World' is frozen in time, space and history, this is particularly the case with 'Third World women' (Mohanty 1988). According to Carby (1983: 71):

> Feminist theory in Britain is almost wholly Eurocentric and, when it is not ignoring the experience of black women at 'home', it is trundling 'Third World women' onto the stage only to perform as victims of 'barbarous', 'primitive' practices in 'barbarous', 'primitive' societies.

Northern feminists have been criticized for universalizing their own particular perspectives as normative, essentializing women in the South as tradition-bound victims of timeless, patriarchal cultures, and reproducing the colonial discourses of mainstream, 'male-stream' scholarship (Carby 1983). What Mohanty (1988) calls the 'colonialist move' arises from the bringing together of a binary model of gender, which sees 'women' as an *a priori* category of oppressed, with an 'ethnocentric universality', which takes Northern locations and perspectives as the norm. The effect is to create a stereotype – 'Third World woman' – that ignores the diversity of women's lives in the South across boundaries of class, ethnicity, and so on, and reproduces 'Third World difference'. This is a form of Othering, a re-privileging of Northern values, knowledge and power (hooks 1994; Ong 1988; Spivak 1990; Trinh Minh-Ha 1989). Mohanty argues that Northern feminism is too quick to portray women in the South as 'victims', to perceive all women as oppressed and as the subjects of power.

This assumption of sameness incurred the resentment of many Southern women at the UN conferences on women in Mexico City (1975) and Copenhagen (1980) and opened deep divisions. Heated debates at these conferences highlighted the profound differences among women across the global divides of North–South and East–West, as well as within regions along class and political lines. The theoretical fallout from these debates is an emphasis on difference as opposed to universalism. The criticisms of black feminists and non-western women living in the North (e.g. Mani 1992), together with input from Northern feminists themselves (e.g. Wilkinson and Kitzinger 1996), has enabled Northern feminism to move on from notions of global sisterhood, to acknowledge differences, deconstruct Othering processes, and celebrate diversity and multiplicity (Flew *et al*. 1999).

Despite this, Narayan (1997, 1998) sounds a cautionary note about the risk of falling into the trap of cultural essentialism. In trying to account for differences between women, seemingly universal essentialist generalizations about 'all women' are replaced by culture-specific essentialist generalizations that depend on totalizing categories such as 'western culture', 'non-western cultures', 'western women', 'Third World women'. The resulting portraits of 'Third World Women', 'African women', 'Indian women', 'Muslim women', 'post-socialist women', as well as the picture of the 'cultures' that are attributed to these various groups, often remain fundamentally essentialist. They depict as homogeneous groups of heterogeneous people whose values, interests, ways of life and moral and political commitments are internally plural and divergent (Narayan 1998: 87–8). The consequence is 'an ongoing practice of "blaming culture"' for problems in Southern contexts and communities (Narayan 1997: 51; see also El Saadawi 1997; Schech and Haggis 2000: 103–4).

The tendency of Northern feminisms to theorize difference in universalizing ways is also problematic. They often fail to acknowledge explicitly that the framing of debates about difference, even in reference to cross-national experiences, is the product of specific Northern experiences of identity formation. Difference, and how feminists and women activists negotiate this, varies both geographically and temporally. For example, whereas Northern feminists emphasize difference, women's movements in African societies have often suppressed difference to build coalitions within their fractured societies (Flew *et al*. 1999: 672). Examining how such dialogues about difference have evolved in specific contexts should encourage feminists in the North to reflect more critically on what historical, social, political and economic conditions have shaped their understandings of difference (see also Benhabib 1999; Felski 1997; Goetz 1991).

Central to a postcolonial critique of gender and development is the problematizing of both the universalist assumptions of Northern feminisms, wherein political goals are assumed to be the same the world over despite vast differences in the everyday realities of women's lives, and the tendency to theorize difference through binaries. Within these binaries the 'Third World Woman' (and previously, the colonized woman) has functioned as the oppressed, backward Other in relation to which an image of a liberated, developed, educated female Northern Self has been constructed (Mohanty 1988; Parpart 1995a, 1995b; Spivak 1988, 1993; Trinh 1989). These representations were not simply confined to development theory and academic circles, but also informed the development industry. As Parpart (1995a: 227) argues:

> Drawing on a long history of colonial discourse which represented Third World women as particularly backward and primitive, development planners continued and even extended the representation of Third World women as the primitive 'other', mired in tradition and opposed to modernity.

This continued to be reproduced in the Women in Development (WID) discourses of the 1970s, reinforcing rather than challenging images of the backward and oppressed Third World woman. While the Gender and Development (GAD) approach of the 1990s was more radical in countering the 'add women and stir' philosophy of WID with a more critical understanding of gender roles and relations in development, it did not challenge images of women in the South as the 'impoverished, vulnerable "other"' [who] . . . need to be saved from poverty and backwardness' (Parpart 1995a: 236).

Postcolonial strategies to dislocate western-centric approaches perceive all knowledge as contestable, in contrast to the hands-off 'respect' of cultural relativists. Certain issues will unite women cross-culturally (e.g. sexist oppression); other struggles, such as those for racial justice or national liberation, might mean confrontation between women. Issues such as veiling in Muslim societies or female genital mutilation, for example, have exposed the divisions between feminists, especially those who see these as normative issues (oppressive of women and thus wrong) and those who demand that they are understood from within the context of the cultures that produce these practices. Postcolonialism seeks to problematize stereotypes and generalizations. For example:

> there is no need to portray female genital mutilation as an 'African cultural practice' or dowry murders and dowry related harassment as a 'problem of Indian women' in ways that eclipse the fact that not all 'African women' or 'Indian women' confront these problems,

or confront them in identical ways, or in ways that efface local
contestations of these problems.

(Narayan 1998: 104)

The challenge is to produce something constructive out of
disagreement, and to combine material concerns and emphasis on local
knowledge with postcolonial and post-structuralist dismantling of
knowledge claims. Ferguson (1998: 95) theorizes this as a new 'ethico-
politics'. She suggests that the problem that Northern feminists need
to confront is that they are located in the very global power relations
that they might aspire to change; hence there is a 'danger of colluding
with knowledge production that valorizes status quo economic,
gender, racial and cultural inequalities'. There is a need for self-
reflexivity, recognition of the negative aspects of one's social identity
and devaluation of one's moral superiority to build 'bridge identities'
across difference. This allows other knowledge to talk back, and
creates 'solidarity between women that must be struggled for rather
than automatically received' (Ferguson 1998: 109). This does not
mean generalizations cannot be made, but it puts the emphasis back
on how they are made. As Schech and Haggis (2000, 113) argue, these
postcolonial feminist approaches are not simply about deconstructing
western feminisms. Rather they provide a more comprehensive project
of re-moulding a conceptual framework 'capable of embracing a
global politics of social justice in ways which avoid the 'colonizing
move'. These concerns have given rise to new feminist thinking
on questions of gender and development, including debates about
globalization, humanitarianism (Ticktin 2011) and alternative
frameworks for thinking about transnationalism (Ramamurthy 2003).

## Postcolonial counter discourses

Postcolonialism allows us to deploy a variety of different resources to
understand the material and cultural aspects of development, as well
as the politics of identity formation and belonging as it varies across
space. Many debates about postcolonialism and development revolve
around the notion of subjectivity – the range of subject positions or
identities that an individual human being as agent or subject mobilizes
or embodies. Postcolonialism seeks to disrupt the notion of a unitary,
self-authoring subject posited by Enlightenment thought and critiques
the 'unreflexive projection of subjectivities as universals' (Werbner

2002: 2). Thus, the universalization of a single image or definition of poverty and poor people is hotly disputed. Postcolonial theory allows us to 'open up the notion of agency', and to 'deepen our understanding of subjectivity by looking at its multiple forms, influences and meanings and opening up the spaces where development's subjects are constructed' (Power 2003: 126). It is not simply marginalization, dispossession and exploitation that form common ground for the making of subjectivities in the contemporary period; rather, there are plural arenas in which economic, cultural and political identities are made. Subjection to the discourses of development is not only about relations of power and domination, but also about resistance and reconstruction. Examples of this resistance and reconstruction are evident in postcolonial film, photography and literature. Analysis of such sources, common in postcolonial studies, imbues development studies with a theoretical richness and openness to information that it traditionally lacks (Sylvester 1999).

## Postcolonial film as counter discourse

As Lewis *et al.* (2012) point out, development has long been a central theme in cinema, often telling the story of contentious relationships between richer and poorer countries and usually from the perspective of western protagonists. Relatively recent examples include: *Beyond Borders* (2003) in which Angelina Jolie plays an aid worker; *Blood Diamond* (2006), which focuses on the complicity of the international diamond trade in civil war in Sierra Leone; *The Hurt Locker* (2008), which depicts bomb disposal units during the Iraq War. Postcolonial issues are thus manifested in film, including the writing of history, nationalism, identity, gender and race. In contrast to Hollywood's treatment of 'development' in its broadest sense, postcolonial films provide a keen insight into and critical commentary on the continual process of political, economic and cultural struggle of power between the divided 'self' and 'Other'. They also represent counter discourses to dominant discourses of development. A notable example is *Bamako* (2006), written and directed by Mauritanian-born filmmaker Abderrahmane Sissako (see Figure 4.2). Set in the courtyard of a mud-walled house in Bamako (the capital of Mali), west Africa, *Bamako* juxtaposes the intimate personal story of an African couple on the verge of breaking up with public political proceedings in which civil society is taking action against the WB and the IMF, whom they

**Figure 4.2   *A poster from the film* Bamako (2006)**

Poster design by KeeScott Screen

directly blame and put on trial for Africa's woes. While the focus of the two institutions has been on 'poverty alleviation', the film argues that they are promoting an economic model that goes well beyond their core missions, and that there is extensive evidence that their policies have exacerbated rather than alleviated poverty in the developing world.

The film arises out of the reality that no court of law exists to call into question the power of the North, and denounces the fact that the predicament of hundreds of millions of people is the result of policies that have been decided by people elsewhere. In addition, one of the witnesses at the trial, Aminata Traoré, refuses to accept that poverty is the main characteristic of Africa by asserting that Africa is a victim of its *wealth*. The film counters images of a continent beset by war and famine by focusing on the will of Africans to fight and their ability to get on with life. As Sissako argues, 'These images represent for me the glance of those who don't have the means to speak out' (www. bamako-themovie.com/about_03.html). The website accompanying the film contains a detailed written critique of the IMF and WB, which attacks the undemocratic nature of both organizations and the deleterious effects they continue to have in Africa (see www.bamako-themovie.com/about_04.html).

Other notable postcolonial African films include those by Sembène Ousmane (see Chapter 2). *La Noire de . . .* (1966) was the first African feature film. *Mandabi* (1968) was the first film made in Sembène's native language, Wolof. *Xala* (1974) was based on Sembène's novel about post-independence corruption and moral decay. *Ceddo* (1977) was heavily censored in Senegal because of its anti-Islam stance. *Moolaadé* (2004), a winner of awards at Cannes, is set in a small African village in Burkina Faso and explores the controversial subject of female genital mutilation. There are many other important African films. For example, Nouri Bouzid's *Nezness* (1992) is set in Tunisia and criticizes both the restrictions associated with what he calls 'the hypocrisy of Islam' and European influence on Arab society. Mostefa Djadjam's *Borders* (1992) confronts the global controversy of refugees from a North African perspective. Bakupa Kanyinda Balufu's *The Draftsmen Clash* (1996) is a political satire about African dictators. Senegalese director Djibril Diop Mambety moves beyond documenting Africa's victimization towards envisioning the continent's recovery. His *The Little Girl Who Sold the Sun* (1999) tells the story of a young disabled girl who triumphs against adversity to become a street vendor. The film is a tribute to the indomitable spirit of Dakar's street children and their capacity for transforming their own lives.

Postcolonial films from around the world reveal different aspects of postcoloniality and the lived experiences of colonized and former colonized peoples. Films by Moufida Tlatli (an award-winning Tunisian Arab film director) include *The Silences of the Palace* (1994) and *La Saison des Hommes (The Season of Men)* (2001) and deal with issues of alterity, subaltern feminism and postcolonialism. Euzhan Palcy's *Sugar Cane Alley* (1984) tells the story of a mulatto boy and experiences of creolization in colonial Martinique. Vietnamese-born Trinh T. Minh-ha's *Surname Viet Given Name Nam* (1989) challenges official culture with the voices of women. The film explores the difficulty of translation, and themes of dislocation and exile, critiquing both traditional society and life since the US–Vietnam war.

A recent example of a Latin American postcolonial film is *Embrace of the Serpent* (2015) by Colombian director Ciro Guerra (Figure 4.3).

**Figure 4.3** *A poster from the film* Embrace of the Serpent *(2015)*

Design by Trigon Film

The film is simultaneously a powerful indictment of the violence of colonialism and celebration of indigenous peoples and disappearing cultures. The film was shot in the rainforests of Vaupés and is described by Guerra as 'an attempt to build a bridge between western and Amazonian storytelling'. It tells the story of two white explorers who travel into the Colombian Amazon 40 years apart in the early and mid-twentieth century. The story is fictional, but draws on the journals of two real-life explorers: German ethnobotanist Theodor Koch-Grünberg and American botanist Richard Evans Schultes. The first explorer in the film, Théodor von Martius, is searching for the healing and hallucinatory yakruna plant, which he believes could cure him of a fatal illness. The second explorer, Evan, is searching for the same plant four decades later because he believes it will help him to dream and heal his soul. The central figure in the film is Karamakate, a warrior-shaman who guides both men. Unlike Hollywood renditions of European encounters with indigenous peoples, the story is told through his eyes.

Karamakate appears to be a lone survivor of his tribe, which he believes to have been wiped out by rubber barons. His memories, or lack of them, flow back and forth through the narrative. With Théodor, he encounters children in a Catholic mission – orphans of the rubber wars – who are being beaten into forgetting their mother tongue. Karamakate tells the boys 'Don't let our song fade away': the loss of language signals the destruction of culture, connection with history and sense of belonging in orate societies. By the mid-twentieth century, the older Karamakate fears that he has become a 'chullachqui': a spectral, empty, hollow version of himself bereft of memories and lost in time. It is through Karamakate that we witness the violence of colonialism and its devastating impact on indigenous people and their cultures.

As Lewis *et al.* (2012) argue, while films that focus on westerners engaging with their own consciences and dilemmas in the face of global inequality and conflict are instructive and can be useful in raising public awareness, they are problematic because of the cost paid in terms of the relative lack of local voices. In contrast, postcolonial films are more oriented towards narratives as told by the people who experience colonialism and its enduring effects in contemporary development. Powerful visual story-telling can come at the cost of historical accuracy or factual detail, but 'cinema plays a role in shaping and reflecting popular perceptions of global development issues [that] in the West cannot easily be ignored' (*ibid.*: 17). Postcolonial films are important counter-narratives to mainstream discourses that can capture the violence of colonialism and the vagaries of development. They are also 'powerful teaching tools for bringing alive and humanizing important, if inherently vexed, global issues' (*ibid.*: 17).

## Photography as counter-discourse

Photography was invented in Europe and it is customary to think of it as primarily a western activity. It was also used to perpetuate imperial discourses and stereotypical representations of colonized peoples (Alloula 1986; Gilman 1985; McClintock 1995; McEwan 2000; Rothenberg 1994). Colonial photographs were often 'studio-fantasies' (Graham-Brown 1988: 40) that say more about Britain in the nineteenth century than they do about the parts of the world that travellers visited and the peoples they encountered. Yet, they conveyed powerful images to the audience and were included for decorative purposes to add colour to already spicy textual descriptions of other peoples and places. However, photography is also a counter discourse and, in addition to critically deconstructing colonial photography, postcolonialism acknowledges the possibility of 'photography's other histories' (Pinney and Peterson 2003). This requires 'looking past' the surface of the image for hidden meanings, and exploring subaltern readings and re-uses of photography. It also requires exploring local modifications of a globalized technology and the ways in which photography can be transformed so that it becomes a tool of local empowerment.

In both Indian and African portrait photography, for example, western aesthetic canons and philosophical conceptions of photography as document and evidence of the past are broken down by subaltern photographic practices that do not document what is, but help invent what is desired or experienced. Similarly, Ghanaian photographer Philip Kwame Apagya's formal portraits in front of commissioned painted backcloths seem to deliberately subvert colonial photography, which also used backcloths (see McEwan 2000), by representing Africans very differently (Figure 4.4). There is no intention to deceive – the images are suspended between realism and a sort of naïveté, both unreal and hyper-realistic. They disturb because the dreams of real African people are foregrounded against a background that praises consumer society,

**Figure 4.4**  *Philip Kwame Apagya's portrait of a Ghanaian woman*

development and technology, while problematizing this as a dream, a fiction, or as vacuous and empty. Apagya's images of smiling African women juxtaposed against food and technology are provocative, not least because this is in direct contrast to how they are usually portrayed in popular and development discourses. They are also disturbing because of the insight they offer into a growing cultural vacuum – development is a dream, empty and materialistic.

Photo-essays have also been used increasingly to create counter-narratives of development. For example, Espelund *et al* (2003: 9–10) contest the representation of Africa as a place of crisis, trauma and victimhood by celebrating the spirit of African peoples, their triumph over adversity and their resilience. Their book is a:

> . . . tribute to the millions of people in Africa who excel themselves and make a positive difference in their communities – individuals who display an ability to survive that is often far more impressive than the achievements made by people in more privileged parts of the world . . . [W]hile the attention [in the news] was on 'the big men' – as one former boy soldier in the book terms the people who rule and make headlines – the women (more often than not) and men who ended up in this book were busy picking up the pieces or laying the foundations for a better future.

These different stories challenge Eurocentric reductions of the global South as simply a space of problems in which development takes place (Rigg 2007; Williams *et al.* 2014). Rather, they contribute to new 'critical cartographies' of development (Silvey and Rankin 2011) focusing on the geographies of everyday lives in the global South, and how the global South is both shaping, and being shaped by, global economic, political and cultural processes.

## Literature as counter discourse

Postcolonial literature re-centres development processes within the lived experiences and consciousness of those subjected to development (Perry and Schenck 2001). Just as the anti-colonial literature of the 1960s gave expression to nationalist and non-nationalist modes of resistance (Chapter 2), so the 'local' and the 'global' can be read in new ways through literature in order to develop new perspectives and find alternative ways of interpreting development practice. Literature as 'a crucial record of subject formation should take its place in an inter-disciplinary, international development debate' (Perry and Schenck 2001: 200), not least because these texts contain important records of the processes of subject formation in a variety of places and spaces undergoing development. Christine Sylvester (2011: 187–8) argues that it is only through the advent of postcolonial analysis that development has been seen 'through the eyes of local people making daily livelihood decisions in situations of conflict, hope, resistance, ambivalence, despair and uncertainty'.

As with film, postcolonial literatures can tell us much about the voices, dialogues, languages and social constructions of post-colonial states and provide an understanding of resistance, which is at once local and transnational. Critical reading of fiction also allows for reflection on the nature of subjectivity in relation to national development. For example, in the paradoxical context of contemporary India, Ghosh (2001: 955) argues that popular fictions 'represent the corruption of the era as well as the desirability of its modernity'. Indian fiction tells the story of the 'deferral' of western modernity in the 'imaginary of the postcolonial nation', providing alternative insights into development issues in India. Ghosh is particularly concerned with gender and Indian modernity (see also Daya 2007) and writes:

> In postcolonial India, the cultural and gendered politics of Indian nationalism can be read through the texts of popular novels where constructions of the 'modern' women in the service of the Nehruvian national development project and its successors are presented for the consumption of the literate middle classes.

These fictions present an image of the liberated, 'modern' Indian woman characterized by processes of westernization. The characters westernize themselves to the extent that they discard their spiritual or traditional roles in favour of the 'development' of the Indian nation. The illiterate, impoverished, peasant woman who might be symbolic of a traditional, underdeveloped India ('out there' and 'back there') is not present in these stories of modern, urban India produced for consumption by English-speaking, educated, middle-class women. As such, these texts are partial representations of contemporary India and reveal the 'false' promises that all citizens could be provided with the necessities of life and the desirable symbols of modernity. Thus, imaginative literature can advance understandings of development and how it is received, experienced and articulated in different places such as India. It also provides a means of understanding struggles over the meaning and practice of development.

Novels can capture the complexities of life in developing countries and the fact that the choices they face are far more difficult than any GNP per capita statistics can capture (Raghuram *et al.* 2009). As Sylvester (2011) argues, the postcolonial refocus on people rather than grand trends is strengthened by its use of 'data sources' that development analysts would never use: novels, testimonials, drama and poetry. Economic or sociological statistics fail to describe and account for poor decisions taken by post-colonial governments and by international organizations involved in development. In contrast, novels that focus on daily life under pressure rather than development per se, with the ineffectual apparatuses of economic and political development present only as a backdrop to the stories, can place the errors of development centrally within the stories. Novels are able to focus attention on what is neither discussed, nor even anticipated, in debates about effective, sustainable, empowering or modernizing development.

Sylvester exemplifies her argument with reference to Chimamanda Gnozi Adichie's *Half of a Yellow Sun* (2007), a novel depicting the Biafra War of the 1960s from the perspective of those engaged in the Igbo nation's audacious freedom break from Nigeria. The story tells of the hopes and aspirations of the Biafrans and of their violent repression by Nigeria which, with the aid of the former colonial power, Britain, starved the Igbo into submission. Sylvester demonstrates how Adichie captures the hypocrisy of development discourses that homogenize people like the Igbo as 'stick figure people'. Rather than advocating welfare safety-nets or other forms of insurance (which in the West have improved the lives of poorer peoples), the main goal of development for poorer peoples in the global South has always been greater self-reliance: 'Yet here was Biafra trying to be self-reliant but running up against the sovereign state that had mistreated the Igbo, a state that was retaliating with its principal friends in ways designed to de-develop the dissident area' (Sylvester 2011: 198). In other words, while political economies continue to work to disempower and, in some cases, de-develop the most marginalized peoples in the global South, development interventions continue to insist that problems would be solved if those people were more self-reliant: the deficiency is with them, not with the systems that disadvantage them. Similar arguments are now being advanced to deal with the fact that the world's poorest people are also the most vulnerable to the negative consequences of climate change. The answer to this problem is that these people become or are helped to become more *resilient*. As MacKinnon and Derickson (2012) argue, the discourse of resilience is problematic because it is an ecological concept that is conservative when applied to social relations, it is externally defined by state agencies and expert knowledge, and understanding resilience at the scale of individuals or places is misplaced because the processes that shape resilience operate primary at the scale of capitalist social relations.

Postcolonial novels are valuable in countering the discourses of self-reliance and resilience, and in challenging the authority of outsiders to diagnose the problems of developing countries (discussed further in Chapter 5). *Half of a Yellow Sun* is a postcolonial novel not simply because it tells a critical story of post-colonial development, but because of the voice through which the story is narrated. The various experiences of the Biafran war are told through the eyes of three protagonists: Ugwu (a village boy from Opi who later becomes a servant), Olanna (a university professor) and Richard (a writer). Richard is English and wants to write an account of the war. Running through the novel are snippets of a book being written by one of the characters and the reader is left to make assumptions about who the author is until the very end of the book. It comes as a surprise that the author is revealed to be Ugwu. When asked about the authorship of these snippets, Adichie replied: 'I wanted a device to anchor the reader who may not necessarily know the basics of Nigerian history. And I wanted to make a strongly felt political point about who should be writing the stories of Africa . . .' (see http://chimamanda.com/books/half-of-a-yellow-sun/the-story-behind-the-book/).

Postcolonial literature is important in learning how to think about the relationship between development and individuals rather than about development aggregates, and to shift the focus of development away from the 'overweeningly statistical' (Sylvester 2011: 2002) that has tended to dominate development studies. Sylvester's point is that development interventions have consistently been made by western experts who have no idea of what people in the global South need or desire. In contrast, novels and other cultural outputs are better able to capture some of these needs and desires:

> If we need vacations, perhaps so do they. If we need gyms and parks, perhaps that is what individuals dream of there too. We in development studies do not really know. And it is time we did, by

moving our relations of analysis and our practices, reaching towards
writings and situations and people and their economies of movement.

(Sylvester 2011: 203)

Sylvester argues that 'We could do worse than advocate a campaign
of reading postcolonial stories as part of development theory, training,
and practice' (*ibid.*). We learn from novels that people in crisis are
already extremely self-reliant and able to make life and death decisions
every day. For example, Nigerian writer Adaobi Tricia Nwaubani's *I
Do Not Come to You By Chance* (2009) tells the story of Kingsley who
returns to Nigeria with a university degree, but struggles to find work.
His only option is to take part in the email scams industry targeted
at extracting money from wealthy westerners. The novel is both a
critique of highly uneven global development that gives rise to this
industry and an account of the ingenuity (in contrast to stereotypes
about Nigerian criminality) of those who struggle to sustain
livelihoods in the mainstream economy.

Other examples of how people in the global South respond to crisis
and change might include the growing body of contemporary poetry
written by indigenous peoples that engages with climate change. Craig
Santos Perez, a native Chamoru (Chamorro) from the Pacific Island
of Guåhan (Guam), refers to this poetry as indigenous ecopoetics,
which expresses the 'idea of interconnection and interrelatedness of
humans, nature, and other species; the centrality of land and water
in the conception of indigenous genealogy, identity and community;
and the importance of knowing the indigenous histories of a place'
(Santos Perez, in Russo, n.d.). Indigenous ecopoetics demonstrates
how native writers 'employ ecological images, metaphors, and
symbols to critique colonial and Western views of nature as an empty,
separate object that exists solely to be exploited for profit'. It also
're-connects people to the sacredness of the earth, honors the earth
as an ancestor, protests against further environmental degradation,
and insists that land (and literary representations of land) are sites of
healing, belonging, resistance, and mutual care' (*ibid.*). Perhaps more
importantly, poetry is a means by which indigenous and subaltern
peoples speak and hold those perpetrating injustices to account.
Kathy Jetnil-Kijiner, a Marshall Island poet, received international
acclaim for performing at the UN Climate Change Summit in New
York in 2014. Her poem about the threat of sea level rises addressed
to her baby daughter – 'Dear Matafele Peinam' (see www.youtube.
com/watch?v=L4fdxXo4tnY) – highlighted the devastation already

being caused by climate change to small island communities like the Marshall Islands (see also Milman and Ryan 2016). It called out those responsible: the 'greedy whale of a company sharking through political seas', the 'backwater bullying of businesses with broken morals', the 'blindfolded bureaucracies'. It articulated a mother's promise to her daughter that she will fight, 'we will all fight', for her future. It reduced hundreds of global leaders attending the Summit to tears. And it perhaps helped influence the important agreement at the Summit to close the gap between carbon emission reduction pledges and actual emission cuts, and the shaping of the new legal agreement on climate change that was approved in Paris in December 2015.

Poetry and novels may hold some of the keys for responding to climate change and for forging collective action. They help us to understand the nature of the challenges people in poorer countries believe they face and how they respond to these challenges. As Sylvester (2011: 203) argues:

> It is their thinking . . . that we need to grasp, not the choices economists dream up for them in rational choice models. What better way to begin the process of insight into individual . . . choices than to read about these challenges in novels about postcolonial settings?

As discussed in Chapter 5, in recent years important drivers of development like the World Bank have begun to acknowledge the importance of understanding how people think in order to create solutions to global problems, although their methods for accessing this thinking are still quite removed from Sylvester's propositions.

## Transforming development

Literature can be used to challenge conventional notions about the nature of knowledge, narrative authority and representational form. Lewis *et al.* (2005) argue that fiction that deals with development issues is of value, is a valid form of development knowledge, and should be taken seriously. They argue that the line between fact and fiction is often blurred, since, on the one hand, all knowledge is subjective and all interpretation somewhat rhetorical and, on the other hand, fiction from its earliest incarnations has often sought to teach. Thus, if policy documents can be understood as 'stories', then fiction can also be read as part of the stock of knowledge about development processes and responses. In addition, fiction can sometimes work better than academic or policy-related texts in representing central issues relating to development. It frequently reaches a wider audience, and thus can have greater influence.

This chapter has explored how stereotypical representations of the South have reinforced the sense that it is both spatially and temporally distanced from the North. However, this belies the fact that contemporary African, Asian, Middle Eastern and Latin American cities are increasingly part of global networks linked by new telecommunications technologies. Hybridity is a feature of many of these places and spaces, where modernity and development are defined and consumed in different ways according to everyday realities and practices. As Power (2003: 139) argues, 'similar transformations can be made to the ideological spaces of development if postcolonial theories and debates are engaged in more widely by those interested in development studies'. Postcolonialism enables us to re-emphasize relations with other cultures as an integral part of studying development and allows us to challenge the authority of development discourses and the way they construct the world. This means deconstructing development and the language of development theorists. It means questioning the construction of development knowledge and the power relations that exist between development theorists and practitioners and the peoples and places they seek to study and represent. It means exploring alternative and counter discourses in the texts of subalterns. It also means attending to 'how places in the non-West differently plan and envision the particular combinations of culture, capital, and the nation-state', rather than assuming they are 'immature versions of some master western prototype' (Ong 1999: 31) (see Chapter 5).

## Summary

- Western-centrism pervades many systems of ordering the world; the pictures painted are often partial and based around 'negative' truths which generalize about and stereotype countries and peoples in the South.
- Development ideas, discourses and practices have often stigmatized economic, political and cultural difference as a way of justifying the advantages one region has over another and legitimizing interference and interventions.

- In too many instances, the South is represented as a world of poverty and violence, death and decay, erasing a sense of the specificity of places and spaces. Northern media and development discourses are central in this.
- The power of these discourses is that they are often internalized and people in the South come to see themselves in these terms. In this way, poverty is understood as something that only technicians in the North can solve, rather than people in the South themselves.
- The 'Third World' and 'Third World chaos' are seen as threats to the prosperous North; the main discourse and ideology of Northern intervention in the South is 'development', even though the logic of this is increasingly questioned.
- Postcolonialism seeks to remove western negative stereotypes about people and places from such discourses. It challenges us to rethink, for example, categories such as 'Third World' and 'Third World women', and to understand how location, economic role, social dimensions of identity and the global political economy differentiate between groups and their opportunities for development.
- Counter-discourses increasingly challenge dominant Northern discourses.

## Discussion questions

1. How did imperial discourses shape political interventions in Europe's colonies?
2. Explain how the process of 'worlding' creates the ideas of the 'West' and the 'Third World'.
3. In what ways do colonial images of Africa persist into the present day?
4. What role does discourse play in the 'new imperialism'?
5. Discuss examples of counter discourses and their effectiveness in challenging dominant discourses of development.

## Further reading

Crush, J. (ed.) (1995) *Power of Development* London, Routledge. A collection of essays that conceptualize development as a discourse and explore the language and rhetoric of development texts. Groundbreaking in its day and still an important book.

Doty, R.L. (1996) *Imperial Encounters: The Politics of Representation in North–South Relations* Minneapolis, University of Minnesota Press. An exploration of how labels influence North–South relations (based on case studies of imperial encounters between the USA and the Philippines, and Britain and Kenya), reflecting a history of colonialism and shaping the way national identity continues to be constructed.

Lewis, D., Rodgers, D. and Woolcock, M. (2005) 'The fiction of development: knowledge, authority and representation'. *London, LSE research online*. Available at http://eprints.lse.ac.uk/archive/00000379/. A concise paper that extends this chapter's argument about the representation of development. It discusses how novels from around the world can provide different representations of the South. The list of recommended reading on literary works (in English) on development is a highly recommended resource.

Pieterse, J.N. (2001) *Development Theory: Deconstructions/Reconstructions* London, Sage. Covering issues of Eurocentricism, critical globalism, intercultural transaction, de-linking and post-development theory, and connecting issues of development with critical theory.

Sylvester, C. (2011) 'Postcolonial takes on biopolitics and economy'. In J. Pollard, C. McEwan and A. Hughes, *Postcolonial Economies* London, Zed, pp. 185–204. A call to take seriously postcolonial readings of individual people's thinking at times of crisis, and to see development through the eyes of local people in contrast to policies and practices of development about life in the aggregate. The chapter advocates the reading of postcolonial stories as part of development theory, training and practice to understand the thoughts and actions of people subject to development interventions.

# Useful websites

www.hrw.org/reports/1996/Rwanda.htm Human Rights Watch report on the gendered violence of the Rwandan genocide.

www.bbc.co.uk/music/thelive8event/liveaid/ Useful archival source on Live Aid, Live 8 and the issues and debates surrounding the two events; with useful links to other campaigns, such as Make Poverty History.

www.pinterest.co.uk/bettinello/philip-kwame-apagya/ Selection of Philip Kwame Apagya's photography.

www.ted.com/talks/chimamanda_adichie_the_danger_of_a_single_story A great Ted Talk in which novelist Chimamanda Adichie tells the story of how she found her authentic cultural voice, and warns that if we hear only a single story about another person or country, we risk a critical misunderstanding.

www.youtube.com/watch?v=L4fdxXo4tnY Kathy Jetnil-Kijiner's powerful performance at the UN Climate Change Summit in 2014.

# ⬤5 Critiquing development knowledge and power

## Introduction

As discussed in Chapter 4, the ways in which development is written and analyzed have been subject to intense scrutiny, reflecting the influences of broader philosophical and theoretical debates that have swept through western social and political sciences. More specifically, since the widely acknowledged crisis, or 'impasse', in development studies in the mid-1980s (see Chapter 3), it has been recognized that development is about power – its operations, its geographies, its highly uneven distribution and strategies for achieving it (Watts 1995). The analysis of power is, therefore, central to contemporary critical development studies (Crush 1995; Radcliffe 1999). Much of this analysis has its roots in postcolonial and feminist theories, both of which have had significant consequences for how development is conceptualized.

This chapter explores the ways in which postcolonial critiques challenge the ways in which the dominant discourses described in Chapter 4 come into being. It examines precisely how particular discourses of development come to be dominant. We have seen that knowledge is a form of power and, by implication, violence. It gives authority to possessors of knowledge, and dominant knowledge closes off spaces for the articulation of alternative knowledge forms. Knowledge has been, and to a considerable extent still is, controlled and produced in the North. Moreover, the South becomes the object of Northern knowledge. Northern knowledge forays into the South have been critiqued for being extractive in intent: raw data was sourced and then returned to the metropole for analysis and theorization (Hountondji 2002). For some, the real power of the North lies not in its massive economic development and technical advances, but in its power to name, represent and theorize (Sardar 1999), a fact which postcolonialism seeks to disrupt. While this might overplay to some extent the power of discourse, it nevertheless has significant political ramifications. The *idea* of development has, until recently, enabled the North to appropriate and control the past, present and future of the South. The power of the idea of development is also quite remarkable given the fact that it has endured in an age of growing global inequality. As we saw in Chapter 1, over 1.2 billion people live on less than $1 per day; the average Gross National Income of the richest 20 countries is presently 37 times that of the poorest 20, which represents a doubling of inequality in 40 years; in 2016 the world's richest 62 individuals owned the same wealth as the 3.6 billion poorest (Hardoon *et al.* 2016; see also Seery and Caistor Arendar (2014)), and in 2017 this reduced to eight individuals. As Wainwright (2008) points out, this extraordinary polarization of wealth has occurred in the 'Age of Development', and he asks: How is it that capitalism reproduces inequality in the name of *development*?; How is it that deepening capitalist social relations come to be taken *as* development?

These questions are answered in part by the fact that Northern-based and Northern-dominated institutions are engaged in global development in ways that are both strategic and tactical, promoting and justifying certain interventions and de-legitimizing and excluding others. Power relations are clearly implied in this process, with certain forms of knowledge dominant and others excluded. These institutions generate ideas about development, which are not produced in a social, institutional or literary vacuum, but are part of a development industry that is implicated in relationships of power and domination (Crush 1995; Goldman 2005). A postcolonial approach to development can reveal the apparatuses of power and domination within which development texts and ideas are produced, circulated and consumed. Thus, it is important to explore how development discourse promotes and justifies real interventions with real consequences, and the connections between the words, practices and institutional expressions of development. Of significance are the connections between the words and images that represent the world and the relations of power that order it.

Reflecting on power also necessitates an understanding of the ways in which this is appropriated or resisted. An important aspect of postcolonial approaches is their acknowledgement that local cultures did not passively accept the cultural practices of colonialism. Rather, imperial culture was appropriated by the colonized in resistance to imperialism and in contesting the power of its knowledge. Relationships of power do not go uncontested, and there are multiple possibilities for challenging the authority of certain kinds of development knowledge through the creation of counter-knowledge and other forms of resistance.

## The architects of development knowledge

Development is about ideology and the production and transmission of policies and discourses. It is not simply about financial and material flows, such as aid and investment, but is also about the flow of ideas. Of concern here are considerations of how the key players in international development acquire dominance for their ideologies, how alternatives are de-legitimized and filtered out by global development organizations and agencies, and how this is resisted. Chapter 3 revealed that modernization theory and, more recently, neoliberalism have dominated ideas about global economic development since the 1950s. Critics argue that dominant discourses of development have prevented the formulation of alternatives and creative ways of addressing poverty in particular contexts (Yapa 2002). This is arguably what created the 'impasse' in the mid-1980s – the dominance in policy of neoliberalism despite perceived failures, a widening gap between richer and poorer countries and the failure to posit realistic alternatives. Precisely how these ideas, based on western notions of progression and forwardness, have become so dominant is of significance.

Ultimately, this is a question of understanding how international development works both in theory and in practice. To challenge misrepresentations and myths and to understand more fully how development works, we also need to understand how particular notions of development are diffused, disseminated and popularized. Thus, we need to look at the relationships between the different agencies and organizations involved in development and how they communicate knowledge and ideas between diverse cultural and economic spaces, and across different spatial scales. It is important to understand 'who is allowed to speak in certain kinds of discourses, where does this take place and with what implications for power relations?' (Power 2003: 170). Of course, there is no singular 'Northern' development discourse. There are distinct differences, for example, between the philosophy that underpins the US approach to development and European approaches, which also change over time. Add to this the often profound differences between European countries. These are rooted in cultural, political and historical differences. US development discourse has been shaped by pragmatism, notions of freedom, efficiency and geopolitical responsibility – what Karagiannis (2004: 18) terms 'American suppression of ambiguity'. European development discourse, as articulated through the EU, is widely and erroneously considered an imitation of US discourse, yet displays doubts and self-reflexivity that leave room for resistance and the elaboration of a more 'solidaristic approach' to the relations between North and South (Karagiannis 2004: 18). These discursive differences also produce different development policies and material outcomes. This can be seen in a range of development organizations, as discussed subsequently. However, underpinning all development interventions is what Tania Li (2007) refers to as a 'will to improve', which has shaped landscapes, livelihoods and identities in the South for almost two centuries.

Colonial officials, missionaries, politicians, bureaucrats, international aid donors, experts in agriculture, hygiene, credit and conservation and non-Governmental Organisations (NGOs) of various kinds have all sought development interventions that are shaped by this will to improve. As we shall see, the key players in international development have been the World Bank (WB), The International Monetary Fund (IMF), the United Nations (UN), governments of countries in the global North, NGOs and, more recently, wealthy philanthropists. These are, in effect, trustees (see Chapter 3), whose will to improve is defined by the 'claim to know how others should live, to know what is best for them, and to know what they need' (Li 2007: 4). It remains to be seen whether or not other 'new' players in international development such as China, India and Brazil will reframe development relationships away from trusteeship. It appears that they do not necessarily set out to dominate, but to enhance capacities for action, and their aims are often expressed as benevolent and even utopian. However, there are concerns that their models of development replicate some of the worst aspects of modernization, which saw investment in vast infrastructure projects in poorer countries in return for enormous resource extraction fuelling industrialization and economic development of richer countries. As Li argues, the outcomes of development programmes are not always bad and they sometimes bring changes that some people want. However, the claim to expertise in optimizing the lives of others is a claim to power. 'Improvement' has become almost banal – it goes without remark. A postcolonial approach questions why this is, particularly given that people in the South are often capable of sustaining lives without 'improvement'. Moreover, there is often a gap between what is intended and what is accomplished by development programmes. While they are designed to improve peoples' lives, to better cater for their needs, to rationalize their use of land, to educate and modernize, these intentions are all implicated in sites of struggle and resistance. In focusing on the relationship between power and knowledge, and how this shapes development interventions, postcolonial approaches cast light on why such resistance exists and reveal the limits of 'expert' development knowledge.

## The World Bank and development knowledge

The WB was formally established on 27 December 1945, following the ratification of the Bretton Woods agreement. Based in New York, it was initially focused on post-World War Two reconstruction in Europe. It is still involved in post-conflict reconstruction, reconstruction after natural disasters, and responses to humanitarian emergencies. From the 1960s to 1980 it focused largely on poverty alleviation and from 1968 onwards began to focus primarily on the developing world. From the 1980s and into the 1990s its focus was on debt management and structural adjustment. In recent years it has focused primarily on providing loans to developing countries for development programmes with the stated goal of reducing poverty. Its two official goals up to 2030 are to: (i) end extreme poverty by decreasing the percentage of people living on less than \$1.90 a day to no more than 3 per cent, and (ii) promote shared prosperity by fostering the income growth of the bottom 40 per cent for every country (www.worldbank.org/en/about/what-we-do). However, according to its Articles of Agreement, all its decisions must be guided by a commitment to the promotion of foreign investment, international trade and capital investment, which many critics argue are not necessarily compatible with poverty reduction. The WB helped devise the Millennium Development Goals (MDGs) and the Sustainable Development Goals (SDGs) (discussed subsequently), which aim to eliminate poverty and create conditions for sustainable development.

The WB has been and remains one of the most influential architects of development knowledge. It positions itself as the most powerful 'think tank' on international development issues and claims to be a 'knowledge bank'. It claims to have 'become a clearing house for knowledge about development – not just a corporate memory of best practices, but also a collector and disseminator of the *best development knowledge* from outside organizations' (WB 1999, cited in Power 2003: 184). It uses its own development 'knowledge' to legitimize and rationalize certain positions. Its intentions 'seem to backfire in the most disastrous ways' (Goldman 2005: vii), but it is able to confirm its position as the leading producer of doctrine and knowledge about how countries should develop. Every decade of the WB's existence 'has been marked by both major improvements in the techniques of development and new types of colossal failures'. Its power in disseminating development knowledge and sustaining the development industry means that it is difficult to 'imagine the world today except through the lens of World Bank-style development' (Goldman 2005: vii–viii). In recent years, however, China has become a bigger global lender than the WB and it will be interesting to see if this generates a different global model of development with more just outcomes for the world's poor. Or will familiar patterns repeat? Zambia, for example, is currently heavily indebted to China and unable to pay back its loans. There are fears that as a consequence the national power supplier could be taken over by the Chinese.

Critics of the hegemony of the WB over development knowledge are numerous. One of the strongest criticisms concerns how the WB is governed. While the organization represents 188 countries, it is run by a small number of economically powerful countries, principally the United States, which provide most of its funding, choose the leadership and senior management, and whose interests dominate (Woods 2007). There has never been a non-American president: Jim Yong Kim, a Korean-American, is the most recently appointed and the first non-America-born president. In 2012, *The Economist* observed 'When economists from the World Bank visit poor countries to dispense cash and advice, they routinely tell governments to reject cronyism and fill each important job with the best candidate available. It is good advice. The World Bank should take it'. Alexander (1996) argues that the unequal voting power of western countries and the WB's role in developing countries resemble the South African Development Bank under apartheid, and that it is an architect of global apartheid.

The WB itself has admitted to persistent failure to provide local citizens and stakeholders with the information they need to meaningfully participate in decision-making about the projects affecting their lives. Thus, there are contradictions in its stated aims of empowering people through information and allowing them to participate. The WB produces hundreds of publications each year, but much of its research contributes more to legitimating what it does rather than advancing understanding of the development issues involved. It also withholds a great deal of information because it is seen to interfere with its own deliberations around policy. It sees itself as the disseminator of 'best knowledge', but self-classifies this as such. To the WB, knowledge 'can be traded, exported or imported' (Power 2003: 186), and it sees developing countries remaining 'importers rather than principal producers of technical knowledge' for some time:

> Poor countries – and poor people – differ from rich ones not only because they have less capital but because they have less knowledge . . . Indeed even greater than the knowledge gap is the gap in the capacity to create knowledge . . . Countries that fail to encourage investment in the effective use of global and local knowledge are likely to fall behind those that succeed in encouraging it.
> (World Bank 1999: 1; 2; 25)

The WB thus sees itself as global and local manager of knowledge for development; consequently, it rides roughshod over or simply ignores indigenous knowledge and theory. Even when indigenous communities 'produce' the knowledge, they remain dependent on those who generate the technology to distil and distribute this knowledge. The WB fails to recognize knowledge production 'from below' and thus crowds out alternative theories, ideas and strategies. This arguably creates the conditions for the dominance of neoliberal, economistic approaches to development and the perpetuation of what is, at root, an ideology rather than an evidence-based approach to international development. As the former Chief Economist at the WB, Joseph Stiglitz (2000) argued on resigning his post:

> There never was economic evidence in favor of capital market liberalization. There still isn't. It increases risk and doesn't increase growth. You'd think [defenders of liberalization] would say to me by now, 'You haven't read these 10 studies', but they haven't, because there's not even one. There isn't the intellectual basis that you would have thought required for a major change in international rules. It was all based on ideology.
> (archive.salon.com/news/feature/2000/05/02/stiglitz/index1.html)

In recent years, the WB appears to have embraced the idea that people in developing countries are capable of thinking and that understanding their thoughts is important to development. However, this is not the postcolonial approach advocated by Sylvester (2011) and discussed in Chapter 4. Instead, as the 2015 *World Development Report: Mind, Society and Behaviour* indicates, this is a move towards behavioural economics and the idea that by understanding how poor people think and make decisions they can be 'nudged' into making better decisions. In South Africa, for example, 5 million people have HIV and teenage girls are three times more likely to be infected than boys. The WB's nudge unit found that many girls were choosing to have sex with older men, believing them to be safer sexual partners, when they were in fact more likely to be HIV positive. The team then found that the girls were better able to retain knowledge about which men are more likely to have HIV if they played a computer game about HIV-risk than if they were given information to read. The results of such approaches may well have positive effects, but they rest on the idea that outside experts know 'what is best' for poor people and that the latter can be manipulated into making decisions that outside experts think they ought to make, discounting local knowledge. Moreover, such approaches do nothing to address structural relations of power, such as the patriarchal social relations that make it difficult for poor women in sub-Saharan countries to negotiate safe sex, or the profound socio-economic inequality that binds risky sexual behaviour to economic dependence on men.

Critics have argued that for too long organizations such as the WB have posed as a solution to poverty and inequality, when in fact there is compelling evidence that their policies are often a large part of the problem, resulting in the impoverishment and marginalization of indigenous peoples, women, peasant farmers and industrial workers and a deterioration of economic, social and ecological conditions. Its programmes 'intertwined with other processes and relations, set the conditions for some of the problems that exist today' (Li 2007: 2). One problem is that, until recently, the hegemony of the WB and the other International Finance Institutions (IFIs) has remained uncontested. The WB is still the major player in global development and the most powerful disseminator of development ideology and rhetoric. While there is internal disagreement within the WB (e.g. Stiglitz), and while its neoliberalization of the spaces of development is incomplete and contested in specific places, its development discourses continue to have enormous significance because they turn into concrete realities in the South. Its will to improve persists because of its parasitic relationship to its own shortcomings and failures (Li 2007). The WB needs programmes that are amenable to technical solutions. Thus, while individuals within the organization might have deep ethnographic knowledge of the places in which it intervenes, its lust for parsimony in rendering this knowledge technical means that programmes ultimately fail to address complexity. They address some problems and not others, and they can create new problems. Consequently, WB programmers come under pressure to devise better programmes, rather than to make programming itself and its outcomes the object of analysis.

## The IFIs, globalization and the myth of free trade

The WB is part of a group of IFIs, which form a tier of unaccountable international governance organizations that have a major impact on the economic affairs of all countries. Alongside the World Trade Organization (WTO, formed in 1995), they have perpetuated neoliberal discourses of globalization and so-called 'free' trade that have dominated international development policy since the 1980s. The term globalization has become so banal that it is difficult to analyze what is happening on the ground. It has been 'fetishized' and is now seen as having 'an existence independent of the will of human beings, inevitable, irresistible' (Marcuse 2000: 1). However, as Power (2003: 145) argues, it is important to retain a sense of how globalization actually exists, how it impacts on peoples and places every day, and how these processes 'are related to the will of human beings'. As Ong (2006) demonstrates in the case of East and South-East Asia, governments and institutions have a more decisive role than markets in the successful experiences of development in the global economy.

Globalization is purportedly founded on the free movement of capital and goods. Proponents of free trade, including the IFIs and WTO, adhere to the view that if a country is economically competitive and efficient, it should be able to compete for a share of the global economic pie. The logic is that if poorer countries trade more, greater economic integration and convergence will occur, leading to the disappearance of inequalities. However, in reality:

● Northern economies are protected by trade barriers, in the form of subsidies for their own goods and tariffs on imported goods – the very anathema of free trade. In 2018, the US government imposed tariffs on some imported goods (e.g. steel) as a protectionist measure.

- Countries in the South lose hundreds of billions of dollars because of trade barriers in wealthy countries; at the beginning of this century, for every $1 given in aid and debt relief poor countries lost $14 because of trade barriers (Sogge 2002: 35).
- Economic and military aid packages are used by Northern economies to split potentially powerful alliances amongst countries in the South (e.g. Nigeria and Pakistan at the WTO meeting in Doha in 2001 – see Box 5.1).
- Trade liberalization has many negative effects on the agricultural policies of Southern countries, shifting governmental priorities from the needs of the people to the needs of a volatile international market. Seventy per cent of the world's agricultural trade comes from the South, yet the global agro-food system is skewed against the interests of poor farmers in the South.
- Over the last 30 years, the South's share of food imports rose rapidly, while its share of exports rose by only small margins because of protected Northern markets. This situation continued to worsen into the 2000s, as examples from a 2005 survey reveal:

  – Ghana (1998–2003): rice imports increased from 250,000 to 415,150 tonnes; domestic rice fell from 43 to 29 per cent of the domestic market (2000–2003); 66 per cent of rice producers recorded negative returns and loss of employment; tomato paste from the EU increased from 3,300 to 24,740 tonnes (650 per cent). Farmers lost 40 per cent of the share of the domestic market.
  – Cameroon (1999–2004): poultry imports increased 300 per cent; 92 per cent of poultry farmers dropped out of the sector; 110,000 rural jobs were lost each year.
  – Mozambique (2000–2004): vegetable oil imports (palm, soy and sunflower) increased 500 per cent; domestic production shrank from 21,000 tonnes to 3500 (1981–2002); over a million families have been affected and thousands of jobs lost.
  – Jamaica (1990–2005): increased onion imports have seen the collapse of the industry.
  – Jamaica (1990–2004): dairy imports saw 50 per cent of farmers going out of business; employment has fallen by 66 per cent.
  – Sri Lanka (1981–2005): dairy imports increased from 10,000 to 70,000 tonnes, consuming 70 per cent of the domestic

market. Local production expanded by less than 15 per cent. (Source: www.fao.org/ES/ESC/en/41470/110301/index.html)

- Developing countries saw a 9 per cent increase in overall food import expenditures in 2007. The more economically vulnerable countries were worst affected, with total expenditures by low-income food-deficit and least developed countries rising by 10 per cent. According to one economist, 'the food import basket for the least developed countries in 2007 cost roughly 90 per cent more than it did in 2000. This is in stark contrast to the 22 per cent growth in developed country import bills over the same period' (FAO, 7 July 2007).
- Rising food prices are a global concern and led to riots against high food prices in 2008 in countries such as Haiti, Egypt, Senegal and Cameroon. In 2010, rising costs of basic foods such as bread precipitated the Arab Spring. The UN indicates serious food shortages in many countries of Africa (Lesotho, Swaziland, Zimbabwe, Somalia, Mozambique, Eritrea, Mauritania, Senegal, Liberia, Sierra Leone, Ivory Coast, Burkina Faso and Cameroon). There are serious food shortages in war-torn Afghanistan and Iraq and chronic food shortages in North Korea. Food riots indicate serious weaknesses in food security policies.
- Where countries have succeeded in breaking into western markets, prices have actually fallen for commodities (e.g. bananas and coffee).
- Critics (e.g. Shiva 2001) argue that local economies and food cultures are being destroyed because of the globalization of the food system; 'market totalitarianism' is a threat to sustainability and the survival of the poor and their communities. The role of women, in particular, is disregarded by neoliberalism; they are disempowered, overlooked and their knowledge quite literally stolen by global biotechnology firms. Development is not sustainable in these circumstances (Panos Institute 2001).

The IFIs and the WTO are committed to working more closely with each other to reduce trade barriers. Some countries in the South (e.g. India) have called for a 'development box' – a protected trade area that would exclude or minimize imports to allow poor countries with large numbers of poor farmers to protect and invest in them rather than removing trade barriers – but this has so far been ignored.

What makes the IFIs and WTO so powerful in terms of determining the nature of development ideologies and interventions is that they can globalize their ideologies and set the terms of the debates, choosing only a form of economics that supports their arguments. Thus, while the WB conducts very little research into world income distribution, its causes and consequences, through its globalization discourses it can shape global development agendas. The IMF imposes strict lending conditions on countries in the South, which require structural adjustment and the adoption of neoliberal macroeconomic frameworks. The WTO is the means through which neoliberal ideology and free trade are put into practice on a global scale (Box 5.1). However, as ActionAid (2005) notes, 'No country has ever become a "developed country" by pursuing free trade policies . . .'.

Neoliberalism is essentially a product of specific Northern economic knowledge, but it is globalized in ways that other types of knowledge are not through the power of the IFIs. However, it is important to understand that neoliberalism is adopted differently in different places (Escobar *et al.* 2002; Ong 2006) and the power of the IFIs and WTO to disseminate particular development ideologies does not go uncontested. For example, the Trade Justice Movement (TJM), a coalition of 80 organizations, is concerned with how the benefits of freer trade can be distributed in favour of developing countries. It states:

> everyone has the right to feed their families, make a decent living and protect their environment. But the rich and powerful are pursuing trade policies that put profits before the needs of people and the planet. To end poverty and protect the environment we need Trade Justice not free trade.

> (www.tjm.org.uk/about.shtml)

## Box 5.1  The WTO and global trade

The WTO is the world's largest economic organization comprising 123 member states. It was inaugurated in 1995 and grew out of the General Agreement on Tariffs and Trade (GATT), which commenced in 1948. Like GATT, the WTO conducts negotiations through rounds. The GATT Uruguay Round (September 1986–April 1994) was criticized by NGOs for paying insufficient attention to the special needs of developing countries. Critics – many from the global South – pointed out that representatives of Northern based multinationals played the leading role in the drafting of policy on agriculture and other trade matters. Agreements on intellectual property and industrial tariffs were particularly disadvantageous to countries in the South. The WTO is intended to deal with some of these inequalities by regulating trade between participating countries, and by providing a framework for negotiating trade agreements and for resolving disputes to ensure participants adhere to WTO agreements, which are signed by representatives of member governments.

The Doha Development Round commenced in November 2001 with an objective to lower trade barriers and promote free trade (Das 2006; Newfarmer 2006), but stalled over disagreements between the North (EU, USA and Japan) and the major countries of the South, represented by the G20 developing countries and especially India, Brazil, China and South Africa. Subsequent ministerial and related meetings (Cancún 2003, Geneva 2004, Paris and Hong Kong 2005, Geneva 2006, Potsdam 2007) all failed to reach agreement, primarily over farming subsidies and import taxes in the North and the demand for fair trade in agriculture in the Global South.

The WTO trade talks attract large-scale demonstrations by activists from both North and South (discussed subsequently). The Doha Development Agenda commits countries to negotiations on opening agricultural and manufacturing markets, services and expanding intellectual property regulation. Proponents claimed the new round would make trade rules fairer for developing countries, but opponents alleged that it would expand a system of trade rules that are bad for development and interfere excessively with domestic policies in Southern countries. Negotiations again broke down at Geneva in July 2008, with disagreement over opening agricultural and industrial markets and reducing farm subsidies in the North. Agreement was finally reached on one aspect of the Doha Round – the 2013 Bali Package. This includes provisions for lowering import tariffs and agricultural subsidies, with the intention of making it easier for developing countries to trade with advanced economies in global markets, and the abolition of hard import quotas by richer countries on agricultural products from the global South. As of 2017, the Doha Round remains unconcluded and its future is uncertain. The fact that any final declaration of the WTO must ultimately be confirmed by the US Congress suggests that the impasse between North and South is unlikely to be resolved in the immediate future, and certainly not in the interests of the South (see Panos Institute 2006a, 2006b; Stiglitz and Charlton 2005).

TJM believes that issues of equity, sustainability and the environment, poverty and the plight of the 'losers' from liberalization should be addressed. It has three main aims (www.tjm.org.uk/briefings/TJM policy0105.shtml):

- that governments, particularly in developing countries, should be able to make their own decisions on poverty and protecting the environment;

- export subsidies that undermine farmers in developing countries should be ended;
- legislation against corporate actions detrimental to people and the environment.

Within dominant discourses, globalization and development are often conflated or at least seen as going hand in hand. However, Africa highlights the problems with some of the dominant myths about globalization. As discussed in Chapter 4, several African economies are growing at healthy rates – often around 5 per cent – primarily because they are starting from a very low base. However, in real terms, Africa's share of world trade has declined from 5.5 per cent to 2.5 per cent since the 1950s (Bora *et al.* 2007) and continues to decline. According to the World Economic Forum (2015), in 2013 Africa's share of global trade remained very small, at around 2 per cent, and exports remained highly focused on commodities (fuels and mining products account for over half of sub-Saharan exports, compared with only about 10 per cent for developing Asian and advanced economies). This exposure to commodities renders the region vulnerable to fluctuations in global commodity prices and can also jeopardize fiscal stability within countries. The reason for this, according to the IFIs, is not inequities within the global system, but because Africa is outside of modernity, outside of development and thus outside of globalization. It is relatively rare to read in major IFI documents anything about the role of Africa in globalization, even though the continent was at the centre of economic forces that helped inaugurate the forces of modernity, and is now at the centre of a new strategic battle for resources and influence between the North, on the one hand, and India and China, on the other. Africa has, for many centuries, been connected to global markets and globalization is certainly not new, yet the impression often given, as White (2001) argues, is that this is a 'novel process for Africa, delivered courtesy of the greater dissemination of western progress and enlightenment and the growing (and inevitable) scale and reach of global capitalism' (cited in Power 2003: 146).

Within neoliberal discourses about globalization by the IFIs, the central role of nation states in development is de-emphasized in favour of global governance. The spatial scales and networks of governance that affect 'poor countries' are thus shifted away from their own governments and towards the IFIs. Within these discourses, capitalism is considered universal and inevitable; it has become impossible to imagine social reality in any other way. The collapse of state socialism and communism exacerbated this in the 1990s. As Gibson-Graham (1996: 125, 130) argues, 'In the globalization script . . . only capitalism has the ability to spread and invade'. This assumes that forms of alternative economies fall away because of the universal and overwhelming force of globalization in the passive 'Third World'. The discursive power of the IFIs has ensured that capitalism is seen as natural, as having the ability 'to invade and to take life from non-capitalist sites while non-capitalism is seen as damaged, fallen and subordinated to capitalism' (Gibson-Graham 1996: 125). While movement of capital across state boundaries is supposed to help the poor and promote growth, evidence suggests that it might exacerbate problems (Cobham 2001). For example, in the 1990s, macroeconomic instability in Asia plunged economies into crisis, with ripple effects throughout the global economy. The Central Bank of Thailand spent $9 billion trying to protect the baht and its links to the US dollar. However, the baht still collapsed, with devastating effects on the economies of Indonesia, Malaysia, the Philippines, South Korea, Singapore, Hong Kong, Taiwan, Laos and Vietnam (Rigg 2001).

This crisis was discursively represented in the North as linked to the nature of Asian capitalism and, in particular, to corruption, rather than to the free market policies imposed by the IFIs. Economic crisis led to political crisis, undermining the legitimacy of some states that had achieved successful growth and bringing about the collapse of governments in Thailand, South Korea and Indonesia. The crisis revealed the fragility of development associated with rapid globalization. Some critics saw this as an opportunity to move away from the destructive policies associated with rapid industrialization. Similar circumstances saw the collapse of the Argentinian economy in 2002, with equally devastating effects on the country's poor. A popular uprising saw the overthrow of President Fernando de la Rua. In both cases, critics were vocal in their disparagement of the impacts of US-dominated IFI policies and the erasure of alternatives:

> The economic collapse of Argentina is the latest failure of the one-size-fits-all model that the United States tries to impose on developing countries . . . Argentina followed the IMF model more faithfully than almost any other nation. Its economy was opened wide; its peso was pegged to the dollar. For a few years this sparked an investment boom as foreigners bought most of the country's patrimony – its banks, phone companies, gas, water, electricity, railroads, airlines, airports,

postal service, even its subways . . . But the dollar-peso peg led to an overvalued currency, which killed Argentine exports. And once there was little more to sell off, the dollars ceased coming in, which pulled money out of local circulation. As Argentina tanked, the IMF's austerity program pushed the economy further into collapse . . . If the United States wants the world's support, much less its gratitude, it needs to let emerging economies follow their own paths.

(Kuttner 2002)

Criticism of the deleterious effects of the IFIs and the WTO was not confined to the popular press; Nobel laureates in economics were also scathing. Joseph Stiglitz (the 2001 laureate) observed that the countries that have benefited most from globalization are those that control the terms of engagement and, thus, the dissemination of development knowledge. Powerful economies benefit by taking advantage of the global market for exports, while countries in the South are coerced through IMF dictates into economic and fiscal policies that leave them vulnerable to forces beyond their control. The IMF promoted an ideology that Stiglitz branded 'market fundamentalism', and as both bad economics and bad politics. This ideology was imposed on countries in the South without proper consideration of the potential effects on poor people or in undermining emerging democracies. Similar criticisms were also voiced by Amartya Sen (the 1998 Laureate), who lamented the fact that globalization as promoted by the IMF was concerned only with market relations rather than with enhancing the social opportunities of millions of poor people. Its models worked to the benefit of investors often at the expense of ordinary people, particularly in the South. Yet countries with the highest growth rates, such as South Korea and China, are also those that have resisted much of the IMF model.

Free market ideologies create a borderless world for some; for others there are new spaces of deprivation and disconnection, which are rarely acknowledged in discourses of globalization. Instead, the IFIs continue to see markets as the primary mechanism for growth and poverty reduction and institutions as needing to serve markets more effectively. However, this fails to acknowledge alternative perspectives that institutions work for markets not for people; corporate monopolies of the market and corporate governance failures are also not discussed. Given the corporate scandals in the USA in the early 2000s and the collapse of the global financial markets in 2008 this seems astonishing. The collapse and bankruptcy of energy company Enron in 2001, for example, exposed its reported financial condition as sustained mostly by institutionalized, systematic and creatively planned accounting fraud. The near collapse of the global financial system in 2008 was triggered by profligate lending, excessive risk-taking and the failure of regulation in the banking sector of advanced economies, which lead to a global economic downturn. Interestingly, this collapse was predicted by Raghuram Rajan, Governor of the Reserve Bank of India, who was highly critical of the ultra-loose monetary policies of the IMF and western governments. However, expertise and criticism from the global South are rarely acknowledged: the neoliberal ideology that continues to drive these policies and the nature of international development remains entrenched. This reveals a debate about globalization that is overwhelmingly and uncritically western-centric in its focus on modernity and capitalism, much like the modernization theories of the 1950s. The WB, IMF and WTO still occupy positions of great power in controlling the global economy because they claim to be the possessors of expertise and the architects of 'good governance' that will best enable poorer countries to develop. However, is this how rich countries really became rich? As Chang (2002, 2–3) points out:

> The short answer to this question is that the developed countries did not get to where they are now through the policies and the institutions that they recommend to developing countries today. Most of them actively used 'bad' trade and industrial policies, such as infant industry protection and export subsidies—practices that these days are frowned upon, if not actively banned, by the WTO. Until they were quite developed (that is, until the late nineteenth to early twentieth century), they had very few of the institutions deemed essential by developing countries today, including such 'basic' institutions as central banks and limited liability companies ... If this is the case, aren't the developed countries, under the guise of recommending 'good' policies and institutions, actually making it difficult for the developing countries to use policies and institutions they themselves had used in order to develop economically in earlier times?

Global economic power remains concentrated in the North. Indebtedness, itself a product of modernization theories and policies, allows Northern capitalist economies to 'manage' the periphery through the promise of loans from the IFIs in return for structural adjustment – what Stiglitz famously referred to as 'bribarization' before he was sacked by the WB. It also allows them to continue to extract a kind of surplus, for example, in interest payments, which maintains a relationship of dependency. As one commentator notes in relation to Africa:

> Structural adjustment has helped to tie the physical economic resources of the African region more tightly into servicing the global system, while at the same time oiling the financial machinery by which wealth can be transported out of Africa and into the global system.
>
> (Hoogevelt 1997: 171)

Neoliberal policies have also had disproportionate effects on indigenous communities within former settler colonies, such as New Zealand and Australia. These are often communities already with high unemployment, ill health and poverty. For example, New Zealand's neoliberal economic experiment in the mid-1980s resulted in large-scale redundancies that deeply affected the Maori labour force:

> Disillusioned young people try and make sense of their lives while being put through training programmes to prepare for work in communities where no one is employing. Television imports American culture and educates the tastes of the young for labeled clothes and African American rap . . . New missionaries and traders make their way into the region. Some tribal leaders talk economic development, others talk self-determination. Other tribes have vigorously pursued a corporate ethos and have attempted to turn collective knowledge into corporate asset bases and financial wealth. Most tribes are struggling to take care of themselves. People are still trying to survive. Is this imperialism? No, we are told, this is post-colonialism. This is globalization. This is economic independence. This is tribal development. This is progress.
>
> (Tuhiwai Smith 1999: 96–7)

In communities where many people are reduced to living on welfare benefits because of unemployment, in communities with high rates of heart disease, cancer, respiratory disease, diabetes and suicide, it is difficult to see this as development or progress. It is also a familiar story for indigenous communities in other former settler colonies.

Since the 1990s, there has been growing discontent with the policies of the institutions disseminating neoliberalism. Poverty reduction has made limited headway and neoliberalism has failed in many post-socialist countries. As discussed in Chapter 3, the catastrophic meltdown of the Asian economies in 1997, and crises in Brazil,

Argentina and Russia, produced a 'post-Washington consensus' based on belated recognition that market 'failures' and institutions are important conditioning factors in the outcomes of development. It is recognized increasingly that 'development' is a complex social process, that institutions work differently in each society, and that social relations with institutions and markets are important factors. Despite this, the North is reluctant to accede to demands to resolve the Doha impasse, and the global economy remains dominated by the knowledge produced within and disseminated by the IFIs.

Interestingly, individuals who have accumulated vast wealth in global corporations are now major funders of WB and other development initiatives. For example, the Bill and Melinda Gates Foundation is a partner of the WB's Development Marketplace. With the Rockefeller Foundation, it also funds the Alliance for a Green Revolution in Africa. Founded in 2006, AGRA is premised on the notion that African farmers need uniquely African solutions designed to meet their specific environmental and agricultural needs, sustainably boost production and help them gain access to rapidly growing agriculture markets. It aims to catalyze and sustain inclusive agricultural transformation in African countries by increasing incomes and improving food security for 30 million smallholder households. However, there are questions about whether Africa can afford this proposed 'green revolution' in terms of human health and environmental sustainability. As Thompson (2006: 562) argues, these goals 'require resources that the continent does not have while derogating the incredible wealth it does possess'. Scientists, agriculturalists and African governments all agree that Africa has not remotely reached its agricultural potential, and have advocated policies for food sovereignty. These diverge sharply from the high-tech, high-cost approach promoted by Gates, Rockefeller and AGRA. As Holt-Gimenez et al. (2008) argue, history suggests that green revolution approaches deepen the divide between rich and poor farmers (see Shiva's 2016 critique of the Green Revolution in India). Technology-led transformation degrades tropical agro-ecosystems, increases environmental risk and leads to the loss of agro-biodiversity. They argue that hunger is not primarily due to a lack of food, but rather because hungry people are too poor to buy the food that is available. Without addressing structural inequities in the market and political systems, approaches relying on high input technologies tend to fail. AGRA sees the private sector as the solution, but the private sector alone will not solve the problems. Holt-Gimenez *et al.* (2008) predict that genetic engineering of crops will make sub-Saharan smallholder systems more environmentally vulnerable and lead to farmer indebtedness.

AGRA embraces neoliberalism and accessing markets as the solution to hunger and poverty, but this ignores the many successful agro-ecological and non-corporate approaches to agricultural development. There is little space within AGRA's alliances for peasant farmers to be the principal actors in agricultural improvement and, critics argue, AGRA's resources would be better invested in supporting their struggle for food sovereignty as a means to end to hunger and poverty in rural Africa. As Hursh and Henderson (2011) suggest, neoliberal green revolution policies have caused considerable damage to economic equality, the environment and education, yet they remain dominant because the power elite, who benefit most from the policies, have gained control over both public debate and policy-making. By dominating the discourses of development and the logic of economic, environmental and educational decision-making, neoliberal actors like the Gates Foundation have largely succeeded in marginalizing alternative conceptions.

AGRA was led initially by former UN secretary general, Kofi Annan. This is bizarre given that in 2002, the UN commissioned a survey by an expert panel of scientists, largely from the South, which suggested that a green revolution would not provide food security because of the diverse types of farming systems across Africa. Annan was informed that there is 'no single magic technological bullet . . . for radically improving African agriculture', and 'African agriculture is more likely to experience numerous "rainbow evolutions" that differ in nature and extent among the many systems, rather than one Green Revolution as in Asia' (Thompson 2006: 563). Thus, Annan agreed to head the kind of project his advisers at the UN told him would not work. The proposed green revolution threatens to dislocate farmers to consolidate land for high-tech farming; monocultures will replace biodiversity, chemical fertilizers pose a serious threat to the environment, and the wealth of ecological and farming knowledge among local small-scale farmers could be destroyed for the cash wealth of much fewer large-scale farmers buying all their inputs from foreign corporations. This example illustrates why questions need to be asked about the power of enormously wealthy individuals, like Bill Gates, to implement development programmes through their Foundations. While some of these are undoubtedly positive, such as the Gates Foundation's Women's Leadership Program in post-conflict societies, others are widely held to be inappropriate because they ride roughshod over local knowledge systems, and destroy local sovereignty over development.

## The United Nations

The UN was founded in 1945 to prevent wars and foster dialogue between nations. Its aims are to facilitate cooperation in international law, international security, economic development, social progress and human rights issues. Like the WB, its headquarters are in New York. Almost every recognized state is a member, but it also tends to be dominated by Northern agendas. Economic development has long been a major concern of the organization and, through its Development Goals, the UN is a major architect of development knowledge. In 2000, it adopted eight specific MDGs:

1  eradicate extreme poverty and hunger;
2  achieve universal primary education;
3  promote gender equality and empower women;
4  reduce child mortality;
5  improve maternal health;
6  combat HIV/AIDS, malaria and other diseases;
7  ensure environmental sustainability;
8  develop a global partnership for development.

These goals contained specific targets to be met by 2015:

1  halve the proportion of people whose income is less than US$1 a day;
2  reduce by three-quarters the maternal mortality rate;
3  reduce by two-thirds child mortality;
4  eliminate gender disparity in education (by 2005);
5  begin to reduce the incidence of malaria and other major diseases;
6  halve the proportion of people without sustainable access to safe drinking water and basic sanitation.

(Source: *Development Goals* 2003)

The MDGs were significant for a number of reasons. On the plus side, they made poverty visible and, in contrast to neoliberal capitalism and market fundamentalism, marked a new global commitment to poverty alleviation. However, they also shone a light on persistent problems in international development. First, they revealed how definitions of poverty continue to be defined by Northern-based institutions working through primarily economic models rather than through lived experience. For example, research shows that people labelled as 'poor' may not consider themselves as such (e.g. McIlwaine 2002). Second, they revealed the unevenness of development. Despite some clear successes (China, the Newly Industrialized Countries of Asia, middle income countries in Latin America), they represented an acknowledgement that 50 years of international development had failed to deliver even basic levels of food, shelter, healthcare and education to millions of people in the poorest countries of the South. Third, the UN's own *MDG Report 2015* reveals that the MDGs were at best only partially successful: 'Despite many successes, the poorest and most vulnerable people are being left behind'. The report revealed that gender inequality persists; large gaps exist between poorest and richest households, and between rural and urban areas; conflicts remain the biggest threat to human development; millions of poor people still live in poverty and hunger, without access to basic services. In sub-Saharan Africa poverty has actually increased. There has been a reduction in the number of people living in abject poverty, but this has little to do with the MDGs and can be almost entirely attributed to the growth of China (*The Economist* 2015). The vast majority of developing countries missed most of the MDG targets in 2015, and nearly all African countries missed most of them. However, this is not a sign that poor countries failed, that aid has been wasted, or that donors did not spend the right amount of money. Rather, it reflects the fact that the targets were unrealistic.

After 2015, the UN launched a new set of SDGs, with targets for development up to 2030. The process of agreeing the SDGs has been much more inclusive, with all 193 member states involved and with much greater engagement with constituencies in the global South, including civil society organizations. The SDGs have set 17 ambitious goals with 169 targets, again oriented towards dealing with global development challenges (e.g. poverty, hunger, health, education, gender equality, access to water and energy, economic growth, sustainable cities, climate action and partnership). Unlike the MDGs, the SDGs have been broadly supported by major NGOs. While the MDGs aimed to deal with problems, the SDGs aim to deal with the causes of the problems and, crucially, are about *sustainable* development. They also take into account the inter-linkages between challenges, rather than attempt to deal with each one separately. Importantly, the SDGs also hold governments to account and ensure that the 17 issues highlighted are on political agendas. However, the SDGs have been criticized for being contradictory, because in seeking high levels of global economic growth, they will undermine their own ecological objectives: development by its very nature cannot be sustainable (see Chapter 8). The UN's definition of poverty is still seen as problematic, with critics arguing that below $1.25 per day as a marker for extreme poverty is inadequate for human subsistence, and that the poverty line should be revised significantly upwards. The SDGs also ignore local contexts and promote a 'cookie-cutter' approach to development policies. In addition, like the MDGs, the goals are likely to be unattainable. *The Economist* (2015) estimates that alleviating poverty and achieving the other SDGs will require about $2–$3 trillion USD per annum for the next 15 years, which does not appear to be feasible. Like the MDGs, by setting unrealistic targets the SDGs risk perpetuating discourses of failure in the global South, and sub-Saharan Africa specifically. In turn, this risks erasing the significant progress that is being made in improving the quality of life in these parts of the world. The South Centre – a Geneva-based intergovernmental organization of developing countries that promotes their common interests in the international arena – also fears that SDGs could introduce new conditionalities tied to aid funding for developing countries.

In addition to the SDGs, the UN Development Programme (UNDP) is the largest multilateral source of grant technical assistance in the world. The UN also promotes human development through various related agencies. It publishes the annual Human Development Index (HDI), a comparative measure ranking countries by poverty, literacy, education and life expectancy. As we saw in Chapter 1, it thus plays a key role in defining 'development' and 'underdevelopment'. The UN, like the WB, is foremost in producing discourses that have long represented development in normative terms, as an 'uncontested human good' (Munck 1999: 200). It is thus shaped by a similar will to improve. As we saw in Chapter 4, development organizations have endeavoured to label whole parts of the world as 'underdeveloped', 'backward' or 'poor' in order to create a moral imperative for intervention on their behalf. Power relations set the terms in which issues are discussed and interpreted, and lead to the formulation of development goals, values and objectives in particular spaces and places. This creates certain realities or regimes of truth, determining how a country and a people are represented, which in turn shapes development interventions. For example, when the WB refers to corrupt or weak governance in countries of the South, the antidote becomes 'good governance' based on specific Northern, neoliberal models of democratization and development. Similarly, the SDGs reduce development to a series of measurable indicators, which forecloses other ways of thinking about politics and development.

The moral imperative that justifies interventions by Northern-based organizations is, in part, generated by popular media representations, as discussed in relation to humanitarian crises (Chapter 4), but also by development organizations themselves (see Power 2003: 173). The UN's global moral imperative works specifically in the context of emergency relief and the humanitarian nature of international development agendas. However, it does little to explain and tackle the causes of the inequalities it seeks to alleviate. The inequities of international trade and economics are side-stepped, downplaying the importance of structural processes of politics and economics. Instead, the global moral imperative generates a 'feel good' or image industry that produces knowledge and ideologies, which subsequently shape development policy, norms and aspirations.

## Northern governments

This 'feel good' industry and the will to improve also underpin approaches by Northern government departments, which are also important architects of development knowledge. For example, USAID (Box 5.2) attempts to make a virtue of the relatively paltry proportion of US GDP given to overseas development aid (0.19 per cent in 2014, compared with 1.1 per cent in Sweden, 1.07 per cent in Luxembourg and 0.99 per cent in Norway who top the league in the North; the UAE contributed 1.17 per cent in international aid in 2014, largely to Egypt) (OECD Observer 2014). USAID constructs itself as a caring benefactor of poor peoples around the world, but is quite naked in the fact that aid is not humanitarian assistance alone and is concerned with promoting US foreign policy interests. It invokes discourses of threat to buttress its global moral imperative – the threat of terrorism, the threat to democracy where this is 'fragile' in parts of the South, the threat of poverty, disease, drugs and crime, the threat posed by 'weak' and 'corrupt' governments. These threats draw upon familiar stereotypes to provide the rationale for the US government's targeting of aid in particular places for particular purposes. They thus underpin its global moral imperative, which is concerned with promoting US interests abroad, but couched in terms of humanitarianism and care.

---

## Box 5.2  USAID discourses of aid: humanitarianism as self-interest (sources accessed 29/9/17)

### Doing the right thing

On November 3, 1961, USAID was born and with it a spirit of progress and innovation. November 3, 2011 marked USAID's 50th Anniversary of providing U.S. foreign development assistance *From the American People*. Our workforce and USAID's culture continues to serve as a reflection of core American values – values that are rooted in a belief for doing the right thing. (https://www.usaid.gov/who-we-are/usaid-history)

### Global moral imperative as self-interest

USAID's mission statement highlights two complementary and intrinsically linked goals: ending extreme poverty and promoting the development of resilient, democratic societies that are able to realize their potential. We fundamentally believe that ending extreme poverty requires enabling inclusive, sustainable growth; promoting free, peaceful, and self-reliant societies with effective, legitimate governments; building human capital and creating social safety nets that reach the poorest and most vulnerable. (https://www.usaid.gov/who-we-are/mission-vision-values)

U.S. foreign assistance has always had the twofold purpose of furthering America's interests while improving lives in the developing world. USAID carries out U.S. foreign policy by promoting broad-scale human progress at the same time it expands stable, free societies, creates markets and trade partners for the United States, and fosters good will abroad. (https://www.usaid.gov/who-we-are)

### Discourses of threat and compassion

USAID's efforts directly enhance American – and global – security and prosperity. The United States is safer and stronger when fewer people face destitution, when our trading partners are flourishing, when nations around the world can withstand crisis, and when societies are freer, more democratic, and more inclusive, protecting the basic rights and human dignity of all citizens. By focusing on these two goals, together, we position ourselves to meet the challenges of today while mitigating the risks of tomorrow. (https://www.usaid.gov/who-we-are/mission-vision-values)

Afghanistan and Pakistan have both faced substantial security and governance challenges over the past decade. In countries that are critical to our national security, progress is fragile, but our continued efforts remain vital. (https://www.usaid.gov/where-we-work/afghanistan-and-pakistan)

## Policy into practice: USAID in Latin America and the Caribbean (LAC)

Latin America and the Caribbean continue to have some of the highest rates of income inequality in the world and economies have slowed. Severe, chronic drought threatens lives and livelihoods. Regional progress in health masks inequalities between and within countries. Worsening citizen security, fueled by a violent transnational drug trade, is hindering growth and undermining democratic institutions in parts of the region. Climate change poses risks, especially in Central America and the Caribbean. And some countries are restricting political rights. Economic and political stability in the Western Hemisphere are vital for the United States. Drug trafficking and violence that afflict our southern neighbors can penetrate our borders and impact U.S. communities. Latin America and the Caribbean are also important and growing markets for American companies – a quarter of U.S. exports go to the region. In LAC, USAID helps to make the United States and the Western Hemisphere more peaceful, secure, and prosperous by strengthening the capacity of governments and private entities to combat crime, improve governance, address climate change, and create an economic environment in which the private sector can flourish and create jobs. (https://www.usaid.gov/where-we-work/latin-american-and-caribbean)

In the first decade of this century and in contrast to USAID, the UK government's DfID allied itself much more explicitly to the UN MDGs (especially 50 per cent reduction in global poverty by 2015), but in doing so was similarly promoting a 'feel good' image for UK government policy. As Power (2003) argues, DfID did not actually commit itself in policy to cutting world poverty in half, but it wanted to be part of the 'noble' and 'moral' task of reducing the numbers of people living in absolute poverty. While there was much more discussion about international development under successive Labour governments than in previous administrations, much of this was 'fuzzy talk' (Sogge 2002), pointing to the future and disguising the actual political and economic injustices in the present that DfID seemed unwilling to deal with. Rather, DfID engaged unproblematically with neoliberal discourses of development, allying itself closely to the 'market fundamentalism' of the IFIs. As a consequence, it was criticized for being 'strong on rhetoric and presentation, but weak on the substantive issues raised by policy objectives and on the means to implement them', preferring to align itself with the latest development knowledge and buzz words (environmental conservation and sustainability, gender equalities, participation/empowerment and good governance) (Power 2003: 180). It drew on the neoliberal development orthodoxies of the WB World Development Reports and thus offered nothing new or radical (Mawdsley and Rigg 2003).

In recent years, under first the Tory-led coalition and now Conservative governments, the UK government's international development policy has become more explicitly aligned with that of the US. In 2013, the then Minister for International Development, Justine Greening, asked:

> How do we maintain [our] vital advantage? Well, international development is a practical part of that approach . . . [W]e are stepping up our Global Fund investment in malaria, AIDS and TB with people like Bill Gates. He's not someone who invests for the sake of it and neither am I – we invest, because we both know this pays back. Of course it's good for those countries, but it's good for us too. It can't ever make sense to allow terrorists to flourish overseas, and to reach our shores before we do something about it. It's sensible to tackle these risks at source . . . [W]e're helping developing economies grow faster but can we be smarter about the UK locking into the business opportunities those emerging economies present? Yes . . . So development doesn't just develop their economies, it develops ours too.
>
> (DfID Press Release, 29/09/13
> 'International Development is 100% in our national interest')

Greening's successor, Priti Patel, continued to pursue international development policies that are in the UK's national interests (see Chapter 3). Critics argue that this represents the 'death of development'. For example, the UK's new 'development partnership' with India has generated much criticism:

> Cash earmarked to help people in poor countries will instead be offered to leaders of middle-income giants such as India, China and South Africa, to get them to buy British exports. Look at Patel's announcement last week of a new 'partnership' with India, which

promised 'support for India to boost economic growth, jobs and trade, which will also benefit Britain.

(Aditya Chakrabortty, *The Guardian* 23/08/16)

In a letter to Patel published in *The Guardian* (18/07/16), international development scholar Jonathan Glennie reminded her that 'Aid works best when it is owned by the recipient', but this has arguably never been a concern of any national government involved in development in poorer countries. DfID was in the past praised widely for promoting livelihoods analysis and the adoption of a sustainable livelihoods framework, even though this ignored how rampant inflation, conflict, political instability and unemployment make people vulnerable. However, the current discursive and policy shift towards promoting economic growth and using foreign aid to bolster UK economic interests overseas represents a considerable move away from a concern with poverty and the world's poorest peoples.

The IFIs, development organizations and Northern government departments often fail to recognize that capitalist penetration in the South is not necessarily inevitable. Non-capitalist ways of organizing have been numerous and widespread, for example, through co-operative and communal organizations. Critics believe that more attention should be paid to these rather than to simply debating whether globalization is 'good' or 'bad' for the poor. Places and identities are never completely reshaped by capitalism and development and there is always contestation and resistance. In addition, neoliberalism is taken up in different ways by different regimes, be they authoritarian, democratic or communist (Ong 2006). This suggests room for optimism in conceiving alternative economic forms (Escobar 2001) rather than being blinded by discourses of modernity, development, capitalism, globalization and free trade. Postcolonial approaches might productively engage with such emerging alternative development knowledge.

## NGOs and development professionals

NGOs are significant players in the development process, as critics of certain forms of international development, but also as disseminators of development knowledge in their own right. Crucially, NGOs are not completely independent of governments because they receive funding from and are often co-opted by them. Indeed, some commentators suggest that NGOs are now too deeply enmeshed in the promotion of Northern state interests to provide any kind of alternative (Duffield 2007). NGOs have the potential to provide alternatives to dominant development discourses, but these are very often co-opted and emptied of radical content. One example is the notion of 'social capital', which has been popularized through the role of NGOs. This is posited as a more people focused alternative to top–down, universal and economistic models. Social capital recognizes that capital is not simply found in industry, commodities and land, but also in human skills and capabilities. It refers to the ability of people to work together for a common purpose in groups and organizations. It is based in ideas about multiple stakeholders and partnerships forged by different states, capital and groups in society. However, it often betrays little understanding of how these relationships work or do not work and can shift attention away from inequalities of power (Fine *et al.* 2001). It is also prescriptive – if development fails locally it is because of a lack of the right kind of social capital, which in turn legitimates interventions in local political economies as development agencies seek to build this.

Debates about social capital have become a new orthodoxy among NGOs, revealing some of the problems in the ways in which they disseminate development knowledge. NGOs can operate at local, national or international levels but do not always have much close contact with local communities. Consequently, they are often criticized for imposing foreign solutions and for being poorly monitored, despite being important allies of social movements. The 1980s and 1990s witnessed a proliferation of NGOs, grassroots organizations (GROs) and community-based organizations (CBOs), primarily in response to the retraction of the role of the state in welfare and social provision under neoliberal structural adjustment. This led to the institutionalization and professionalization of development in NGOs, and to further dissemination of development as certain ideas, practices and discourses (Escobar 1992). Critics argue that Northern NGOs have mobilized only compassion for 'children and victims of disaster' and funds from donors without building the 'capacity' that they claim to build, and often seem more concerned with their own survival than with challenging poverty or tackling 'real' development issues. As Townsend argues, 'There is a great deal of *talk* about "participation", "listening to the poor" and "partnership", and the donor organizations are often committed to these goals in principle. But the practice usually falls short of this talk' (cited in

Power 2003: 183). Lara Bezzina (2017), for example, argues that International Non-Governmental Organizations (INGOs) concerned with disabled people in Burkina Faso work through western models of disability and fail to understand the lived experiences of the people they target for development. Since local grassroots disabled people's organizations are heavily heavy dependent on external partner INGOs for funding and support, the resultant development interventions are usually ineffective. INGOs also rely on a dual logic of difference and oneness, which enables them to represent the global poor as different and distant and yet like 'us' by virtue of their humanity (Dogra 2014). Their messaging often relies on colonial discourses while simultaneously erasing colonialism and its legacies as a factor that continues to shape global economic structures, power relations and patterns of poverty and inequality.

Southern NGOs have much less power to influence projects and how they are conceived. In some cases, ideas generated in the South are taken up by Northern NGOs, which modify them to fit a neoliberal agenda, crowding out locally based alternatives and local perceptions of what constitutes development. Micro-credit, which was pioneered by the Grameen Bank in Bangladesh to create economic and social development from below, is one example of this. As Power (2003) argues, Southern NGOs have to be up on the buzzwords to acquire funding and support in partnerships with Northern NGOs, hence notions like social capital are disseminated from North to South. While it is difficult to generalize about NGOs, there are particular ways in which knowledge is traded, discussed and disseminated within this community. It is important to examine how this relates to knowledge of local peoples who are subject to their interventions. Like governments and IFIs, NGOs create and disseminate development knowledge that has material effects. A postcolonial approach requires examination of the processes by which certain concepts – like social capital – assume pre-eminence and shape development interventions. It also requires interrogation of absences or blind spots within development knowledge.

## Biopolitics and development in the South

Until very recently, one blind spot within development studies has been its relationship with biopolitics – 'a form of politics that entails the administration of the processes of life at the aggregate level of population' (Duffield 2007: 5). This relationship is apparent in the aim of states to progressively expand their ability to support life and help their populations realize their optimal productive and reproductive potential. Biopolitics is a necessary condition of liberal modernity. This is because modernization creates a 'surplus population' that either cannot or is unable to adapt to changes wrought by industrialization and other processes of modernization. As Duffield (2007: 9) argues, biopolitics is required because these surplus populations represent a threat to the security of the state: 'development provides a solution to the disorder that progress unavoidably brings: the disruption and redundancy of established livelihoods and trades, the erosion of traditional rights and responsibilities, unemployment and pauperization'. The idea of development is thus the liberal solution for dealing with surplus populations, in contrast to modernity's other solutions: extermination or eugenics.

Where biopolitics has been considered, attention has been focused less on development as a liberal biopolitical solution than on what Sylvester (2006: 66), calls 'pernicious bare life biopolitics that could be termed fascism'. We can think of this as a reference to two specific issues brought into view by postcolonial approaches: (i) those recent post-colonial transitions that have witnessed states inflicting injury, suffering, death and genocide on parts of their populations in the name of development; (ii) the ways in which present-day development activities (e.g. by governments and multinational corporations) continue a colonial, capitalist, biopolitical project of managing, controlling and sometimes destroying subaltern bodies. While development studies has often neglected such issues, postcolonial approaches have begun to explore them within the context of development:

## 1.    Post-colonial transitions

It is only recently that development specialists have begun to critique how some states have persecuted elements of their populations in the name of development. Much of this draws on the work of Giorgio Agamben (1998, 2005), who writes about a politics of exception in which states exempt themselves from their own laws against killing to construct zones (e.g. prison camps) in which individuals are reduced to 'bare life', without basic human rights and hope of justice, who can, therefore, be killed. A state of exception is brought about through the increase of power structures that governments employ in supposed times of crisis. The state or individuals extend power, while diminishing and rejecting questions of citizenship and individual rights. While Agamben uses the examples of Nazi Germany and former US President Bush's 'War on Terror' as states of exception, which are manifested in places like concentration camps and the prison camps at Guantánamo Bay, there are many instances of post-colonial states extending their power in similar ways in the name of national sovereignty and development. The case of contemporary Zimbabwe is one example.

Following independence in 1980, Robert Mugabe's ZANU-PF government became increasingly authoritarian and, after 2000, allowed the legitimate sphere of politics to be supplanted by a politics that turned exceptions into the rule. Sylvester (2006) describes how ageing state elites created a twisted version of the anti-colonial struggle by legitimating parastatal youth militias and calling these (impossibly) 'war veterans'. These militias occupied Euro-Zimbabwean farms and forced owners off without compensation – all white farmers were sinners because they had profited from colonialism and continued uneven development, while the youth militias were 'war veterans' and thus could not be sinners. The government broadened the list of sinners to include black farm workers, who were driven from their homes for the sin of having laboured for white bodies. Homosexuals were persecuted by the state, their sin being that homosexuality is a 'European disease', un-African and a product of colonialism. Courts scaled back the inheritance rights of women. Urban ZANU-PF youth turned on opposition party members and the government expelled foreign journalists on the grounds that they were British agents. As Sylvester (2006: 68–9) argues, Zimbabwe's war against enemy bodies also killed foreign investment, donor aid, agriculture, most industry and tourism. ZANU-PF accepted no blame for economic collapse. In May 2005, *Operation Murambatsvina/Restore Order* (OM/RO) was launched. This was a massive 'clean-up' campaign characterized by demolitions and evictions, designed to clear out the 'filth' of informal housing and trading in Harare, which the government deemed an obstacle to development. In reality, this targeted urban dwellers, whose supposed sin of supporting opposition parties resulted in the burning and bulldozing of homes and market stalls, the removal of their occupants to re-education camps in rural areas, and the destruction of livelihoods for thousands of poor people. It is unsurprising that these conditions and high-level corruption led eventually to a coup d'état in November 2017 that removed Mugabe from power.

Zimbabwe is not alone in reducing elements of its population to 'bare life' so that they can be removed or destroyed in the name of national interest and development. Indeed, Sylvester suggests that post-colonial elites often subsume sovereign exception into statecraft to the continued detriment of their citizenry. These elites are intimately familiar with colonial biopolitics, the development project and, in many cases, violent decolonization. Consequently, they are also familiar with the notion of sovereign exception. Dictatorships in Latin American, African and Asian countries illustrate that when the colonized territory becomes a sovereign state it takes on the normal state entitlement to exception. In other words, it has the right to create a range of people that can be killed by the state for a variety of exceptional reasons. In the past, military dictatorships have justified their rule as a way of bringing political stability for the nation or rescuing it from the threat of 'dangerous ideologies'. In Latin America the threat of communism was often used, while in the Middle East the desire to oppose Israel and, later, Islamic fundamentalism proved an important motivating pattern.

Eritrea is another prime example in which biopolitics, militarism and development have become entwined. Once celebrated as the 'African country that works', Eritrea descended into economic crisis and severe human rights violations. This was due in part to the border war with Ethiopia that began in 1998, but was primarily the result of the 'high modernist' development policies and ideologies of the government, which have intensified militarism, and the controlling and disciplining of human lives and bodies by state institutions, policies and discourses (O'Kane and Hepner 2011). Eritrea has pursued a model of national development that rejects most neoliberal strategies as imperialist, while simultaneously exploiting trends and mechanisms associated with globalization in order to achieve its own nationalist ends. The single-party state has become increasingly oppressive, nationalizing land and forcing young people into either compulsory labour or the military. The consequences of these policies have provoked mass emigration of Eritrean people, many of whom risk their lives to flee through Sudan and Libya and end up in migrant camps or detention centres in places like Calais. In July 2018, Eritrea and Ethiopia formally restored relations, which ought to improve conditions in both countries, but it is likely to take many years to reverse the negative effects of prolonged economic crisis, militarism and severe human rights violations in Eritrea.

One of the almost universal characteristics of a military government is the institution of martial law or a permanent state of emergency. This state of emergency creates the state of exception in which oppression is sanctioned. Another characteristic is the way in which many Northern states turn a blind eye, or tacitly or even overtly support such regimes, purportedly in the interests of their own 'security'. The US involvement with military dictatorships in Argentina, Brazil, Chile, Uruguay, Paraguay, Guatemala and El Salvador in the 1970s, for example, was famously exposed by a former CIA agent (Agee 1975). In some post-colonial states (e.g. Zimbabwe), the exceptions have grown in pernicious effect and the exceptional has become increasingly normal. While Agamben recognizes this as fascism, the international development community refuses to name it as such and 'an international community of wink-wink states endeavours to look the other way; or it complains feebly' (Sylvester 2006: 70).

Patel and McMichael (2004: 236) argue that international political economy becomes the stage on which exceptional and often fascistic local policies are defended as normal by those who see themselves as 'sinned against and unsinning, demonizing – correctly – the imperial apparatuses of control without implicating themselves in its functioning'. For example, in 2005, the Dutch Development Minister, Agnes van Ardenne accused the Ugandan President Yoweri Museveni of planning to retain power by amending the constitution to allow unlimited terms of office, and warned him that international donors would not tolerate such a change (Sylvester 2006). The president retorted that Uganda donates its raw materials to the West, which are processed for profits overseas. In other words, sinned against by the imperial apparatus of uneven development, Museveni implied that a post-colonial head of state cannot be a sinner, while the Dutch minister tried to prevent Museveni's anticipated sins without including her own state's colonial-development sins of the past and present. As Sylvester argues, fascism is always already latently present in the sovereign exception in economic conditions associated with capitalism and colonialism. Moreover, post-colonial elites have inherited biopolitical tools to continue the pattern as a form of anti-colonial revenge and response to real and manufactured development crises. Sylvester (2006: 75) argues:

> Although Zimbabwe today is not genocidal Rwanda of 1994, the state brings global fascist histories home through biopolitics that deliver pain to the very Zimbabweans it once endeavoured to rescue. Overwhelmingly, it is the black subaltern there who cannot speak (out) and increasingly cannot eat. How do we help her when the pernicious

state would like to harm her and increasingly prohibits even NGOs from getting close to her?

## 2.    Present-day development biopolitics

Critics argue that many present-day development activities continue a colonial, capitalist, biopolitical project that has always contained within it the latent traits of global fascism (Patel and McMichael 2004). As we saw in Chapter 2, colonial government worked by managing and controlling colonized bodies. Colonized people were made to cover their bodies, subject their bodies to western forms of hygiene, fill their bodies with western knowledge, move their bodies through slavery and displacement to different lands (both overseas and within the colonial state), use their bodies for slave and wage labour, and fight other bodies in the name of the colonial state. Moreover:

> Bodies that seemed too 'other' to fit on the approved colonized/
> development line could suffer assault and death through holocausts of the
> sort Australians conducted in Tasmania, the cavalry led in the US West,
> and the Spanish and Portuguese unleashed in the American colonies.
>
> (Sylvester 2006: 68)

This body disciplining has continued into post-colonial contexts and in this respect development differs little from colonialism. Development relies on and exemplifies a 'deployment of disciplinary technologies at the level of the individual' (Patel and McMichael 2004: 237), of which there are countless contemporary examples. For example, thousands of residents were forcibly evicted in Beijing and any who dared to protest were imprisoned in illegal 'black' prisons' to make way for development projects associated with the 2008 Olympics. Similar forced evictions, in some cases leading to violence, were carried out in Rio de Janeiro's favelas in the year leading up to the 2016 Olympics. Many already poor people lost their homes, and in some cases their freedom and livelihoods, to make way for a two-week mega-event considered more to be economically more important. Millions have been displaced to build the Three Gorges Dam on the Yangtze River, now considered by many in China to be an environmental disaster. Over 300,000 people living in Delhi's biggest slum were evicted in 2004 under a government plan to transform the banks of the Yamuna river into a tourist and leisure centre in defiance of citizens' rights (Figure 5.1). The humanity of poor people is erased; they are literally

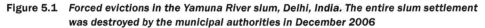

Figure 5.1 *Forced evictions in the Yamuna River slum, Delhi, India. The entire slum settlement was destroyed by the municipal authorities in December 2006*

Source: Parivartan Sharma/Reuters for Housing and Land Rights Network, New Delhi

erased from areas targeted for development that will benefit the affluent, and those who resist are subject to other forms of discipline, such as imprisonment.

There are also countless biopolitical stories to be told about the complete disregard for people's safety in the South by national and multinational corporations based in the North. Amongst these, the sale of unsafe contraceptives by the US government for mass consumption in the South during the late 1970s is notorious. The dumping of Depo-Provera – an injectable contraceptive not approved for use in the USA – on the South was particularly controversial. This was justified on the grounds that 'overpopulation' in the 'Third World' and the risks of dying in childbirth were so much greater than in the USA, so the use of almost *any* contraceptive was justified, even if the side-effects were severe and the death rates unacceptable (Ehrenreich *et al.* 1979). The explosion of the Union Carbide factory in Bhopal, India in December 1984 exposed hundreds of thousands to deadly toxins, killing approximately 20,000 to date and continuing to affect thousands more who suffer from breathing difficulties, cancer, serious birth defects, blindness, gynaecological complications and other problems. Authorities had tried and failed to persuade Union Carbide to build the plant away from densely populated areas, but the company refused because of the expense that such a move would incur. The company, now owned by Dow Chemicals, still refuses to accept responsibility and to compensate affected communities. Coca-Cola has been heavily criticized for causing extreme water shortages in developing countries where supplies are already scarce. Evidence from War on Want (2006) appeared to show that Coca-Cola has had a serious impact in communities in several Indian states and in Latin America. Researchers uncovered areas in Rajasthan where farmers have been unable to irrigate their fields after Coca-Cola established a bottling plant, and have revealed similar problems in Uttar Pradesh. The forced closure of a Coke plant in the southern Indian state of Kerala, after it was alleged to have contaminated local water, is well documented (see, for example, www.indiaresource.org/campaigns/coke/2003/continuingbattle.html) and has been followed by other plant closures in India. In 2015, the Tamil Nadu government withdrew permission to allow Coca-Cola to build a plant in the area due to farmers' fears that doing so would deplete their groundwater resources and there have been widespread protests against Coca-Cola throughout India (Figure 5.2).

**Figure 5.2**    *Students protesting against Coca-Cola, Varanasi, India 2014*

Source: Citizens News Service

As these examples demonstrate, the bodies of subalterns can be destroyed, poisoned and experimented upon in the name of development in ways that are unacceptable in the North, though this is increasingly being resisted. The promise of postcolonial development studies lies in its abilities to 'follow bare life politics into its hideouts, and into our texts and toolboxes as well, searching with undeflecting and nuancing eyes' (Sylvester 2006: 75–6). The onus is on development studies to better see and address the problematic biopolitics of the contemporary world, particularly where these are bound up with notions of development. It is also important, therefore, to consider the relationship between development and biopolitics not simply as latent fascism but also as the *liberal* solution to surplus populations.

Development emerged in the nineteenth century as a solution to the potential disorder created by European industrialization, imperialism and the ending of slavery. As discussed in Chapter 3, it involved exercising trusteeship over surplus life (Cowen and Shenton 2006). In the contemporary period, development is again bound up with issues of security. As Duffield (2007: 227) argues, development embodies an urge to protect and develop others less fortunate than ourselves, which is both noble and emancipatory. However, this is transformed into 'a liberal will to govern through the assertion of an educative trusteeship over life that is always experienced as somehow incomplete'. The question for Duffield is whether development's urge to protect and emancipate (its will to improve) can be separated from its current association with security and emergency. A dichotomy exists in which Northern states protect themselves through providing social protection to their surplus populations, while in the South essential needs are seen as being met through self-reliance. Development as 'betterment is not concerned with extending levels of social protection similar to those enjoyed in Europe to, say, the peoples of Africa'. On the contrary, 'essential needs are seen as being met naturally from self-reliance' (Duffield 2007: 218). One solution is to create alternative development through rejecting self-reliance (which has arguably morphed into resilience in contemporary discourses) in favour of providing protection through social insurance for the world's surplus population. However, this places the impulse to protect and improve in the hands of the state, thus further privileging its role and repeating many of the current and historical problems of state-centred approaches to international development. Freeing this impulse from the state is desirable, which is why the 'hopeless enmeshment' of the NGO movement with state interests is considered by some to be a 'tragedy' (Duffield 2007: 231).

Duffield's solution is important in highlighting the potential of postcolonial politics. He sees social movements as one possible solution since they promote a sense of global solidarity. Such solidarity foregrounds the mutuality and reciprocity between provider and beneficiary. It constructs the grievances and problems of geographically, socially and culturally distanced others as deeply intertwined. Differences are acknowledged but similarities become the basis for shared concerns and political action. Social movements render global solidarity political:

> Distant struggles are common points of departure that collectively problematize the over-arching, anti-democratic and marginalizing effects of global neoliberalism, whether as struggles against hospital closures in mass consumer society or the ruination of pastoralist livelihoods beyond its borders. In this respect, it minimizes attempts to divide and striate humankind either according to measures of development and underdevelopment or those of culture.
>
> (Duffield 2007: 233)

The political strategies of transnational social movements thus speak to a postcolonial politics within development since, rather than educating the poor and marginalized, 'it is more a question of learning from their struggles for existence, identity and dignity and together challenging the world we live in' (Duffield 2007: 233–4).

## Transnational resistance and theorizing back from the South

Northern development discourses can be considered tyrannical because of their tendency to take over domains other than the economic domain originally ascribed to them (Karagiannis 2004: 138). Development is inherently 'political'. However, postcolonial approaches that open up the ambiguities of development discourses also entail a spirit of opposition, particularly opposition to accounts of development that see no future against its hegemonic power and take no account of the creativity, agency and resistance of subaltern peoples on developing countries. The exercise of power necessarily gives rise to resistance, negotiation and appropriation and this can be seen both historically and contemporary political movements.

Historically, the imposition of western notions of progress and development was resisted by indigenous peoples from the moment of the first colonial encounter. For example, the European colonization of time (Chapter 3) that underpinned the global expansion of capitalism met with much resistance. In both South Africa and Australia there was constant resistance to time discipline, including: the destruction of symbols of colonial time, such as mission bells; 'walkabouts' or leaving colonial space-time for traditional practices; and everyday subversion, such as lateness and foot-dragging. Where they worked in their favour, colonized peoples also re-appropriated European time practices, like negotiating work contracts, or insisting on the Sabbath as a day of rest. Nanni (2012) documents how indigenous and colonialized peoples made a mockery of supposedly objective western time by pointing out its irrationalities. The imposition of the northern European four-season agrarian model on Australia looked even more absurd when compared with Aboriginal knowledge, which more accurately represents natural cycles by identifying six or more annual seasons and multi-year seasons that can span as many as 28 years. Throughout the southern African colonies, the monthly salary payment was treated with great suspicion by African labourers, who observed regular lunar 28-day months. Their frustrations at the randomness of Gregorian 28-, 30- and 31-day months undermined its claims to be logical and rational, and the labourers often insisted on the use of indigenous time-practices to reckon payment schedules. This is not to downplay the violence of imperialism, which by destroying indigenous time destroyed and de-legitimized indigenous culture. Rather, it is to recognize indigenous agency and resistance in the formation of colonial cultures – to recognize, in the words of Frantz Fanon, that 'colonialism must accept the fact that things happen without its control, without its direction' (cited in Nanni 2012: 210).

Although not enough is known about interculturation of ideas about time, colonization was not simply a unilateral imposition of western ideas on other people. Pre-colonial cultures also helped define colonial time as hybridized time. Nanni argues that, for example, some settlers in Australia were receptive to Aboriginal astronomical knowledge that lined up with the European understandings of constellations. In South Africa, missionaries sometimes worked through the Xhosa calendar because it had twelve months and isiXhosa names for months are still used in the Eastern Cape. And European time never completely eradicated indigenous time. Despite decades of French colonization,

Bourdieu was still able to write about the Kabyle peasants in early 1960s Algeria as having:

> An attitude of submission and of nonchalant indifference to the passage of time which no one dreams of mastering, using up, or saving . . . Haste is seen as a lack of decorum combined with diabolical ambition. The clock is sometimes known as 'the devil's mill'; there are no precise meal-times; the notion of an exact appointment is unknown; they agree only to meet 'at the next market'. A popular song runs: 'It is useless to pursue the world, No one will ever overtake it'.
>
> (in Thompson 1967: 58–19)

Despite the violence of imperialism and colonialism, African cultures remain to a large extent polychronic rather than linear – people do several things at the same time. This is often bemoaned as 'Africa time' (or, in Latin America, 'mañana'), but what westerners think of as a lack of punctuality or a lax attitude about time may also be understood as a different approach and method in managing tasks, events and interactions. Lives are not linear and people tend to manage more than one thing at a time rather than in a strict sequence, including personal interactions and relationships. In many places colonized by Europe, an emotional time consciousness has persisted and still resists and frustrates western mechanical time consciousness. This creates responsibilities towards collectives, rather than to an individual, which have to be juggled and, while frustrating to westerners, these different time cultures are rooted in maintaining deep social networks and responsibilities.

In some places there are now active attempts to decolonize European time. In 2014, the clock on Bolivia's Congress building was altered to reverse the numbers and turn the hands counterclockwise, instead of clockwise, to reflect the country's location in the Southern Hemisphere. Bolivia's president of Congress, Marcelo Elio explains, 'Clocks are an evolution of the sundial, and in the northern hemisphere a sundial's shadow runs clockwise, while in the southern hemisphere it moves counterclockwise — making the modern clock a representation of light in the northern hemisphere' (*New York Times*, June 15 2014). He called the clock's alteration 'a clear expression of the de-colonization of the people'.

This may be an overstatement, but it captures how the colonization of time is not simply a history of imposition, but of resistance, negotiation and interculturation. As Nanni (2012: 180) puts it: 'Time, like colonialism in general, was not coherently imposed, nor coherently resisted; it functioned, rather, as one of the many sites of cultural exchange, tension and negotiation which came to characterize nineteenth-century relations between colonizers and colonized' (p. 180). Having spent most of the nineteenth century trying to eradicate it, it is perhaps ironic that it is now 'westerners who dream of going "walkabout", leaving the temporal matrix of modernity behind' (Manktelow 2014). The Slow Movement – a resistance to the commodification of time and the pace of modern life – has become a cultural trend in the global North. The Slow City movement began in Italy in 1999 linked to the slow food movement and is now spreading more widely across Europe and North America. According to Carl Honoré, author of *In Praise of Slowness*, slowness is about 'savouring the hours and minutes rather than just counting them. Doing everything as well as possible, instead of as fast as possible. It's about quality over quantity'. This is a refusal of a commodified view and a different valuing of time. It coincides with a sometimes problematic romanticization of 'indigenous time', but the similarities with Aboriginal views of 'whitefella time' are notable as the following poem, transcribed from the Northern Territory in the 1950s, illustrates:

> White man properly different kind,
> Watch on wrist, then wind . . . wind . . . wind;
> Pull up sleeve and look all day,
> Look when work and look when play. . .
> My man listen and then him say,
> 'No time sleep', and 'No time stay',
> Too much hurry down the street,
> Terrible place when, 'No time eat'.
>
> (in Nanni 2012: 227–8)

In a similar vein, development organizations like the WB exercise power in scripting development programmes, but these are often contested and rejected on the ground in the South. Development organizations are often unable to change the social forces and state practices that shape development. This resistance is not always positive and might be a product of, for example, despotic or corrupt state practice or social forces that sustain caste and class inequalities. However, when development programmes fail, people mobilize on the basis of rights. As Li (2007) argues, they understand clearly the relationship between defects in programmes and their own insecurities, which are seen as attacks on their ability to sustain their own lives.

This mobilization reveals the limits of both 'expert' development knowledge – programmes fail to deliver intended outcomes – and of the power of development organizations – the WB is not omnipotent because its programmes generate resistance. Yet the WB is successful in setting the parameters for how North–South relations are understood. It is thus incredibly powerful, even if its projects fail or are rejected and resisted. Postcolonialism is not simply a critique of development discourses and their emanation in policy and practice, as outlined above, but also creates possibilities for rethinking development, for considering resistances to dominant discourses of development and alternatives to them, and for interrogating how 'we' speak and write about the South.

## Anti-globalization and the beginnings of resistance

Global protests against globalization are now commonplace, especially around the WTO meetings – the first large-scale protest occurred at the Seattle meeting in 1999 – and the annual meetings of World Economic Forum (WEF) conferences, where up to 1,000 international companies meet at what some call a 'Business Olympics' for the global corporate elite. At the heart of these protests is debunking the myth of free trade and opening up the 'ideological space' of development in search of alternatives. Anti-globalization thus has much in common with postcolonial critiques of dominant development discourses and ideology. For example, the World Social Forum (WSF) is a global coalition of various protest organizations that originated in Brazil's civil society organizations and has its headquarters in in Porto Alegre. The slogan of the first forum in 2001 was: 'Another world is possible'. It attempts to globalize resistance by building a coalition between erstwhile localized and fragmented movements, bringing together, for example, farmers from East Africa, fisher folk from India, trade unionists from Thailand and students and activists from around the world. It is a counterpoint to the WEF and aims at promoting a 'post-capitalist' world through the creation of alternative forms of communication, practical economics and social and political organization. It sees this as a long-term process, beginning with land reform, mechanisms to democratize control of finance capital, redistribution of wealth through social security and public re-appropriation of public resources such as water, knowledge, seeds and medicines.

In response to the increasing significance of global protest, particularly against the development policies of the IFIs, some politicians in the North are beginning to acknowledge that world trade is not in fact a solution to poverty and global inequality. However, generally these problems are still seen as distant and removed from the key political issues in the North: 'Elite western policy-makers seem to regard the growing income equality gap as they do global warming. Its effects are diffuse and long term and fears of political instability, unchecked migration flows and social disruption are regarded as alarmist' (Wade 2001: 80).

In the late twentieth century, European policy shifted from accepting responsibility for colonialism and its after effects – reflected in paternalistic models of development as aid – to a more forgetful politics in which post-colonial countries are responsible for their own fate (Karagiannis 2004). Unfortunately, this prevents the emergence of a new politics of solidarity between North and South that might lead to a different development ethics. However, people in the South are beginning to show their belief in alternatives to neoliberalism, and are now seeking to support alternative sources of knowledge and pursue strategies of 'constructive disengagement'.

The 1990s witnessed the rise of social movements around the world in response to the hegemony of neoliberalism and the failures of development. These diverse movements are fighting for political autonomy and cultural, ecological and economic survival and include squatter movements, women's groups, human rights organizations, workers' rights organizations and youth groups. These are often global networks of resistance, such as the WSF, *Via Campesina* (an international peasant union formed in 1992, uniting farmers, rural women, indigenous groups and landless peoples in campaigns on food sovereignty, agrarian reform, credit and external debt, technology, women's participation and rural development) and the Occupy movement (an international social and economic justice movement, which began in New York in 2011 and campaigns under the slogan of 'We are the 99 per cent' to draw attention to the fact that the world's richest 1 per cent have expanded their wealth considerably since global financial crisis in 2008). As one Indian protest movement puts it:

> It's not about the First and Third World, North and South. There is a section of the population that is just as present in the US and in Britain – the homeless, unemployed people on the streets of London – which is also there in the indigenous communities, villages and farms of India, Indonesia, the Philippines, Mexico, Brazil. And all those who face the backlash of this kind of economics are coming together – to create a new, people-centred world order.
>
> (Patkar, quoted in Power 2003: 209)

In promoting solidarity networks, popular 'participation' and 'empowerment' approaches, these movements are also challenging Northern assumptions and aim to democratize development (see Chapter 6). In some cases, the resisters have come to power. For example, in 2005, Evo Morales became the first indigenous leader to be elected president of Bolivia and was re-elected in 2009 and 2014. Morales has sought new avenues of social development, arguing that indigenous people have not benefited from the country's development, that foreign companies have too much control of Bolivia's resources, and that the benefits now should flow to the nation and especially its poorer communities. What is interesting, then, is how the resisters to western-style development are also shaped by a will to improve when they attain power. Given the recency of such political changes, the similarities and differences in how this will to improve is articulated *within* and *from* the South remain as yet unexamined, but this is important and fertile terrain for postcolonial development studies. These political changes have also coincided with the rise of indigenous theory and subaltern knowledge.

## Indigenous theory and subaltern knowledge

Traditional knowledge was often seen as an obstacle to development. However, with the rise of post-development and postcolonial critiques and the failure of development interventions to reduce global poverty rates and rising inequality, the universal applicability of western scientific knowledge and developmentalism has been fundamentally questioned. In fact, some scholars have suggested that western science is also an indigenous and/or local knowledge, since it is localized in the institutions of the North. The difference between western and other knowledge is that the former gained global pre-eminence through the processes of colonialism and neocolonialism (Escobar 1995a; Mohan and Stokke 2000). As Radcliffe (2017: 2) argues, the ethical and political context for knowledge production is fragile and contested. As a strand of decolonial theory, indigenous theory focuses attention on the specificity of the disempowerment of indigenous peoples, seeks political alternatives to the colonial present, and examines the material and political costs of settler colonialism (*ibid.*). Indigenous theory and subaltern knowledge are also now seen in some quarters as critical to sustainable and balanced development, as being more attuned to local conditions and contexts, and more inclusive of local people, offering them a voice within the development process (see Chapter 6).

The significance of indigenous knowledge and the need to consult local people about its development was raised as long ago as the 1980s (e.g. Chambers 1983), but it is only relatively recently that this has been given any credence in mainstream development circles. As Briggs and Sharp (2004) argue, there has been some previous engagement with local knowledge, but only at a technical rather than a conceptual level. In addition, indigenous knowledge was allowed to offer technical solutions only if it accorded with, rather than challenged, the dominant scientific and/or development world view. In recent years, however, indigenous knowledge has been appropriated into mainstream development thinking. As with many new ideas, this appropriation is not unproblematic – a liberal embracing of the concept of indigenous knowledge has arguably led to a loss of its radical and political elements, while disguising the continued dominance of western knowledge and power.

One example of this is the WB's appropriation of indigenous knowledge. As Briggs and Sharp (2004) point out, within WB lexicon, indigenous knowledge has been abbreviated to 'IK', emphasizing the view of indigenous knowledge as a technicality and hiding the deeper ways of knowing behind the neat sign. In 1998, the WB published its 'Indigenous knowledge for development: a framework for action' (see www.worldbank.org/afr/ik/ikrept.pdf), in which it argues for the importance of paying attention to the knowledge base of the poor. However, the framework lists mostly technical and discrete knowledge that can be identified (e.g. herbal medicine). It does not acknowledge the significance of dealing with deeper knowledge that might underlie technical knowledge and inform the world view of local people. Such knowledge might contain local understandings of decolonization, social justice, gender relations, familial responsibility, human–environment relations and so on. Moreover, the report is clear that indigenous knowledge should not threaten the established order: 'IK should complement, rather than compete with global knowledge systems in the implementation of projects' (World Bank 1998: 8). Existing scientific knowledge (and thus western experts) still defines what indigenous knowledge is useful and worthy of dissemination. In other words, the appropriation of IK is incommensurate with the political and analytical aims of indigenous theory, which seek decolonization. This is not an engagement with knowledge from the global South that challenges the colonial legacies of development or responds to a broader responsibility to make knowledge differently (Jazeel 2017).

Briggs and Sharp also caution against some other approaches to indigenous knowledge, which can lead to a freezing of traditional cultures and ways of knowing. They draw on Silvern's study of the Ojibwe tribe to demonstrate how this played out in the context of native Americans' use of natural resources (Silvern 1995, cited in *ibid.*: 669). The Ojibwe have traditional rights to fish the lakes of northern Wisconsin, but the methods they currently use are seen as problematic by many other residents in the area. The grounds for complaint are that, rather than using 'traditional' methods such as birch bark canoes, pitch torches and spears, the Ojibwe now use motorized boats, battery-powered lamps and metal harpoons. The knowledge of conservation that underlies these technological advances are seen as irrelevant; for the Ojibwe's claims to indigenous knowledge to be recognized their methods of fishing would have to remain unchanged from some point in the nineteenth century, when they were considered 'traditional'. In this example, indigenous knowledge only counts as such if it is related to a 'traditional' culture that has been frozen in time. Claims to indigenous knowledge that might also be 'modern' are seen as a contradiction. A false binary between indigenous and western development knowledge is easily reinforced, yet the two are not necessarily always disengaged. However, accounts of the politics of knowledge in development studies fail to adequately address how subaltern knowledge is translated and used in development policy (McFarlane 2006: 36).

Other problems relate to particularism – an assumption in studies of indigenous knowledge that each situation is unique and that every development struggle is therefore localized and specific. The problem with this is that it can be at the expense of a global vision (Spivak 1988: 290); those structures that operate at global levels to reproduce inequality are conveniently overlooked by too myopic a view of the local. It is important therefore, not to divorce indigenous knowledge from power relations operating at multiple scales and to acknowledge that it cannot be understood outside of a critical analysis of economic, social, cultural and political contexts. As discussed in Chapter 2, Spivak emphasizes the need to challenge the easy assumption that the postcolonial scholar can recover the standpoint of the subaltern, or the development specialist to recover the standpoint of people in the South. Simply appropriating indigenous knowledge and theory does not equate with challenging the dominance of Northern development discourses. Rather, there is a need for a rigorous interrogation of the politics of speaking and writing that inflect development discourses (see Chapter 6).

Decolonizing development knowledge requires interrogating how 'deep colonization' by Eurocentric ideas continues to marginalize indigenous perspectives (Howitt and Suchet-Pearson 2006). As Sarah Hunt (2014: 28) argues, 'it is not enough to talk about Indigenous ontologies without addressing how we come to understand what Indigeneity and ontology themselves mean'. She argues that western ontologies of Indigeneity are not the same as Indigenous ontologies, and thus 'making ontological shifts in the types of geographic knowledge that is legible within the discipline requires destabilizing how we come to know Indigeneity', as well as the representational strategies used in engaging with it. Decolonizing development knowledge also requires acknowledging that Eurocentric development discourses and practices reflect highly problematic assumptions about the relationships between people and between people and their surroundings. As Howitt and Suchet-Pearson (2006: 323) argue: 'failure to challenge these assumptions risks re-imposing colonial power relations on groups who make different sense of the world'.

## Contesting hegemonic knowledge: postcolonial feminisms

As discussed in Chapter 4, the power of western feminism to speak for women elsewhere has not changed since colonial times and came under scrutiny during the late 1980s. The critique from women in the South is a precursor to, and instructive of, broader postcolonial challenges to the dominance of western development knowledge. Trinh (1989) described the exclusionary tactics of western feminism that make the concerns of 'Third World women' 'special' because they are not 'normal' but Other, and because they are not written by white women. She wrote (1989: 82):

> Have you read the grievances some of our sisters express on being among the few women chosen for a 'Special Third World Women's Issue' or on being the only Third World woman at readings, workshops and meetings? It is as if everywhere we go, we become someone's private zoo.

White academics in the North are empowered both economically and socially to make women in other cultures the object of their investigations, when the reverse is often neither possible nor feasible. For example, Sittirak (1998: 119) describes her experiences as a Thai woman studying in Canada:

> Officially, there are no regulations to prevent me from expiating Canadian or any other ethnic groups. However, like many other 'international students' who received scholarships from development projects, it implicitly seemed that we 'should' focus on our own issues in our homes. That is the way it is. At that moment, I did not question as to why a Thai student had to focus on Thai issues, while Canadian students had much more academic privilege and freedom to study and speak about any women's issues in any continent from around the world.

Spivak, in particular amongst postcolonial theorists, has taken gender and development to task for the 'matronizing and sororizing of women in development' (1999: 386) and for representing women in the South as oppressed by 'second-class cultures' (1999: 407). By wishing to come to the aid of oppressed 'sisters' in the South, feminist intervention from the North replicates the colonialist civilizing mission. And, by assuming women's solidarity as a basis from which to speak for all women, it both disavows the obstacles to gender equality in the North and ignores the historical, cultural and socio-economic differences between women in different parts of the world.

The consequence of these criticisms is that the presumed 'authority' and 'duty' of Northern academics to represent the whole world has increasingly been questioned both within and without its ideological systems (Duncan and Sharp 1993). As discussed in Chapter 4,

Northern feminists have increasingly begun to recognize that international feminism is constituted by a multiplicity of voices, including those of women in the South, which demands the recognition of difference and the multiplicity of axes and identities that shape women's lives. Greater emphasis is now placed on the 'positionality' of the researcher in relationships of power. This is much more than a question of being culturally sensitive or 'politically correct', and requires a continual and radical undermining of the ground upon which one has chosen to stand, including, at times, the questioning of one's own political stance (Duncan and Sharp 1993). Black feminists and women in the South are fighting for spaces in which to articulate their own demands and shape their own political agendas. Furthermore, marginalized women are resisting their representation by elite women from within their own cultures, many of whom are now located within the Northern academy. There are clear overlaps, therefore, with new social movements that are directly resisting the imposition of Northern development norms.

Constant reflection on the creation and production of knowledge remains important. As hooks (1990: 132) argues, 'if we do not interrogate our motives, the direction of our work, continually, we risk furthering a discourse on difference and otherness that not only marginalizes people of color but actively eliminates the need of our presence'. As we have seen, Spivak (1988) takes this argument a step further, arguing that the subaltern cannot speak, since s/he is required to be imbued with the words and phrases of western thought in order to be heard. Therefore, the subaltern cannot be heard because of the privileged position that academic researchers and development consultants from the North occupy. This leads to 'epistemic violence' – the erasure, trivialization and invalidation of ways of knowing the world that fall outside of western languages and philosophies. The subaltern is thus always 'caught in translation, never truly expressing herself, but always already interpreted' (Briggs and Sharp 2004: 664). In response, postcolonial feminisms acknowledge not only the situatedness of knowledge, its cultural specificity and, therefore, its partiality, but also pay greater attention to 'unlearning' privilege as loss (Spivak 1990, 1993).

As discussed in Chapter 3, this unlearning is an ethical response to the conundrum posed to Northern researchers and practitioners by epistemic violence. In terms of educational opportunity, citizenship and location within the international division of labour most academics are privileged.

Privileges based in race, class, nationality, gender may have prevented us from gaining access to Other knowledge, not simply information we have not yet received, but the knowledge that we are not equipped to understand by reason of our social positions. Spivak's 'unlearning' of privilege involves working hard to gain knowledge of others who occupy those spaces most closed to one's privileged view and attempting to speak to those others in a way that they might take us seriously and be able to answer back. Postcolonial feminisms, therefore, allow for competing and disparate voices among women, rather than reproducing colonialist power relations where knowledge is produced and received in the North, and white, middle-class women have the power to speak for their 'silenced sisters' in the South (Amadiume 1997).

Spivak also outlines a formulation of ethics in which she posits the ethical relation as an embrace between parties who learn from each other, which has implications for thinking about hearing and writing tactics. This embrace is not the same as wanting to speak for an oppressed constituency, since the problem is not that the subaltern cannot speak but that she is not heard. As Jacobs (2001: 731) puts it, 'It is very common nowadays for the postcolonial politics of not speaking for the other to override an alternative postcolonial politics of listening to the other'. The latter only cease being subaltern when, to use Gramsci's terms, they become organic intellectuals or spokespeople for their communities. Such a change will not be brought about by intellectuals attempting to represent oppressed peoples or by merely pretending to let them speak. However, interactions between academic and non-academic researchers in disparate locations can generate new languages and social representations that can become 'constituents of alternative social visions and practices', as well as 'enabling new political identities and initiatives' (Gibson-Graham, 2002: 108). Despite the problems of the inevitable partiality of the privileged academic view, recognizing the effectiveness of knowledge 'creates an important role for research as an activity of producing and transforming discourses, creating new subject positions and imaginative possibilities that can animate political projects and desires' (Gibson-Graham 2002: 105; 1994).

## Postcolonializing development knowledge?

For the development researcher, feminist and postcolonial ethics of unlearning and learning to learn are not about speaking for an individual or group, but developing new positions through interactions between researchers and people in different locations. It is an ethical imperative that points towards transformation, recognizing a condition that does not yet exist, but working nevertheless to bring that about (McEwan 2003). The politics of development would then emerge as provisional and constantly under review. It is important, however, that postcolonialism does not become another colonizing discourse, yet another subjection to foreign formations and epistemologies. This requires greater sensitivity to the relationship between power, authority, positionality and knowledge. Briggs and Sharp (2004) provide a method for framing this ethico-politics by distinguishing between liberal and radical politics. They argue that there must be a radical attempt to engage indigenous people and knowledge, rather than a liberal attempt that integrates views into pre-given positions and falls short of dealing with the many different kinds of indigenous knowledge and ways of knowing.

In terms of transnational development networks, postcolonial politics and ethics involve reconciling the role of transnational networks in local political economies of development, with attention to the ways in which subaltern knowledge is deployed (McFarlane 2006: 45). At stake are the ways in which subaltern knowledge is translated and used in development, and also how it is mediated as it travels and is reshaped. It involves considering how learning takes place between actors in transnational development networks and what kinds of politics and practices emerge through the exchange of people and information (McFarlane 2006). Such an approach goes beyond a concern with whether NGOs or donors are accountable to subaltern groups to consider the extent to which such agents can and do listen to subaltern individuals and groups as well as what they do with what they hear. Central to this is the search for a more equitable dialogue. A postcolonial approach to development knowledge requires 'alertness to the different kinds of knowledge and spatialities produced by different actors, and the ways in which some of these become dominant, while others are marginalized or abandoned' (McFarlane 2006: 46). Concepts such as 'mistranslation' (Gupta 1998) and 'misreading' (Said 1984) are useful for tracing these more relational spaces of development. Thus, although postcolonial theory does not solve the problem of underdevelopment, its concern with subaltern voices and knowledge can provide a much-needed counter to the Eurocentrism of much writing on development.

This chapter has explored how institutions, governments and organizations are often the most powerful architects of development knowledge, which is shaped by the will to improve, and informs development policy and practice in the South. This will to improve is neither necessarily malevolent nor benevolent. Nor should the power of development organizations be overstated, since they often struggle to impose their will through programmes and are often faced with resistance. In addition, we should not forget that individual scholars and academics are also authors of development knowledge. We saw in Chapter 4 how dangerous it is to assume that 'we' can encounter the South, and especially the 'Third World subaltern', on a level playing field. Interactions towards and representations of people in the South are inevitably loaded, shaped by favourable historical and geographic positions, and material and cultural advantages resulting from imperialism and capitalism. Producing knowledge about the global South inevitably involves relationships of power. Those who claim to be able to represent the South are in privileged positions of power, able to speak for (represent politically) and speak about (portray) the subaltern subject, who, as we have seen, is thus denied space in which to speak herself. This power dynamic is not simply a problem for Northern development specialists, but also for those members of the professional classes within the global South or the Third World diaspora – what Spivak terms the 'native informant'. The latter is also in a position of relative power in relation to the subaltern:

> Being postcolonial or 'ethnic', according to Spivak, does not necessarily or naturally qualify one as Third World expert or indeed subaltern; in fact, valorizing the 'ethnic' may end up rewarding those who are already privileged and upwardly mobile, at the expense of the subaltern.
>
> (Kapoor 2004: 631)

For Spivak, the dangers in valorizing the 'ethnic' or 'indigenous' are that this new-found 'nativism' is a form of reverse ethnocentrism in which the 'native informant' is complicit in the spread of a 'new orientalism' (Spivak 1993: 56).

Many people working in the field of development might argue that we represent the South so as to know it better in order to be of help. Yet scholars such as Spivak, Said and Escobar argue that things are never this innocent – knowledge is always imbricated with power, so that knowing the South is also about discursively framing, disciplining, monitoring or managing it. Spivak, in particular, has criticized the way in which Northern researchers further their own personal and institutional interests through fieldwork and data collection in the South. For her, this is a form of imperialism where the South provides resources for the North. Knowledge production is, in effect, cultural imperialism, to which there are two elements. The first is 'benevolent first-world appropriation and re-inscription of the Third World as an Other' (Spivak 1988: 289), where information is extracted from the South to serve the purposes of researchers in the North. The second is the privileging of theory (Spivak 1988: 275), where researchers transform the 'raw facts' gathered from the South into 'knowledge'. The field is thus the repository of data and the academy is centred as the site for value-added theory – subalterns tell stories and researchers theorize for them. For example, social movements do politics but Northern academics formulate 'new social movement theory' (Kapoor 2004: 633). For Spivak, the Third World is worlded on the basis of a theory/practice binary, which belies the fact that the production of theory is also a practice. In this practice, subaltern stories are not seen as sophisticated theory and, again, the subaltern voice is not heard.

Development 'knowledge' frequently travels from wealthier, more powerful countries in the North, it generally travels one way, and it travels as a solution rather than as a basis for learning (Mawdsley *et al.* 2002). This is not simply an epistemic divide, but a material and institutional one (see Jones 2000), which underpins an implicit tendency to view the South as a mix of countries that knowledge travels to rather than from. This is entrenched by the organization of knowledge production in the Euro-American academy. Euro-American norms continue to dominate the social sciences and knowledge production more broadly. This perpetuates the assumption that 'we' have nothing to learn from 'them' since the only kinds of knowledge that can be taken seriously by the Euro-American academy are those that conform to its particular formats of writing, citation and history. Moreover, Enlightenment principles still shape understandings of what counts as 'proper' knowledge production, with many specialisms required to 'detach morality and political interest from proper scholarly research' (Appadurai 2000: 14). This differs from many intellectual contexts in the South, where the role of the public intellectual has been central and where contributions are not necessarily conditioned by professional criteria of criticism, publication and dissemination. As McFarlane (2006) suggests, it is imperative not to insist that overseas scholarship conform to Euro-American precepts in order to be taken seriously and to develop ways of internationalizing social science research.

In terms of what this means for development studies, Maxwell and Riddell's (1998: 28) insights are useful:

> There is one route I think we should not take, which is that each of us should try to merge all our work into one, covering North and South . . . Instead, people who specialize on the North or South will continue to do so, but should make new efforts to learn from each other, to explore common problems brought on by convergence, and perhaps to develop new theory together. The best place to start might be with specific topics, like public works, food policy or participation – indeed, with the meaning and measurement of terms like 'poverty' and 'social exclusion'. This will enable collaboration to be built inductively, from the bottom-up.

At the core of such approaches is a spirit of openness to different conceptions, practices and modes of knowledge production, based in an ethic of respect and critical reflection. As Slater (1997: 648) argues:

> In a world increasingly configured by global connectivity, a strong case can be made for posing the significance of another three Rs – respect, recognition and reciprocity. If our geopolitical imagination in the field of knowledge is going to be open, nomadic, combinatory, critical and inquiring, it can displace the hold of Euro-Americanist thought and find ways of learning from the theoretical reflexivity of different writers and academics from other worlds and cultures . . . Mutual respect and recognition must include, if they are to be of any meaningful ethical value, the right to be critical and different on both sides of any cultural or intellectual border. Reciprocity and dialogue can only emerge if there is a will to go beyond indifference and historically sedimented pre-judgments; to engage in analytical conversations with each other in ways that can make the outside part of the inside and vice versa and has the potential to engender mutually beneficial encounters.

## Summary

- Postcolonial approaches engage critically with how development is disseminated and how certain development knowledge becomes hegemonic.

- A focus on the relationship between power knowledge reveals how one particular doctrine – neoliberalism – continues to dominate, reinforcing and exacerbating spaces of exclusion and disconnection which might be considered neocolonial.
- The IFIs that perpetuate neoliberalism have failed to achieve their targets of poverty alleviation and development. Critics believe that economic globalization has brought the planet to the brink of environmental catastrophe, unprecedented social unrest and an increase in poverty, hunger, landlessness, migration and social dislocation.
- The growing sense of a need to reform the Bretton Woods institutions implies challenging and confronting the normative and discursive nature of western-centric development. A new and different global architecture is required for producing and sharing knowledge about development; postcolonialism reveals how dominant discourses have prevented us from seeing alternatives.
- A postcolonial approach also reveals resistances to hegemonic forms of power: international and transnational resistance have raised important questions about the need to democratize development institutions, the nature of twenty-first century world development and the dissemination of development knowledge that in turn feeds into policy and intervention.
- International feminisms, the WSF and the Occupy movement have emerged as important alternative international 'assemblies' for the discussion and networking that is taking place between peoples and grassroots organizations around the world.
- There are powerful commonalities between anti-development activism and postcolonial theory (including postcolonial feminisms). Both share similar rejections of dominant development discourses and similar understandings of the problems in neoliberalism. Both contest the monopolization of knowledge about development by IFIs and insist that other ways of knowing are possible.

## Discussion questions

1 How do postcolonial approaches help us understand how development knowledge is disseminated?
2 How do postcolonial approaches help us understand how development knowledge is translated into development policy on the ground?

3   Why are the IFIs so powerful in determining development interventions and outcomes in the South?
4   Explain the significance of globalized resistance to neoliberalism.
5   Discuss the similarities in postcolonial, feminist and post-development critiques of development knowledge.
6   What alternatives does postcolonialism offer in rethinking development knowledge?

## Further reading

Goldman, M. (2005) *Imperial Nature. The World Bank and Struggles for Social Justice in the Age of Globalization* London and New Haven, Yale University Press. Path-breaking ethnographic study of the workings of the WB that critiques its propensity for exacerbating problems it seeks to solve, its ability to tame criticism and its dominance of development.

Hoogevelt, A. (2001) *Globalization and the Postcolonial World: The New Political Economy of Development* London, Palgrave. Accessible critique of the impacts of globalization on development in the South.

Kapoor, I. (2004) 'Hyper-self-reflexive development? Spivak on representing the Third World "Other"' *Third World Quarterly* 25, 4, pp. 627–47. Essay on the politics of representation and a helpful discussion of (mis)readings of Spivak's 'Can the subaltern speak?'.

Li, T. Murray (2007) *The Will to Improve. Governmentality, Development and the Practice of Politics* Durham, NC., Duke University Press. A brilliant analysis of the will to improve using ethnographic methods and drawing on research in Indonesia.

Mohanty, C.T. (1988) 'Under western eyes: feminist scholarship and colonial discourses' *Feminist Review* 30, pp. 61–88. Agenda-setting essay critiquing western feminism and representations of 'Third World' others.

Patel, R. and McMichael, P. (2004) 'Third Worldism and the lineages of global fascism: the regrouping of the global South in the neoliberal era' *Third World Quarterly* 25, 1, pp. 231–54. Interesting critique of neoliberalism that focuses on the biopolitics of development.

## Useful websites

www.actionaid.org ActionAid International.

www.bond.org.uk British Overseas NGOs for Development (BOND): List of 'Gleneagles +1' reports (see specifically www.bond.org.uk//campaign/gleneagles.htm).

https://www.gatesfoundation.org/ Bill and Melinda Gates Foundation, Global Development Program.

www.ictsd.org International Centre for Trade & Sustainable Development (ICTSD).

www.odi.org.uk/wto_portal/index.html Overseas Development Institute, WTO portal.

www.oxfam.org/en Oxfam International.

www.southcentre.org South Centre.

www.unicef.org/publications/ UNICEF publications.

www.viacampesina.org Via Campesina (International Peasants' Movement).

www.globaljustice.org.uk/ Global justice and World Development Movement.

https://fsm2016.org/en/ World Social Forum, Montreal 2016.

# 6  Agency in development

## Introduction

The history of development as both knowledge and a form of
intervention also implies North–South relationships constituted
through agency and passivity. The creators of knowledge and those
intervening are assumed to have agency; those receiving knowledge
and being 'developed' are assumed to be passive. As Linda Tuhiwai
Smith (2012: 1) argues, this is deeply problematic for those formerly
colonized people who are rendered objects of this knowledge and of
development interventions:

> It galls us that Western researchers and intellectuals can assume to
> know all that it is possible to know of us, on the basis of their brief
> encounters with some of us. It appalls us that the West can desire,
> extract and claim ownership of our ways of knowing, our imagery,
> the things we create and produce, and then simultaneously reject the
> people who created and developed those ideas and seek to deny them
> further opportunities to be creators of their own culture and own
> nations.

Chapter 3 traced some of the continuities and discontinuities between
the colonial and post-colonial eras, in particular the shift after
1945 towards humanitarian assistance. Of enduring significance to
development studies is the fact that the 'cures' for the 'ills' of recipient
countries in need of humanitarian assistance were first forged through
the experiences of colonization and imperialism. These forms of
humanitarianism, with their roots in nineteenth century imperialism,
have been dismissed by many postcolonial critics as little more than
a legitimating screen for imperial domination. This is not to deny
the fact that there were debates about the nature of humanitarianism
during imperialism and many who opposed it (Gandhi 1998: 27),
so any mapping of the idea of development during imperial times

needs to acknowledge these differences. What is important is that humanitarianism allowed those peoples who were seen as the object of humanitarian concern to be appropriated into the consciousness of the would-be benefactors. In other words, humanitarians in the North put themselves in positions to speak authoritatively about the sufferings of others in the South, who were deemed incapable of speaking for themselves.

This also encouraged the formulation of prescriptive principles aimed at relieving the suffering and making improvements to the lives of the sufferers. While we might think of this humanitarianism in positive terms, it is rooted in a priori assumptions about where agency and passivity reside. Positioning the humanitarian in the North as capable saviour invariably perpetuates notions of the incapable victim in the South. Moreover, as Lacquer (1989: 188) argues, this sense of responsibility for distant strangers may never be able to escape the imposition of 'epistemological sovereignty over [their] bodies and minds'. The preceding two chapters have demonstrated how this sovereignty has been established and perpetuated, specifically through development discourses and practice and also through biopolitics. As Power (2003: 138) argues, it is precisely the establishment of this sovereignty over the minds and bodies of distant others that can and must be challenged through postcolonial approaches to development. This chapter focuses more specifically on agency, resistance and possibilities for marginalized and subaltern peoples, who are often the subjects and objects of development (Paolini 1997), to be the agents of development.

A significant aspect of postcolonial theory that has emerged out of cultural and literary criticism is its 'celebration of the particular and the marginal', which envisages peoples of the South 'carving out independent identities in a de-Europeanized and hybrid space of recovery and autonomy' (Paolini 1997: 84). This emphasis on the marginal and on 'carving out' identities has important implications for development studies. It involves acknowledging that development is never experienced uniformly across the South. As Radcliffe (1999: 84) argues, 'cosmopolitan jet-setters in São Paulo live one kind of development while women in sub-Saharan Africa walking for hours to collect water experience a completely different kind of development'. The question is, how do we listen to and learn from this difference, and how do we acknowledge and understand the agency of people in these places in development thinking and practice? One important aspect of this involves creating space within the Northern academy and the spaces of development practice to listen to people living in the South whose voices are usually not heard, ignored or co-opted in development policy and practice. Decolonizing development would then go further to create the conditions to enable the self-determination of indigenous and subaltern peoples, acknowledging their rights to define development in their own terms.

The agency/passivity binary is also intertwined with discourses that rendered the South as the primitive, backward, stagnating, underdeveloped Other, in contrast to the dynamic, progressive, modernizing North. In order to challenge these notions, this chapter begins by illustrating that rather than being 'out there' and 'back there', the developing world is integral to 'modernity' and 'progress' in the North, contributing directly to the economic wealth of countries in the North through labour and resources. In addition, postcolonialism attempts to hear the voices of the marginalized, the oppressed and the dominated through a radical reconstruction of history and knowledge production that seeks to account for the agency and resistance of peoples subjugated by both colonialism and neocolonialism. This is relevant in challenging the notion of a single path to development and demanding acknowledgement of a diversity of perspectives and priorities.

The politics of defining and satisfying needs is a crucial dimension of current development thought, to which the concept of agency is central. Postcolonialism prioritizes questions such as: who voices the development concern, what power relations are played out, how do participants' identities and structural roles in local and global societies shape their priorities, and which voices are excluded as a result? The overlaps between postcolonial, grassroots development and participatory approaches, which attempt to overcome inequality by opening up spaces for the agency of people in global South contexts, are significant. However, poverty and unequal access to technology and other resources make this increasingly difficult. Therefore, the chapter concludes with an examination of the links between postcolonial approaches and new forms of agency and resistance.

## Breaking down binaries between worlds

Postcolonialism invokes an explicit critique of the spatial metaphors and temporality employed in development discourses. The latter have a tendency to designate countries of the South as both spatially and temporally distanced from Northern modernity and progress; the 'Third World' is both 'out there' and 'back there'. In contrast, postcolonial perspectives insist that the 'other' world has been and is 'in here' and that ideas about modernity and progress are not forged solely in the North. Assumptions of temporal and spatial distance are erroneous when one considers both the historical and contemporary connections between North and South. The two have always existed in relation to each other, connected through complex (rather than unidirectional) flows of people, ideas, commodities, languages and cultures. Scholars have pointed out that the origins of modernity in Europe are profoundly intertwined with the conquest of the Americas that began in 1492. The origins of European industrialization and global capitalism can be traced directly back to 1610 and the beginning of Colombian Exchange, which saw the enormous transfer of human populations, plants, animals, cultures, technology and ideas between the Americas and Europe that was triggered by colonialism, extraction and accumulation. Therefore, the South has been and is integral to 'modernity' and 'progress' in the North, contributing directly to the economic prosperity and development of Northern countries through labour and resources (Box 6.1). As discussed in Chapter 2, this agency is also apparent in the ways in which the modalities and aesthetics of the former colonies have partially constituted languages and cultures in the North.

It should also be acknowledged that Asian, African and Latin American modernities have long had trajectories that are not necessarily determined by the West. These modernities give shape to everyday life, provide diverse and distinctive ways of comprehending the world, fashion identities and influence responses to present-day conditions. For example, as Comaroff and Comaroff (2012: 118) argue, the African continent has generated iconic cultural forms such as popular Christianity, mass-mediated musical modes and cinematic genres such as the Nollywood straight-to-video movie industry. They claim that such creativity is 'at once productive and destructive in flouting, repudiating, remaking, European templates' (*ibid.*) and it need not be compared with Europe, since it has its own genius. Similar arguments could be made about Latin American and Asian creativity and cultural diffusion: for example, religious syncretism in Latin America, which predates European colonialism and provided a mechanism for absorbing forms of Christianity and African animism from the colonial period; or the diverse popular and fusion music in South East Asia countries that originates in traditional local musical models and incorporate aspects of global musical styles. These are not outcomes of alternative modernities, but vernaculars, just as Euro-modernity is a vernacular. In paying attention to these vernaculars as expressions of agency within modernity and development, postcolonialism attempts to re-write the hegemonic accounting of time (history) and the spatial distribution of knowledge (power) that constructs countries in the South as removed and separate from trajectories of development in the North.

Thinking about North and South in a relational way – drawing out the historical and contemporary connections that are constituted through flows of people, resources, cultures and so on and focusing on commonalities, connectedness and hybridity rather than difference and separateness – is important because it also points towards breaking down the associated binary of active (Northern) agent/passive (Southern) recipient. If the South has been and is active in Northern development, then it becomes more difficult to imagine development that is simply delivered to or imposed by the North on a passive South. For example, migrant labourers from the South have recently been depicted not only as crucial agents in Northern economies, but also in development in their own countries through their return of financial remittances to their families and communities.

A recent World Bank study (2016) reports that officially recorded remittances to the South were US$441 billion in 2015, a dramatic increase from US$31.2 billion in 1990 and US$167 billion in 2005. It also reports that the true size of remittances, including unrecorded flows through formal and informal channels, is believed to be significantly larger. Remittances constitute the second-largest capital flow behind FDI and are now three times the amount of Official Development Assistance (ODA) (which it overtook in 1995) and for some time have been the fastest growing type of financial transfer to countries of the global South (DfID 2006). In 2015, the top recipient countries of recorded remittances were India, China, the Philippines and Mexico. As a share of GDP, however, smaller countries such as Tajikistan (42 per cent), the Kyrgyz Republic (30 per cent), Nepal (29 per cent) and Tonga (28 per cent), were the largest recipients. High-income countries are the main source of remittances: the United States is by far the largest, with an estimated $56.3 billion in recorded outflows in 2014, followed by Saudi Arabia, Russia, Switzerland,

Germany, the United Arab Emirates and Kuwait. Some development scholars have seen this as a new, bottom-up way of financing development for countries in the South, with migrant labour the key driver in localized development, including providing for household economies, funds for schooling and so forth. Generally, there has been a focus on the positive dimensions of remittances and to celebrate migrants as active economic agents whose risk-taking altruism boosts economic growth in both the global North and South (Brown 2006). This is important, but it should not be forgotten, however, that while remittances challenge perceptions of the passivity of people in the South in development, this is neither unproblematic nor without debate.

---

## Box 6.1 The role of the South in Northern development: examples past and present

### Cadbury: UK philanthropy and African slavery

The story of Cadbury, one of the largest manufacturers in twenty-first-century Birmingham, is told at Cadbury World, part of the chocolate factory in the Bournville district of the city. Visitors are presented with the story of the long association of Cadbury with ethics and progressive social policy, the antislavery position of its Quaker founders and the building of the model village of Bournville, which housed company workers at the end of the nineteenth century in exemplary conditions. However, the darker side of the Cadbury story is not conveyed, specifically its association with slavery in west Africa at the beginning of the twentieth century. The philanthropic endeavours of the Cadbury family would not have been possible without the huge profits the company made at the end of the nineteenth century and, in turn, these profits would not have been possible without the company's complicity in slavery, which endured well beyond the formal abolition of European slavery.

Although Portugal had abolished slavery in all of its colonies by 1870, it allowed contract labour. African workers could sign agreements of their own free will by which they committed themselves to five years' labour at a set wage. In reality, the workers were coerced – they were, in all but name, slaves. Portugal continued to ship forced labourers from Angola to its islands of São Tomé and Príncipe, 150 miles off Africa's west coast. Repatriation was virtually impossible and the death rate was as high as 12 per cent. Forty thousand slaves produced the cocoa beans that Cadbury had been buying since 1886. Between 1901 and 1908, the two islands supplied half of Cadbury's cocoa beans. The British government turned a blind eye to Portuguese slavery because it was trying to recruit labourers from its African colonies to work in South Africa's gold mines. The Cadbury family first learned of slavery in Portuguese Africa in 1901 but, despite knowledge of the brutality and very high death rates on the islands, continued to purchase the products of slavery until the company was publicly exposed by journalists in 1909 (Satre 2005). Cadbury defended its reputation by claiming to be using quiet diplomacy to put pressure on Portugal, but coerced labour continued in Portuguese Africa until independence in 1975. Thus, the success of one of Britain's most famous manufacturers, one of the biggest employers in Birmingham, and one of the more socially progressive businesses at the beginning of the twentieth century, was incontrovertibly bound up with the exploitation and inhumane treatment of workers in Africa. The ability to provide exemplary housing for the families of workers at Bournville was directly related to the profits accumulated through the simultaneous exploitation of African workers who provided the cocoa.

## Africa, India and the British economy

Between 1990 and 1997, the number of people coming into the nursing profession in Britain fell from 18,980 to just over 12,000. Nurses recruited from abroad accounted for 26 per cent of the 16,000 nurses registered in 1997, and five years later that figure had grown to 43 per cent of the registered total of 37,000. Many of these nurses migrated from the Philippines, South Africa and India. This trend of filling staffing gaps in the UK's National Health Service with overseas workers has continued. According the UK government, in 2017, over 18,000 members of staff in the NHS are from India, over 15,000 are from the Philippines and over 11,000 are from African countries (principally Nigeria, Zimbabwe and Ghana).

(http://researchbriefings.parliament.uk/ResearchBriefing/Summary/CBP-7783)

Recruiting from the global South was controversial, since poorer countries were investing resources in training professionals who then left for the relatively higher salaries in the UK. The UK government imposed stricter criteria, but during the 1990s a significant number of migrant workers from the South were effectively propping up the National Health Service and, despite stricter immigration rules, they are likely to continue to play an important role when the UK exits the European Union. In 2014, 14 per cent of professionally qualified clinical staff and 26 per cent of doctors were non-British. The 2016 referendum decision to leave the EU has caused a 90 per cent reduction in nurses applying to work in the UK from EU countries. In response, in June 2016, the government relaxed an NHS language test for overseas workers in attempt to deal with the resulting staffing crisis. The delivery of a universal system of high quality health care is effectively dependent on trained migrant workers, many of whom are from the global South, because of a skills deficit in the UK economy (http://society.guardian. co.uk/health/comment/0,7894,1545536,00.html; see also Mackintosh *et al.* 2006). Meanwhile, most foreign computer workers entering the UK in 2005 were immigrants from India. More than eight out of ten of the 22,000 overseas IT workers came from India, suggesting that multinational companies are recruiting staff in low cost countries to fill skills gaps in the high cost UK market. Despite increasingly anti-immigration government policies, this trend has continued and, in 2016, Indian nationals accounted for 57 per cent of the total skilled work visas granted (53,575 of 93,244). The highest number of these remain in the IT sector (42 per cent), followed by professional, scientific and technical activities (19 per cent). Ironically, low-skilled IT jobs (such as call-centre work) continue to be shipped to India by the UK service sector (see Kofman and Raghuram 2006)

## Mexico and the US economy

The US economy has become increasingly reliant on migrant labour (legal and illegal), particularly from Mexico. The Mexican foreign-born population grew by 104 per cent during the 1990s, from 4.3 million to 8.8 million persons overall and is currently over 11 million. As the Mexican foreign-born population grew in the 1990s, it became an increasingly important part of the US labour force. While Mexican immigrants were 2 per cent of the US labour force in 1990, by 2000 this had nearly doubled, with Mexican immigrants accounting for 4 per cent of the US workforce. In crop agriculture, fruit and vegetable manufacturing, animal slaughter, landscaping and car servicing Mexican workers were more than one-fifth of the entire workforce. In 2002, the American Immigration Law Foundation reported the following findings (Paral 2002):

- *Mexican workers are integral to US economic growth.* While other immigrant groups also perform essential worker jobs, the size of the Mexican population makes its impact on the US economy more quantifiable.
- *New jobs have not required advanced education.* [Nearly 43 per cent of all job openings by 2010 required only a minimal education, at a time when native-born Americans are obtaining college degrees in record numbers and are unlikely to accept positions requiring minimal education.]

The low-wage sectors of the US economy (agriculture and services) are thus heavily reliant upon migrant labour from Mexico. There has long been a widespread debate in the USA about the significance of economic migrants to its economy, particularly in low-wage, labour-intensive sectors, and their status in relation to US citizenship. In March 2006, proposed legislation that would make it a felony to be in the United States without proper papers and a federal crime to aid illegal immigrants provoked massive street demonstrations that brought major cities to a standstill, with marches of over 500,000 in Los Angeles and 300,000 in Chicago. Many carried banners and wore T-shirts declaring 'We Are America' (see http://www.nytimes.com/2006/03/27/national/27immig.html). That many illegal immigrants pay taxes and are currently serving in the US armed forces in Iraq and Afghanistan intensified the debate.

A motivating factor behind President Trump's current proposed policies – including the construction of a new US–Mexico border wall, more border patrol agents, and stricter deportation policies – is a belief that immigrants are stealing job opportunities from American workers. As Trump said in July 2015, 'They're taking our jobs. They're taking our manufacturing jobs. They're taking our money. They're killing us'. However, research in August 2017 by the Brookings Institute revealed:

- The impact of immigrant labour on the wages of US native-born workers is low and undocumented workers often work the unpleasant, back-breaking jobs that native-born workers are not willing to do.

- Immigration is tied to positive economic growth and innovation in the US, and is especially important for areas that are experiencing a decline in domestic migration and as the nation's population gets older and fertility remains low.

https://www.brookings.edu/blog/brookings-now/2017/08/24/do-immigrants-steal-jobs-from-american-workers/

Some scholars view remittances as manifestations of global inequality (de Haas 2005; Wimaladharma *et al*. 2004). On the basis of evidence from low-paid migrant workers in London, Datta *et al*. (2007) challenge the notion that remittances represent a panacea with regard to development funding, and a win-win situation for all concerned. They highlight three problems with this notion:

- The link between migration and poverty reduction is misplaced, given that migration rarely involves the poorest sectors of society and is much more likely to occur from middle-income countries such as Mexico or India.
- The costs and drawbacks of viewing remittances as the new development mantra from the perspective of migrants are much less explicitly acknowledged. While there is some recognition of the costs at the global and national levels, the costs to migrants themselves in terms of the sacrifices they have to make are largely ignored (Datta *et al*. 2007: 47; Pratt 2012). Remittances are invariably generated by underpaid, exploited and often excluded migrants. These sacrifices are both economic and emotional in nature, with many migrants underestimating the costs of living in the North and the difficulties involved in being separated from family and friends. The growing anti-immigrant sentiments in advanced economies, which ignore the economic benefits of migration and have fuelled populist responses such as Trump's wall, the UK's exit from the EU and its government's creation of a 'hostile environment' for non-UK citizens, can also make migrant work an unpleasant experience. And the appalling working conditions of migrant workers in Gulf countries are currently under increased scrutiny, with attention focused on Qatar and its abuse and exploitation on migrant workers from Bangladesh, India and Nepal in its preparations to host the 2022 FIFA World Cup (Amnesty International 2016).

- However attractive policies to promote remittances may appear to national governments in the North and South or to IFIs, the reality is that they absolve them of the responsibility to develop policies that address the structural inequalities that lead to South–North migration in the first place. It also ignores the fact (as the cases of Mexican migrants in the USA and IT workers from India in the UK illustrate) that the labour markets of the North are often functioning only because migrants are doing the jobs that natives often refuse or do not have the skills capacity to do.

Therefore, while migrants are acknowledged as economic agents, it is important not to celebrate uncritically their role as the generators of remittances and the impetus behind development from below. Many of these migrants toil for low wages in labour markets that often discriminate against them and live in societies that invariably exclude them (Datta *et al.* 2007). While they may perceive themselves as the 'lucky' ones or be perceived as such by those who remain at home, this should not erase the reality that these migrants are parts of the engine that is driving an inequitable global system and an unethical shift in global development policies.

What migrant workers do provide, however, is evidence of agency in development from within countries of the South. The South is not removed, separate or disconnected, either temporally or spatially, but is integral to development in the North, with migrant labourers also fuelling development in the South through remittances. Where development discourses of the 1950s and 1960s sought to force the South into a map of 'globally normative patterns' (Power 2003: 139), postcolonialism emerges to disrupt conventional geographies with accounts of 'contact zones' (Pratt 1992), of North–South 'interaction' and to focus instead on the hybridity (mixing) of cultures and identities. Postcolonialism thus has much in common with indigenous politics and wider mobilizations by subaltern peoples, and the challenges these currently pose to notions of development, particularly in former settler colonies such as Canada, the USA, Australia and New Zealand.

There is a very long history of resistance by African-descended and indigenous peoples against colonialism, racism and exploitation, but in recent years these mobilizations have become more visible and politically effective. Indigenous and subaltern peoples have protested and occupied, created their own institutions, engaged with the UN, the WB and NGOs and created their own media to demand land rights, political and cultural recognition, autonomy, environmental protection and rights to language, knowledge and culture practices (Cupples and Glynn 2014). As Larsen (2006: 312) argues, indigenous territories 'are contested spaces bound up in (post)colonial ideologies of otherness, while also serving as sites of representation for alternative agendas of cultural and economic production'. Indigenous and subaltern politics around the world has been important in reconceptualizing territories as 'flexible, fluid spaces where all varieties of unexpected political alliances, cultural productions and resource-management strategies can be spawned' (Larsen 2006: 312). The decolonizing mobilizations emerging in these spaces are challenging dominant understandings of development, changing the terrain in which development is conducted, asserting alternative ways of knowing and being in the world, fashioning effective political resistance and defining alternative indigenous projects for development. In particular, they have brought dominant understandings of time and space into question. Economic restructuring in former industrialized countries has destabilized the conventional metaphorical role of the 'future' in modern, meta-narratives of social and economic progress (Harvey 1990). Several authors have recounted the deep disillusionment within indigenous politics with the idea of future improvement through industrialization. Consequently, alternative understandings of non-industrial values for land and resources based on historical practices are being advanced (Hayter 2003), which may produce alternative trajectories of development within indigenous territories (Larsen 2006). This in turn forces us to rethink the very spatiality of development and development discourse, and to reconsider the meaning and theorization of the relationship between peoples and places.

## Problematizing passivity and partnership

Postcolonial approaches have begun to provide greater insights into the Eurocentric nature of much of development policy and praxis and to highlight problematic representations of the passivity of people in the South. However, few critiques have explored detailed case studies, particularly on how discourses of passivity shape policy in development agencies. One exception is Maria Eriksson Baaz (2005), who examines the discourses, identities and politics of Scandinavian donor agents in Tanzania. Baaz's interviews with development workers reveal how their identities are constructed in opposition to an African 'Other'. Neither identity is completely stable – the African 'Other' is sometimes depicted in romantic terms, but more often is seen through a derogatory lens as passive, corrupt and dangerous. Of critical importance is that the issue of identity underpins how development aid is planned and negotiated. Baaz's analysis is at the level of NGOs and donor agencies and adds empirical weight to the more abstract arguments about discourses of development outlined in Chapter 4. In particular, she demonstrates that:

> The image of an open, trustworthy, organized and committed Danish development worker Self in opposition to an implicit (and . . . explicit) image of the Tanzanian partner as unreliable, uncommitted and disorganized . . . cannot be understood simply as insignificant words describing the Self and Other. These meanings are manifested in development practice.
>
> (Baaz 2005: 2)

Despite this, when understood through Bhabha's notion of hybridity (see Chapter 2), development interventions cannot be conceptualized as a simple mimetic process whereby powerless 'receivers' adopt Northern values. The operation of power and the dominance of the North within the development industry cannot be neglected. However, recognizing hegemonic structures and power inequalities also allows for recognition of hybridity in which Northern values are mediated, and oftentimes challenged and remoulded by actors in the South, in the constant creation of new diversities.

While hybridity exists in practice in relations between North and South, stereotypes around a 'passive/active' dichotomy continue to play a particularly vital role in the formation of identities in the context of development aid. Passivity and lack of responsibility are associated with the partner in the South; activity and responsibility define the donor and the Northern development worker. This perceived opposition is held up as one reason for the apparent development 'gap' between North and South. As discussed, this has a long history dating back to the colonial period and remains particularly dominant in the post-colonial view of Africa in the media and popular cultural representations. In the context of the development aid relationship, unequal power and co-operation is characterized by conflicting goals; where partners resist this is perceived as non-commitment and passivity. However, as Baaz (2005: 121) suggests, images of a passive Other and active Self must also be understood in relation to the idea of aid-dependence, which occupies a central role in Northern discourses of partnership.

Within this discourse, the partners in the South are presented as spoilt by aid and as people who prefer things delivered to them rather having to work for them. The idea of aid dependency in development discourses remains powerful, to such an extent that when development partners in the South demand economic support, refuse to participate in meetings without being paid an allowance, or refuse to contribute with labour or funding, their actions are interpreted through the lens of passivity and dependence (Baaz 2005: 125). Running alongside these representations are stereotypes of cultural differences, which depict partners in the South as corrupt, unreliable and untrustworthy in contrast to the transparency, reliability and trustworthiness of the Northern development worker. In practice, as Baaz argues, these discourses create a relationship of mistrust.

Power (2003) poses this as a question of trusteeship, arguing that the colonial and neo-colonial notion of trusteeship has been replaced by an equally problematic (as Baaz's case study testifies) discourse of partnership. As discussed in Chapter 3, the notion of trusteeship has underpinned development from the Enlightenment period. Under colonial administrations, trusteeship revolved around the mission to civilize Others, to strengthen the weak and to give experience to 'childlike' colonial peoples who required supervision and guidance. Agency resided with the trustees; the childlike Other remained passive within colonial notions of development. These notions were reinvented after 1945, wherein the UN and the IFIs became the new trustees who aimed to promote the welfare of 'natives' and 'advance' them towards self-government and development. More recently, the language of trusteeship has largely been abandoned in favour of discourses of partnership, precisely because of colonial connotations, but it reappears implicitly in much of postwar international development:

> The distinguishing feature of the late twentieth century question of development is that trusteeship, the integral of the nineteenth century doctrine of development, has been renounced as the source of action toward development. From the immediate historical experience of formal imperialism and its end, it is easy to see why trusteeship should be renounced when the goals of development for post-colonial Third World states are explicated by people who are neither of the state nor the Third World.
>
> (Cowen and Shenton 1996: 446)

At the heart of this is the fact that international development is still very much something that is defined and enunciated by the North:

> Just as in colonial times, the frameworks and strategies of development are authored outside the country concerned and are grounded in foreign (neoliberal) ideologies. Just as colonial states sought to govern their territories and administer their peoples, so the IFI's agenda of 'good governance' today seeks to discipline the realm of local politics, to prescribe the kinds of political change that are possible in the South.
>
> (Power 2003: 132)

The language of partnership is thus a means by which development agencies attempt to counter accusations of neocolonialism (Cooke and Kothari 2001), but whose development policies are still problematically defined by similar relationships of power. Until recently, this was central to the UK government's development strategies after 1997 (Department for International Development 1997, 2000a, 2000b). However, the 2015 White Paper on International Development (HM Treasury/DfID 2015) dispensed altogether with the language of partnership and instead repeatedly declares the UK to be driving global development initiatives in the UK's national interest (Noxolo 2017: 2). Rather than being concerned with tackling poverty and building long-term partnerships with developing countries, the 'national interest' (which translates into a focus on security, crisis and emergency) provides justification for the diversion of UK aid spending away from the jurisdiction of DfID and into the jurisdiction of the National Security Council. As Noxolo (2017: 2) argues, UK government development policy is remarkable in that it 'absents people in the Global South almost entirely' to focus on 'challenges' for the international community and 'investment opportunities' for the UK economy. Thus, as discussed in Chapter 5, when former Secretary of State for International Development, Priti Patel, referred in 2016 to a 'strategic partnership' with India, (www.gov.uk/government/news/priti-patel-hails-strategic-partnership-between-the-uk-and-india), national interest was to the fore. When subject to scrutiny, these policies expose the problems inherent within discourses of partnership and the ways in which they simultaneously mask and perpetuate the active/passive binary within development. They also mask the 'disturbingly colonial' shifts in UK government policy (Noxolo 2017: 2).

A different example of the problems with partnership is the New Economic Partnership for Africa's Development (NEPAD), a partnership of 54 African countries formed as part of the African Union. This is an initiative formulated by African states that seeks to build on local and global interdependencies by acknowledging the importance of partnership and co-operation in international development. The then South African president, Thabo Mbeki, was a major driver of the initiative, requesting an annual commitment of US$64 billion in aid, loans and investments at the June 2002 G8 summit. NEPAD aims to ensure Africa ends its 'marginalization' from international capitalism and seeks to develop a 'new framework of interaction with the rest of the world . . . based on the agenda set by African peoples through their own initiatives and of their own volition, to shape their own destiny' (NEPAD base document, cited in Bond 2002: 1). While this appears to be a partnership with terms set by African states and driven by African agendas, critics argue that this is far from being a locally authored strategy. It gives the impression that it reflects and arises from Africa's rich traditions of political struggles, but in reality, according to critics, NEPAD is driven by the interests of the North, and is very much based on an unequal partnership in which the North still has the power to determine development trajectories in the South. Consequently, many African intellectuals and social movements have been united in opposition.

Patrick Bond (2002: 1) is one critic of NEPAD who has been vocal in his opposition:

> [I]s it a sell-out of Africa's legitimate aspirations for social, environmental and economic justice? . . . Has anybody or have any organisations aside from a few ruling elites and their international capitalist allies and backroom technocrats been party to its authorship? . . .
> 'We do not want the old partnership of a rider and a horse', Mbeki insisted in mid-June when Libyan leader Muammar Gaddafi criticised NEPAD for its obeisance to 'former colonisers and racists'. . . .
> Africa's progressive movements and intellectuals are uniting in anger mainly because Mbeki's plan surrenders so much terrain to the international structural power relationships which are responsible for Africa's last quarter-century of social dislocation, economic austerity and deindustrialisation, ecological degradation and state fragmentation.

NEPAD has attracted numerous other criticisms, which Bond aptly summarizes:

● It evolved 'under conditions of smoke-filled-room secrecy', driven by the then US President Clinton and UK Prime Minister Blair, the G8, the IFIs and international capital and excludes civil society organizations. As a result, the plan denies the rich contributions of African social struggles in its genesis. Instead, it empowers 'transnational corporations, Northern donor agency technocrats, Washington financial agencies, Geneva trade bureaucrats, Machiavellian Pretoria geopoliticians and Johannesburg capitalists, in a coy mix of imperialism and South African subimperialism' (Bond 2002: 1).

- It is based on the promotion of failed neoliberalism and free market economic policies (especially the 'Washington Consensus' model of economic development rejected by civil society organization). The logic is: Africa is marginalized within the globalization process, which constitutes a serious threat to global stability, so greater globalization is required. In contrast, leftist and women's movements in Africa argue that the continent's continued poverty is a direct outcome of excessive globalization, because of the 'drain from ever declining prices of raw materials (Africa's main exports), crippling debt repayments and profit repatriation to transnational corporations' (Bond 2002: 1).
- It advocates privatization, especially of infrastructure such as water, electricity, telecoms and transport, which will fail because of the insufficient buying power of most African consumers.
- More insertion of Africa into the world economy will worsen fast-declining terms of trade, given that African countries are reliant on cash-crop and minerals production whose global markets are glutted.
- Grand visions of information and communications technology are hopelessly unrealistic considering the lack of simple reliable electricity across the continent.
- Instead of promoting full and immediate debt cancellation, the NEPAD strategy is to support the WB's Poverty Reduction Strategy approach and the Highly Indebted Poor Country (HIPC) debt relief initiative. Yet even the WB concedes its HIPC plan has failed to make Africa's foreign debt 'sustainable':

Malawi's worsening starvation, due to a famine amplified when the country's grain stocks were sold thanks to International Monetary Fund 'advice' to first repay commercial bankers, is emblematic, and so extreme that even that wretched country's leaders are publicly blaming the IMF. When it comes to other financial flows, speculative 'hot-money' investments in emerging markets such as South Africa have harmed not helped the vast majority. And most foreign loans over the past thirty years have detracted from local capital accumulation, because they have allied corrupt African state elites with foreign bankers who drain the continent by facilitating capital flight. NEPAD calls for more of each . . .

(Bond 2002: 1)

- It appeals to all the peoples of Africa to 'become aware of the seriousness of the situation and the need to mobilize themselves in order to put an end to further marginalization of the continent and ensure its development by bridging the gap with the developed countries'. Bond sees this as hypocrisy, since elite rulers in Africa are the only ones to have benefited from globalization and live in luxury (at great distance from the masses). When progressive Africans do attempt to 'mobilize', they are nearly always met with repression.
- It is embedded in Northern free market ideology, but the solution to Africa's problems is more likely to be found in Africa's own cultures and values, for example, the South African Freedom Charter (1955) and Reconstruction and Development Programme (1994), which have instead been superseded by neoliberalism. Ironically, NEPAD does not acknowledge the potential significance of African agency in development knowledge and policy.

While NEPAD has received some acceptance from those initially very critical, and has in general become less controversial as its programs have become more concrete, it has also been criticized by some of its initial backers. For example, Senegalese President Abdoulaye Wade accused NEPAD of wasting hundreds of millions of dollars and achieving nothing. It remains a relatively poorly resourced framework and, while the aim of promoting greater regional integration and trade among African states is welcomed, the neoliberal macroeconomic principles NEPAD endorses remain contested.

Partnership was also central to the UN's Millennium Development Goals (MDGs): Goal 8 aimed to develop a global partnership for development. In a 2012 report on the progress with MDG8, however, the then Secretary-General, BanKi-moon highlighted significant problems in meeting this goal:

> [T]he Task Force has had difficulty identifying areas of significant new progress and for the first time there are signs of backsliding ... The waning of support for the global partnership for development

may be understandable in the context of a protracted economic and
financial crisis.

(United Nations 2012: ix)

The nature of the 'partnership' envisaged by the MDGs reveals
the operation of power relations within international development.
For example, Target 8.A: 'Develop further an open, rule-based,
predictable, non-discriminatory trading and financial system' indicates
that 'partnership' in the MDGs equated with liberalism, which in
turn meant submission to economic rules already established by
economically powerful countries in the North; Target 8.B: 'Address
the special needs of the least developed countries' with more generous
ODA could be said to represent yet more empty promises, given
that over 50 years of similar promises prior to the MDGs did little to
address the needs of such countries; Target 8.E: 'In cooperation with
pharmaceutical companies, provide access to affordable essential
drugs in developing countries' illustrates that 'partnership' with
corporations that enjoy an abusive monopoly was prioritized; Target
8.F: 'In cooperation with the private sector, make available the benefits
of new technologies, especially information & communications'
establishes that partnership was subject to the condition that the private
sector benefits. Samir Amin (2006) referred to this 'partnership' as
'doctrinaire neoliberalism'. He accused it of advocating extreme
privatization aimed at opening new fields for the expansion of capital,
the private appropriation of agricultural land, corporate advantage
through maximum deregulation, uncontrolled opening up of capital
movement and the forbidding in principle of states from interfering in
economic affairs. None of this historically has worked to the advantage
of poorer countries and, as critics like Amin argued, partnership in
these terms resembled neo-colonialism.

Despite criticisms, partnership remains a key goal of the successor
to the MDGs – the Sustainable Development Goals (SDGs) – and
thus of international development policy up to 2030 (Figure 6.1).
For example, Goal 17 of the SDGs aims to 'Strengthen the means of
implementation and revitalize the global partnership for sustainable
development'. It states that the SDGs 'can only be realized with a
strong commitment to global partnership and cooperation' (www.
undp.org/content/undp/en/home/sustainable-development-goals/goal-
17-partnerships-for-the-goals.html). This is intended to be a more
ambitious version of MDG-8, but arguably suffers from the same key
defect. The question remains: what kind of new Global Partnership

**Figure 6.1** *The Sustainable Development Goals*

United Nations icon

model is it advocating? The Financing For Development (FFD) SDG document stresses a more diverse range of actors and sources of cash than the previous reliance on aid: Foreign Direct Investment, domestic resource mobilization, ODA, philanthropy and South-South cooperation. However, it is short on concrete commitments, least of all from the OECD countries. As Pogge and Sengupta (2016: 89) argue, 'the world's most powerful agents – affluent states, international organizations, multinational enterprises – are once again shielded from any concrete responsibilities for achieving the SDGs'. They argue that given their wealth and influence, these agents ought to be taking the lead in providing the resources needed and in implementing systemic institutional reforms to address the root causes of poverty. They also argue that the targets for SDG17:

> should have specified the concrete responsibilities of the affluent states ... If the world's most influential agents had been held sufficiently accountable for what they owe toward making sustainable development work, the concepts of partnership and universalism would have been more meaningful, rather than what they are now likely to become: a smokescreen for extreme global inequalities.
>
> (*Ibid.*: 89–90)

As Pieterse (2001a: 17) argues, partnership – such as that outlined in
NEPAD and the UN MDGs and SDGs – serves an ideological role
in the neoliberal policy framework and 'papers over contradictions
and the roll-back of government'. It makes naïve assumptions about
the power and power relations that exist between partners. Baaz
(2005) illustrates that these partnerships are never equal. In essence,
partnership is often fancy window dressing, but does not disguise the
fact that creditors still tell recipient nations how their development
should proceed and requires outside approval and endorsement.
Agency still resides in the North. As Maxwell and Riddell (1998:
264) argue in their criticism of DFID's prioritization of business and
commercial involvement in development partnerships:

> We know how best to achieve development . . . we know how you
> should alleviate poverty . . . either you accept the approaches we think
> are right for you or you will not qualify for a long-term partnership
> with us . . . if you do not accept our view of development then we will
> not provide you with aid.

These are just a few examples of how governments and development
agencies in the North have not really moved away from the long
history of unequal relations of power and representation that
characterize conditionality debates. Partnerships are generally
'intrinsically one-sided' (Slater and Bell 2002: 346) and often work
to the detriment of countries in the South. There is little to suggest
that the commitment to 'partnership' in the SDGs will be any
different. Furthermore, despite a more concerted attempt to include
representation from poorer countries in the drafting and agreement
of the SDGs, there is still an enduring refusal in international
development policy to learn from the global South. For example,
Fan and Polman (2014) argue that the SDGs are not ambitious
enough because they fail to acknowledge the real progress being
made in developing countries and what might be learned from this.
There is an urgent need for greater emphasis on eliminating hunger
and under-nutrition and, if organizations like the UN were willing
to learn from the experiences of countries China, Vietnam, Brazil
and Thailand, this could be achieved earlier than the SDG target
of 2030. Fan and Polman suggest three pathways to achieving this
goal. The first, which has been successful in China and Vietnam,
is agriculture-led development with market reforms, but sustained
investment in human development. The second, modelled on Brazil,
is social protection where macro-economic reforms parallel pro-poor

social spending. The third, adopted by Thailand, is a combination of agriculture-led development, market and macro-economic reforms, with investment in human development and pro-poor spending. According to Fan and Polman, all four countries are on track to eliminate hunger by 2025, which puts them well ahead of the UN SDG target. Despite this, initiatives like NEPAD and the SDGs continue to advocate free market, neoliberal solutions to global development problems, despite there being little evidence of these working and plenty of evidence to suggest they may, in fact, worsen inequality. This suggests both a poverty of imagination at the core of international development and wilful neglect of development knowledge and agency in the global South.

## Poverty of imagination and erasure of alternatives

The failure to acknowledge that agency within development might also reside in the South not only reinforces the hegemony of Northern theories, models, practices and policies, but it also closes off the possibility that there might be alternatives. One effect of this has been to continue to privilege Northern 'expertise' in responding to humanitarian crises, despite the fact that local knowledge and practices are often as, if not more, effective (see Box 6.2). Another effect has been that much of the South followed the Washington consensus, embracing policies that have been directed at a 'quixotic quest for higher GDP' (Stiglitz 2006). A further effect has been the neglect of alternative models of economic development, despite these often being more successful than neoliberal models in other parts of the world. The strongly Confucian model (based on a notion of human morality and good deeds) adopted in South Korea, for example, saw it become one of the rapidly industrializing Asian 'tigers' during the 1960s. The UNDP (1996) held up South Korea as one of only a few countries to have developed in an equitable manner, maintaining the most equitable income distribution of any economy in the world. However, while this might be of interest to economists, it is given little relevance within development economics beyond the South East Asian context since this was seen as fundamentally rooted in Korean culture. Thus, while neoliberalism, which is rooted in western cultural values, travels because through the hegemony of western knowledge it can be claimed as universal, the Korean model remained a localized, particularized model for success. It follows that adherents to the Washington consensus had little to learn from Korea because its economic progress was inherently tied to its cultural values – notions of an orderly and hierarchical society, and the significance of family, equity and reciprocity that underpin a profit-sharing model of capitalism. Agency within development thus remains with those institutions, organizations and governments in the North that advocate and universalize neoliberal models of development.

---

## Box 6.2 'People's Science' and dealing with Ebola in West Africa

As discussed in Chapter 4, an outbreak of the Ebola virus in Guinea, Liberia and Sierra Leone in 2014 prompted near hysteria in the international community and a new Live Aid campaign, with experts predicting that millions would be infected within months. The mounting panic in Europe and North America saw the reassertion of familiar tropes about West Africa: Africans were unwilling to break with traditional customs and irrational superstitions that were making the spread of the disease worse. Funeral practices that insisted on the ritual washing of the deceased were threatening the region with an epidemic of biblical proportions. However, as Paul Richards (2016) argues, while hysteria broke out in the global North, West Africans were quietly enacting calm, considered and deeply rational responses in affected communities, which meant that the doom-laden predictions failed to materialize. Richards suggests that money, technology, drugs, vaccines and robot nurses mattered much less in bringing the disease under control than indigenous knowledges and the role of local chiefs. Ebola declined in Sierra Leone's rural south-east much earlier than in the north-west, despite receiving less aid and technical assistance. Here, far from being in a fatal grip of superstition and tradition, villagers recognized the risks of their funeral practices and adapted them accordingly, while also respecting the sacred dimensions of funeral practices. In contrast, the response by the World Health Organization was heavy handed in ordering 'safe' burials that were ignorant of local expertise and common sense, and insensitive to these practices, which generated friction with local chiefs and governments. Richards demonstrates that the humanitarian response to Ebola was most effective in those areas where it supported local initiatives and that it obstructed recovery when it ignored or disregarded local knowledge. There are wider lessons for development in cases such as this.

---

The story of Chinese development is similarly underplayed within development studies and it is only recently that economists have begun to argue that dominant Northern models might have something to learn from the Chinese example. As we saw in Chapter 4, popular and academic representations of China are often negative, depicting China as a threat to global economic security and focusing on human rights abuses, environmental destruction and a dictatorial political regime. There are undoubtedly many serious challenges facing China in all these areas (as there are similar challenges in many Northern contexts) but, significantly, China has recently achieved the world's largest reduction in poverty (around 500 million people have been lifted out of extreme poverty in less than 30 years). As we saw in Chapter 5, the success of the MDG to halve extreme poverty rests almost entirely on China's achievements. While some would claim this has been achieved because China has liberalized its economy and embraced globalization, it could be argued that it is China's understanding that market economies cannot be left unfettered and unregulated (in contrast to neoliberal models) that has brought about this transformation.

China recently adopted its 12th and 13th five-year plans (2011–2020), setting the stage for the continuation of one of the most remarkable economic transformations in history, while improving the wellbeing of almost a quarter of the world's population. As Stiglitz (2006) argues, 'never before has the world seen such sustained growth; never before has there been so much poverty reduction'. Stiglitz suggests that the key to China's enduring success has been its almost unique combination of pragmatism and vision. Unlike much of the rest of the developing world that has followed the Washington consensus and neoliberal models of development, China has sought sustainable and more equitable increases in real living standards. The government realizes that it has entered a phase of economic growth that places enormous and unsustainable demands on the environment, which will eventually compromise living standards. Thus, the plans since 2006 have each placed great emphasis on the environment.

Rural economies in China are growing rapidly and the pace of urban development is exceptional. While this has reduced poverty, inequality has been increasing throughout the twenty-first century, with growing disparities between cities and rural areas, and coastal regions and the interior. The 2006 WB *World Development Report* acknowledged that inequality, not just poverty, should be a concern. China's successive plans have set out agendas for tackling the issue head on. Central to the 2006–2011 plan were programmes for achieving a more equitable and harmonious society. This included eradicating extreme urban poverty through the payment of a government subsidy (*dibao* – a monthly minimum income guarantee) to urban dwellers and beginning to tackle rural poverty through the provision of pensions. China also recognized that future growth would need to be based more on domestic demand than on exports, requiring increases in domestic consumption. Strengthening social security, especially pensions and social insurance programmes, and public health and education aimed to reduce social inequalities, increasing citizens' sense of wellbeing, and promoting consumption by increasing consumer confidence and decreasing the tendency to prioritize savings over spending (Stiglitz 2006). The 2011–2020 plans focus specifically on sustainable and equitable development with the aim of addressing rising inequality and more sustainable growth. China recognizes that equitable wealth distribution is required for increased consumption, and the need for social infrastructure and social safety nets. Its policies are aimed at bridging welfare gaps between countryside and cities and encouraging people to share the fruits of economic growth. Its aim is to become a 'moderately prosperous society'. This rests on President Xi Jinping's promise of January 2018 to lift a further 43 million people out of extreme poverty within three years. The focus is now on rural areas, mainly in the west and south of the country, where the focus is on poor individuals, and on drawing up specific plans tailored to individuals, rather than simply helping poor places to develop in the hope that wealth will trickle down to the poorest. This will not eradicate poverty in China, nor will it necessarily reduce the gap between richest and poorest, but it will likely eradicate extreme poverty as defined by the Chinese state.

China's state-driven and state-managed path to development allows it to at least plan to mitigate against the polarizing forces of industrialization, marketization and globalization. In addition, while concerns are expressed about the lack of democracy and human rights, some economists now acknowledge that China's more centralized governance structure gives it an ability to act in the face of growing inequality and global environmental challenges where the decentralized governance structures in Northern countries render them virtually powerless to act. In the case of greenhouse gases, Stiglitz (2006) suggests that while the USA says that it cannot afford to do anything, China's senior officials have acted more responsibly. The state has imposed new environmental taxes on cars, petrol and wood products, using market-based mechanisms to address its and the world's environmental problems. It remains to be seen if pro-growth arguments will be advanced to counter strong social and environmental policies, which would not only fail to deliver growth but would threaten China's vision of a harmonious future. Stiglitz suggests that the only way to prevent this is open discussion of economic policies to expose fallacies and provide scope for creative solutions to the challenges facing China. He also argues that 'most people outside China do not fully appreciate the extent to which its leaders . . . have engaged in extensive deliberations and consultations as they strive to solve the enormous problems they face', and contrasts this to some western democracies, where decision-making is surrounded by 'excessive secrecy' and confined to 'a narrow circle of sycophants'.

China is significant because, in contrast to neoliberal models, its development is not premised on the notion that market economies are self-regulating; rather, they need to be actively managed to ensure that their benefits are shared widely. The current Chinese model thus attempts a difficult 'balancing act that must constantly respond to economic changes' (Stiglitz 2006). It also demonstrates that alternatives to dominant modes of neoliberalism are possible and, in contexts outside of the North, might be more effective in both delivering economic development through marketization and managing the detrimental impacts on societies and environments. The Chinese model is by no means without its problems, particularly concerning human rights (there have also been problems of corruption in the *dibao* system), but it does highlight the existence of agency outside of the North and points to the possibility of alternative models emerging from developing world contexts.

## Shifting power in international development?

China's emergence as a global economic power has also coincided with a decline in the global authority of western economic models following the financial crisis in 2008. This was triggered by profligate lending in the US subprime mortgage market and excessive risk-taking by banks, which developed into an international banking crisis and the worst global economic depression since the 1930s. Not only has the moral authority and claims to expertise of the North been damaged by the crisis, but advanced economies no longer lead global economic growth and the growing economic significance of some countries in the global South is also being reflected in shifts in political power (Sidaway 2012; Raghuram *et al.* 2014). While before the crisis 'emerging' and 'developing' economies were 'lagging behind', the IMF has found that between 2007 and 2014 they accounted for 69 per cent of global

GDP (Hasmath 2015; IMF 2014). This shift was also signalled by the extension of the Group of Eight (G8) – the eight highly industrialized nations of France, Germany, Italy, the UK, Japan, the USA, Canada and Russia – to the G20. The G8 (now G7 after the suspension of Russia following its annexation of Crimea) formerly held an annual meeting to foster consensus on global issues like economic growth and crisis management, global security, energy and terrorism. In other words, eight countries, which prior to 2008 comprised 14 per cent of the world's population, 60 per cent of the world's GDP and 72 per cent of the world's military spend, formed an exclusive club through which to influence matters of global significance. The G20 – which includes Argentina, Australia, Brazil, China, India, Indonesia, Mexico, Saudi Arabia, South Africa, South Korea and Turkey – grew in significance following the Washington Summit of November 2008, which was called in response to the global financial crisis. In September 2009 world leaders from the group announced that the G20 would replace the G8 as the main global economic council of wealthy nations. Writing in the *Observer*, Elliot argues:

> [T]he [Washington] summit effectively sounded the death knell for the exclusive club of rich nations represented by the G8. The G20 includes all the major developing nations – China, India, Brazil and Indonesia – as well as energy-rich nations such as Russia and Saudi Arabia. As far as global governance is concerned, the G20 is the future, the G8 the past.

Recent years have seen greater inclusiveness in other global development institutions. The World Food Programme now has more inclusive structures and processes; the UN's Development Cooperation Forum (DCF), established in 2007, is more inclusive than its predecessor, the OECD-DAC. The BRICS countries (Brazil, Russia, India, China and South Africa) are now recognized as a major driver of global development (Mawdsley 2010). At a summit in Delhi in March 2012, the leaders of BRICS states not only launched a critique of western financial hegemony, but they also set out an alternative vision for a new world order of development anchored not by the WB and IMF, but by the Bank of the South (Prashad 2012). The BRICS are also beginning to 'colonize the metropole' (Comaroff and Comaroff 2012: 123): witness Brazil's dominance of the global biofuel economy, the Hong Kong banking sector's role in the creation of new financial markets, Chinese and Russian investment in the UK property market, and the reach of Indian auto and steel industries in

Britain. Meanwhile, the IMF and WB have seen some changes in voting and management structures, but radical rebalancing of power has been resisted by 'traditional' power holders. Consequently, new agreements, networks, platforms and forums are being formed outside of the IMF/WB to influence global development, including the BRICS Forum, India-Brazil-South Africa (IBSA), the Forum on China–Africa Cooperation, the India–Africa Summit Forum. These platforms are concerned principally with diplomatic alliances and agreements, trade and investment, and responding to the failure of the WB and IMF to reform. Recent years have also seen an increase in regional and bilateral trade agreements between developing countries. Increasing South–South cooperation raises the prospect of increasing irrelevance of trade with advanced economies and under the regulation of western-dominated institutions. The growing role of regional banks (e.g. African Development Bank; Inter-American Development Bank) may begin to challenge WB hegemony. Alternative institutions, funding streams, power centres, practices and cultures are eroding the former hegemony of the global development institutions (e.g. WTO) and the 'traditional' bilateral donors (see McEwan and Mawdsley 2013; Mawdsley 2012).

These significant shifts in power and the increasing agency of actors in the global South are beginning to be felt in the agreements that emerge from global governance. Until recently, Northern theories and models dominated international efforts at 'sustainable development'. In June 2012, the UN Rio+20 Earth Summit met in Brazil with the objectives of securing renewed political commitment for sustainable development, assessing the progress and implementation gaps in meeting previous commitments (dating back to the original Rio Summit in 1992), and addressing new and emerging challenges. Its major themes concerned how to build a green economy to achieve sustainable development and lift people out of poverty, and how to improve international coordination by building an institutional framework for sustainable development. The outcome was profound disagreement between the UN, which wanted a 'green economy roadmap' with environmental goals, targets and deadlines and developing countries, who wanted new 'sustainable development goals' to protect the environment, guarantee food and energy provision to the poorest, and alleviate poverty. The failure by wealthy countries to acknowledge that a shift from sustainable development to 'green economy' would erode social development provoked enormous protests from social movements and resistance from countries in the global South. Consequently, as Clémençon (2012) argues, the compromise that was reached in the final report – *The Future We Want* (www.un.org/en/sustainablefuture/) – attempts to set out a grand vision for addressing global challenges of sustainable development (economic, social and environmental), but only reiterates promises made elsewhere. It also fails to lay out a coherent 'roadmap' or to define binding targets with specific deadlines. While this might be an unsatisfactory outcome, it reflects a changing political reality in international negotiations, with developing countries playing a more assertive role in pushing poverty eradication as the overarching priority. The question remains, however, as to whether these macro-level shifts will be reflected at grassroots levels of development and will benefit those most in need.

There are also questions about the role of emerging economies as development actors in other global South countries and whether their influence is likely to be positive. One example concerns land acquisition, sometimes referred to as land grab. The geography of land grab has changed: it was once the most significant feature of European colonialism, but today Brazil, India and China are among the top ten countries investing in land overseas (Anseeuw *et al.*, 2012). Yet the picture is complicated by that fact that India is also among the top ten countries in which land is being acquired by foreign investors. A postcolonial approach to understanding these shifts seeks to comprehend how those affected experience them, what new political communities might be emerging within the realm of development in response, what claims they might be making and how they are participating in an agentic civil society. Global elites and the super wealthy can now be found across the North and South; most the world's poor are now located in middle income countries. In addition to rethinking cartographic ontologies and geopolitics of knowledge, we might also ask: What new class alliances are being made possible through this phase of globalization?; How do they differ from those of the colonial period (Raghuram *et al.* 2014); and, What new forms of agency and resistance are emerging?

## New forms of agency and resistance in development

In response to the problematic and unequal nature of relationships between North and South, postcolonial approaches challenge the notion of a single path to development and demand acknowledgement of a diversity of perspectives and priorities. The politics of defining and satisfying needs is a crucial dimension of current development thought, to which the concept of agency is central. In recent years, questions have become increasingly important within development practice, such as: Who voices the development concern?; What power relations are played out?; How do participants' identities and structural roles in local and global societies shape their priorities?; and, Which voices are excluded as a result? In some cases, subaltern actions, which have largely been ignored in elite decision-making vis-à-vis the expansion of global capitalism in former colonies, may provide alternatives in development practice.

One example of this is Ian Yeboah's (2006) case study of urban water privatization in Ghana. This suggests that the attempt to privatize water provision stalled because the underlying rationale and the means through which it was implemented contradicted the interests of Ghanaians and favoured the interests of global capital. The move to privatization was driven by Eurocentric development practice, with which the Ghanaian state was willing to comply because of its dependence on overseas sources of funding and because the psyche of elites, politicians and decision-makers in Ghana is essentially shaped by Eurocentrism. This crowded out subaltern Ghanaian voices and agency and at great cost. Furthermore, Yeboah suggests that the opposition to privatization was hardly based on subaltern interests. Rather, 'it was a classic example of anti-development that suggested people-centred development and human rights, not the *actions* of subalterns in finding culturally-specific ways of solving *their* water problems' (Yeboah 2006: 63, emphasis added). Consequently, and in the absence of state solutions, those privileged by wealth and geographical location have been able to develop post-traditional solutions (e.g. combining pre-colonial underground wells with modern electric pumps), to ensure that they protect their own water supplies. Meanwhile, the vast majority of those not similarly privileged by class and geography still face severe hardship in accessing safe drinking water.

Yeboah's study is a good example of the possibilities of grounding the nexus between postcolonial theory and development practice. This is notoriously difficult to achieve (see McEwan 2003; Power 2003; Robinson 2003a; Sylvester 1999), but extremely important for development. The example of Ghana's water privatization demonstrates the importance of subaltern voices and agency in understanding of the polemics of development, identifying the Eurocentrism of much development practice and highlighting the need to investigate further the possibilities of subaltern alternatives. As Sylvester (1999) suggests, grounding the nexus of postcolonial theory and development practice implies listening to and concerning ourselves with the welfare of the subaltern. It also highlights the importance of creating space in which subalterns can help define the nature of the problem and the type of development intervention required to solve it. Thus, postcolonial approaches attempt to overcome inequality by opening up spaces for the agency of people in the South. Of course, poverty and unequal access to technology make this increasingly difficult. Academics in the South, for example, rarely have the same access to books and technologies of communication as their Northern counterparts. However, despite this, the work of academics in formerly colonized countries has led to an important questioning of authorization and authority.

There is a long tradition within critical development studies of attempting to create spaces in which people in the South can claim both voice and agency. These attempts have not necessarily been named as 'postcolonial' approaches, and very often have their roots in participatory, feminist and post-development approaches. They do, however, have much in common with the postcolonial imperative of hearing the subaltern speak and acknowledging their agency, and they are worth considering here in terms of how they relate to postcolonial critiques.

## Lessons from postcolonial feminisms

Concerns with subaltern voices and agency came to prominence in feminist development studies, where the requirement to understand the lived experiences of women in the South acquired increased urgency from the 1970s onwards. It is now recognized that women in the South often face multiple challenges, including poverty, unemployment, limited access to land and credit, legal, social and cultural discrimination, sexual abuse and other forms of exploitation and violence. While these challenges might resemble those faced by all women, the specificities of history, political economy and culture make them differentially oppressive to women in the South. In Spivak's words they are 'doubly subaltern'. Gender and development approaches have acknowledged these realities, while resisting portrayals of women in the South as victims. Rather, as active agents in development they meet these challenges and are active in confronting and overcoming them, often in remarkably creative and empowering ways. Development, then, is not something that is 'done to' women in the South; rather, the latter play an active role in contributing to the discourses and practices of development.

In addition, postcolonial approaches have forced a move away from totalizing discourses and a singular feminism – based upon the vantage point of white, middle-class Northern feminists, which failed to acknowledge the differences between women – towards the creation of spaces where the voices women from the South can be heard (see Chapter 5). There are now well-established debates concerning representation, essentialism and difference, which have made researching and writing about gender relations and 'women', especially outside one's own cultural milieu, an incredibly complex topic. In the face of sustained criticism, many Northern feminists are now acutely sensitive to the intersections of power with academic knowledge and their privilege in relation to 'Other' women, and are developing more ethical ways of researching and writing (Burman 1995; Madge 1993; Radcliffe 1994, 2015). A postcolonial narrative of identity formation has been used to create a new politics of representation, which sees subjects as fractured and mutually involved in the construction of identities. Robinson (1994), for example, attempts to displace the privileged fixed position of the researcher, to deconstruct the dualism between 'self-researcher' and 'other-researched', and instead to find a 'third space', where mediations of meanings and interactions of interpretations become the object of investigation. Crucial to this is a recognition that the subjectivities of both researcher and researched are mutually constructed through the research process, and that meanings and interactions are also mediated, as is knowledge itself. Transforming the research process in this way involves recognizing that the researcher does not have unilateral control over the research process and the need to 'speak with' rather than to or for the people with whom one is engaged in research. As Spivak (1990) argues, 'speaking with' people from other places and cultures involves openness to their influence and the possibility of them 'speaking back'. It also links to broader notions of breaking with Eurocentric concepts of development and finding other ways of knowing and being, which are discussed further in Chapter 7.

Postcolonial feminisms, in the abstract at least, allow for competing and disparate voices among women, rather than reproducing colonialist power relations where knowledge is produced and received in the North, and white, middle-class women have the power to speak for their 'silenced sisters' in the South. The challenge for those working through gender and development approaches is to continually turn this into practice.

## Lessons from participatory approaches

There are some overlaps between feminist approaches and participatory methods, which, although still often problematic and not always particularly successful, are at the very least notable in their attempt to engage with issues of agency. Participation is often related to grassroots development and has been a strategy of NGOs and development organizations since the 1990s. According to the WB, 'participation is a process through which stakeholders influence and share control over development initiatives and the decisions and resources which affect them' (in Otzen *et al*. 1999: 6). Similarly, as Onibokun and Faniran (1995: 9) argue, 'Community/public/ citizen participation is the act of allowing individual citizens within a community to take part in the formulation of policies and proposals on issues that affect the whole community'. More radical definitions, however, not only emphasize community involvement in the processes of local development, but also demand that social development lead to the empowerment of community members. This involves social change to bring about improved living standards within the community and is especially significant to women, empowering them to liberate themselves from oppressive social structures and to create development structures that work for everyone.

Participation in development planning can encompass several dimensions, including appraisal (understanding local communities and their conceptions of development and other processes), agenda setting (involving local communities in decision-making, defining development needs and policy responses), efficiency (involving local communities in projects) and empowerment (building self-awareness and confidence in communities and individuals). It aims to recognize and support greater involvement of local people's perspectives, priorities, knowledge and skills as an alternative to donor-driven and outsider-led development. Participatory Rural Appraisal (PRA), pioneered by Robert Chambers (1994), is now widely used by NGOs, often working in partnership with community-based organizations (CBOs). This attempts to account for local understandings and knowledge rather than to simply impose Northern understandings and solutions. Similar approaches are also being used in Participatory Urban Appraisal. Central to such approaches is the importance of local communities setting agendas and deciding on priorities, with the intention that local people acquire decision-making power and influence.

One example of radical participation approaches is Reflect (Regenerated Freirean Literacy through Empowering Community Techniques):

> . . . a diverse and innovative approach to adult learning and social change, used by over 350 organizations in 60 countries. It has been used to tackle a wide range of issues, from peace and reconciliation in Burundi, to community forestry in Nepal and holding government accountable in El Salvador.
>
> (www.actionaid.org.uk/323/reflect.html)

Originally piloted by ActionAid in the mid-1990s, the key idea behind Reflect is to merge the pedagogical and political philosophy of Paulo Freire (1972; see Chapter 2) with the techniques of PRA. This combines the technocratic expression of participation with a more theoretical, political and radical approach. It is also theoretically informed by gender and development thinking, and seeks to develop women's capacity to take on participatory roles at community level and beyond. It establishes participation by:

- engaging participants in dialogical discussions of their socio-economic problems;
- using visual graphics (maps, calendars, matrices and diagrams) or drama, story-telling and songs to capture social, economic, cultural and political issues and to structure and illustrate the discussion;
- highlighting keywords from discussions, which form the basis for literacy development;
- participants devising means of solving the problems by identifying action-points to be addressed either by Reflect groups or higher-level organizations.

The Reflect approach links adult learning to empowerment and, therefore, strengthens the voices of poor people in education decision-making at all levels. It creates a democratic space where people can analyze issues for themselves. It is also a basis for mobilization. Although Reflect projects are diverse, they all focus on enabling people to articulate their views: the development of literacy and other communication skills is closely linked to the analysis of power relationships and the active engagement of people in wider processes of development and social change. While members of a Reflect circle learn the basics of literacy, they also learn how to access information or demand services more effectively. Reflect circles often strengthen people's dignity and self-confidence, as well as having an impact on improving resource management, health practices, children's education, local community organization and civic life.

ActionAid began using Reflect with 15 dalit (lower caste) women's groups in the Saptari District of Nepal. After an initial analysis of the caste system and the situation of women, an action plan was developed to counter discrimination based on forming an organization to create group strength. Eight dalit sanghams (or women's groups) were formed and linked together. This gave dalit women a powerful voice at district level. The dalit movement decided to abandon their traditional jobs as dictated by the caste system. In particular, they refused to dispose of the dead carcasses of animals. This was taken as a conscious stand against caste-based exploitation and their refusal to do the work was publicized. There was an immediate backlash from the local high-caste community who refused to serve dalits in local shops. The dalit sanghams planned a careful and united response – using both the law and the media creatively to get their way. Early success with this campaign gave the dalit movement momentum, generating confidence in and commitment to the movement. New members were attracted and before long a national-level dalit movement had emerged. Its strength is rooted in the sanghams and Reflect circles, and means that each victory gives rise to new campaigns, sustaining the momentum of change. The dalit movement in Nepal continues to campaign on issues as varied as education, land reform and citizenship (www.actionaid.org.uk/index.asp?page_id=719).

The results of Reflect have often been impressive, with genuine transformation taking place in gender relations, community–state relations, and between age groups within communities (Hickey and Mohan 2005). Participants report self-realization, increased participation in CBOs, and increased community level actions. Women and Reflect facilitators have become key resource people for the communities. Reflect is inextricably linked to citizenship formation, in that it focuses on people's ability to participate in civil society, to effectively assert their rights and assume their responsibilities.

Beyond these radical approaches, participation has rapidly become institutionalized over the last 20 years in the discourses, if not the practice, of many mainstream development organizations. This is primarily because it promised a new approach that would give 'the poor' more voice and choice in development. Where it is deployed through radical approaches (e.g. ActionAid's use of Reflect) its effects in harnessing the agency and voice of subaltern groups and individuals have been striking. However, all too often participation has been appropriated in non-radical ways. Consequently, Cooke and Kothari (2001) refer to it as the 'new tyranny' in development. As Willis (2005: 104–5) argues, it tends to ignore:

● the time and energy requirements for local people to participate;
● the heterogeneity of local populations and the power relations within them, meaning that 'community participation' does not always involve all sectors of a population;
● that simply being involved does not necessarily lead to 'empowerment';
● that focusing at the level of communities can lead to a failure to recognize wider structures of disadvantage and oppression.

Participation has been taken on board by organizations and institutions such as the WB, but often in problematic ways that do not challenge or engage with existing practices and power relations. People become 'stakeholders' and the goal of participation is subsumed in the rhetoric and delivery of 'good governance' (McEwan 2003). Andrea Cornwall (2006) highlights the culturally specific notions of democracy and governance that are embedded in the various permutations that discourses of participation have taken, arguing that the roots of participation can be found in colonial approaches and are thus not as radical as some critics assume. She questions whether the limited institutional recipes purveyed by Northern development agencies can ever produce the benefits that are claimed of participation in development. Participatory approaches, therefore, are often compatible with top-down planning systems, but have not really heralded profound changes in prevailing institutional practices of development (Mosse 1991). They have largely been unable to give voice and agency to subalterns in the South. As Hickey and Mohan (2005) argue, participatory approaches are most likely to succeed:

- where they are pursued as part of a wider radical political project;
- where they are aimed specifically at securing citizenship rights and participation for marginal and subordinate groups;
- when they seek to engage with development as an underlying process of social change rather than in the form of discrete technocratic interventions.

Moreover, the Reflect example is evidence that radical adult education initiatives must operate in relation to a social movement to be effective in the long term.

## Social movements as a site for postcolonial activism

Progressive social movements are now viewed by several theorists as a means of realizing potential for radical change – they are, in effect, a means for translating the aims of postcolonial theory into practice. This move towards social movements has been particularly influenced by post-development (Escobar 1995b). Although they are often characterized as a form of resistance to development, they are effective in using identity-based forms of participatory politics to extend the boundaries of citizenship to marginal groups. Indeed, Hickey and Mohan (2005) argue that some movements are better understood as being located within a critical position vis-à-vis the ongoing project of modernity rather than being 'postmodern' alternatives to development.

As discussed in Chapter 5, the starting point for many contemporary social movements is a critical resistance to the forms of exclusion and exploitation that have resulted from the broad processes of neoliberal globalization and forms of statist and corporate development. For example, anti-dam movements such as the Narmada Bachoa Andolan (NBA) in India, have opposed both the actual project of development undertaken by the Indian state and the ideological representation of 'development' that underpins it (Routledge 2003; see also Omvedt 2004, 2005). This represents a direct challenge to the moral legitimacy of the state regarding its contract to protect and develop its citizens

(Routledge 2003: 259) and a form of contemporary subaltern resistance aimed at resisting the detrimental effects of globalization (Box 6.3). Other forms of resistance are aimed directly at statist forms of exclusion, such as the dalit and anti-caste movements in India (Omvedt 2006). These seek to contest and oppose the equation by the dominant castes of Indian tradition with Hinduism, and Hinduism with Brahmanism (the religion of ancient India that takes its name both from the predominant position of its priestly class, the Brahmans, and from the importance given to Brahman, the supreme power). This equation works to exclude lower castes from large parts of Indian society. Within this system, the lower castes are 'untouchables' or outcasts who are discriminated against, particularly in rural areas, and prevented from doing all but the most menial of jobs. In response, dalit movements, like those in Nepal, have nurtured alternative traditions, which question this way of looking at Indian society and its history.

Across the global South, the defence of place by subaltern social movements is often constituted as a rallying point for both theory construction and political action. Thus place-based struggles by indigenous peoples concerning issues such as environmental destruction, resource extraction, water privatization and the destruction of livelihoods might be seen as multi-scale, network-oriented subaltern strategies of localization. In such cases, subaltern groups are consciously building translocal solidarities as a means of achieving local aims and ambitions. Of equal significance are struggles over land rights claims and resurgences in localized political identities based around indigeneity. In recent years, social movements have become increasingly significant transnationally as well as domestically. By building new links among actors in civil societies, states and international organizations, such movements multiply the channels of access to the international system for groups that might otherwise be marginalized (Keck and Sikkink 1998). This is especially important and influential in areas of environmental and human rights, where local contestations can be connected into international advocacy networks. These networks become political spaces in which 'differently situated actors negotiate social, cultural and political meanings of joint enterprise' (Keck and Sikkink 1998: 3), making international resources available to new actors involved in domestic political and social struggles. In facilitating the voice and agency of previously marginalized peoples, such networks are of significance to postcolonial politics.

## Box 6.3  Social movements and environmental justice: Ogoniland, Nigeria

Environmental justice is one area of politics that erodes the distinctions between North and South and in which there are growing global coalitions. Environmental injustice refers to the inequitable or uneven distribution of environmental 'bads' or 'goods'; the solution is distributive justice – an equitable distribution of 'bads' and 'goods' so that the poor are not always targeted. A notorious example of the impact of environmental 'bads' on poor people is Ogoniland in southern Nigeria.

Before his execution on 10 November 1995, Ken Saro-Wiwa brought to international attention the exploitation of Ogoni land and people (a population of 500,000 in the Niger Delta) by multinational oil companies (Shell and Chevron) in partnership with the military dictatorship, led by General Sani Abacha. Saro-Wiwa and his compatriots were executed, despite an international outcry, on the grounds of political and economic treason. Saro-Wiwa was campaigning against environmental destruction – gas flaring and crude oil spills that have polluted the Ogoni's drinking water and farmland. The poorly maintained oil infrastructure frequently collapses and Shell blames the local people, accusing them of sabotage. In the 40 years up to 2001, over 4,000 oil spills were recorded in the Niger Delta.

The discovery of oil in 1957 has not brought prosperity to the people of Ogoniland. The spoils of the industry have been shared by the military dictatorship and the oil companies, while the majority of Nigeria's 130 million people live in poverty. The country has earned $300 billion from oil exports, but the average per capita income is $290 (less than a dollar a day). Living standards in Ogoniland have fallen since the British left in 1960. Resistance began in 1990 with the founding of the Movement for the Survival of the Ogoni People (MOSOP) and the drafting of the Ogoni Bill of Rights by Saro-Wiwa, which focused on the right to self-determination, the right to a share of resources coming out of their land and the right to control their environment. In response to the Ogoni uprising, between 1993 and 1998 2,000 Ogoni people were murdered by the military, and many women and girls were raped and mutilated. A hundred thousand people were displaced. Saro-Wiwa's execution saw Nigeria suspended from the Commonwealth in 1995. However, the British government called for Nigeria to be readmitted in 1997 and the oil embargo was never implemented.

The bulk of Nigeria's oil is exported to the UK and USA. Shell's operations continue to devastate the environment – Nigeria leads the world in gas flaring and Shell's activities are seen as a major contributor to global warming. Poverty has deepened as fishing and farming ecologies are ruined. Despite elected civilian governments being in power since 1999, military era decrees still control oil production and successive presidents have turned to the US for naval ships and weapons to enable the government to contain environmental and human rights protestors, which it depicts as 'criminals and restless youth'. As one critic argues:

> Royal Dutch Shell and other associated-[multinational oil corporations] (e.g. Chevron Corporation) . . . have taken more than $30 billion from Ogoniland, leaving behind ecological devastation,

destitution, environmentally induced illnesses, and a shorter life
expectancy among the people.

(Adeola 2000, in Maples 2003: 241)

The Ogoni protests are a combination of environmental and social injustice that
highlights the interconnectedness of environmental concerns with other social and
political concerns. For the Ogoni, environmental justice is not simply a matter of
development but also a matter of survival. The movement has become internationalized,
with Greenpeace and Amnesty International calling for a boycott of Shell. In response,
the Nigerian government has been more stringent in demanding that Shell abide by
better environmental directives; the support of international protest movements was
vital in securing this concession. In 2015, Shell agreed a $84m (£55m) settlement with
residents of the Bodo community in Ogoniland for two massive oil spills in 2008 and
2009. Globalization often causes the problems, but can also lead to effective, globalized
resistance. Disempowered and exploited minorities like the Ogoni people were able to
act and brought their cause to a wider, international public through media exposure, court
battles, links with powerful transnational protest movements and international pressure
groups to bring about some measure of environmental justice. However, in 2011, a UN
report said the Ogoniland region could take 30 years to recover fully from the damage
caused by years of oil spill. Meanwhile, the area remains one of the most polluted
inhabited places in the world. In October 2017, MOSOP once again appealed to the
international community claiming that that Shell had forcefully commenced the laying
of pipelines in Ogoniland with military backing and without the consent of the Ogoni
people (https://www.pmnewsnigeria.com/2017/10/19/mosop-accuses-shell-resuming-oil-
exploration-ogoniland/, accessed 5/1/18). The decades-long struggle is set to continue.

Progressive social movements seek to promote a more equitable
relationship between North and South and a new global order based
on universal human rights protection, greater financial equity,
recognition of labour and women's issues. The conceptual link
between the environment and notions of rights is leading to legislation,
at both national and international levels, which seeks to protect this
relationship between people and their environments, recognizing
the links between livelihoods, development and environments. In
Mexico, the Zapatistas have campaigned against free trade agreements
and the patronage mode of politics, as well as the state's relegation
of indigenous peoples as inferior citizens. Based in the Chiapas
region and led until recently by Subcomandante Insurgente Galeano
(previously known as Subcomandante Marcos, Figure 6.2), they
advocate politics of anti-free trade and pro-democracy based on
Mayan principles of reciprocity. They have successfully used the
internet to cause panic on global markets in a struggle to expose the

**Figure 6.2   Subcomandante Marcos and the Zapatista Movement, Mexico**

Source: AP/Gregory Bull/Vice News

re-colonization of Mexico's poor by international economic interests.
Zapatista politics are not about civic inclusion of a marginalized
people *per se*, but about redefining citizenship from an indigenous
perspective, expressed as a call for 'A political dynamic not interested
in taking political power but in building a democracy where those
who govern, govern by obeying' (cited in Patel and McMichael 2004:
248). The Zapatistas seek to forge alliances with other indigenous
groups seeking decolonization and postcolonial resistance around
the world (Lunga 2008; Hiddleston 2009). In 2007, they hosted
the Intercontinental Indigenous Encounter, which brought together
indigenous peoples to mark '515 years since the invasion of ancient

Indigenous territories and the onslaught of the war of conquest, spoils and capitalist exploitation' and to share experiences. In 2016, in a break with their tradition of rejecting Mexican electoral politics, the movement agreed to select a candidate to represent them in the 2018 Mexican general election. In May 2017, María de Jesús Patricio Martínez – a Mexican and Nahua woman – was selected to stand. Perhaps drawing from the success of indigenous movements elsewhere (e.g. Bolivia – see below), this may signal a change in approach from resistance only towards establishing autonomy as indigenous people and citizens of Mexico.

Other social movements around the world are fighting for political autonomy, cultural, ecological and economic survival, and justice. Brazil's Landless Workers Movement (MST) is the largest and most powerful social movement in Latin America, with an estimated 1.5 million landless members organized in 23 out of 27 states (see www. mstbrazil.org/?q=about; Patel and McMichael 2004: 249–50). It argues that democratic transition has not led to democratic transformation or to the emergence of substantive forms of citizenship. It advocates for and works towards not simply obtaining its share of both land and political power within Brazilian society, but the fundamental transformation of the structures of power within Brazil. The MST carries out long-overdue land reform in a country mired by unjust land distribution. In Brazil, 1.6 per cent of the landowners control roughly half (46.8 per cent) of the land on which crops could be grown. Just 3 per cent of the population owns two-thirds of all arable land. Between 1985 and 1996 rural unemployment rose by 5.5 million, and between 1995 and 1999 a rural exodus of 4 million Brazilians occurred.

Since 1985, the MST has peacefully occupied unused land where they have established co-operative farms, constructed houses, schools for children and adults and clinics, and promoted indigenous cultures, a healthy and sustainable environment and gender equality. It has won land titles for more than 350,000 families in 2,000 settlements as a result of direct action, and 180,000 encamped families currently await government recognition. Land occupations are rooted in the Brazilian Constitution, which says land that remains unproductive should be used for a 'larger social function'. Land seizures under the slogan: 'Occupy! Resist! Produce!' lead to the formation of co-operatives, which involve social mobilization that transforms economic struggle into political and ideological struggle. Democratic decision-making is practiced to develop co-operative relations among workers and alternative patterns of land-use, financed by participatory budgeting to cover social and technical needs. The MST's success lies in its ability to organize and educate. Members have not only managed to secure land and, therefore, food security for their families, but also continue to develop a sustainable socio-economic model that offers a concrete alternative to globalization that puts profits before people and humanity. It has pioneered the production of staple foodstuffs for the Brazilian population at large (with a formal outlet through the national Zero Hunger Programme), filling a significant gap left by agro-export priorities. It has also ranged itself against corporate sovereignty, declaring in May 2003 that fields planted with transgenic crops by large farms would be burned.

Other examples of social movements include the Assembly of the Poor in Thailand, a rural movement protesting their exclusion from the 'Asian miracles' about which the IFIs are so enthusiastic. They campaign for the direction of development to be determined by and for the benefit of the, people – in other words, to be truly participatory. Similarly, Vandana Shiva in India is a well-known activist campaigning against ecological destruction and the disempowerment of rural women. All movements remain committed to building solidarity out of respect for diversity, aim to democratize development, and are thus significant for re-theorizing agency in development.

The globalization of social movements, through alliances such as the World Social Forum (Chapter 5), allows for the construction of enduring networks of relationships among diverse civic and cultural initiatives, forging an alternative organizational and discursive space to that occupied by corporate globalization. As Patel and McMichael (2004: 250) suggest, previous anti-systemic social movements worked to reform or institutionalize countervailing power within institutions or societies. The problem with this was that they privileged the universalist themes of modernity, which crystallized in the statist project of development and which is now the target of this new sensibility that challenges a singular, reductionist vision of development. It is important, then, that global justice movements not only work to reform and transform existing institutions, but that they also support alternative models that lie outside of and are not paralyzed by the logic of economic reductionism. By linking to broad global campaigns and networks regarding the environment and human rights, new social movements appear to have found a way of 'relating the universal and the particular in the drive to define social justice from the standpoint of the oppressed' (Harvey 1993: 116). In so doing, as Hickey and Mohan (2005) argue, they have articulated a mode of political action capable of imagining and generating alternative development futures not only for its immediate constituency, but also for a broader community of dispossessed and marginalized peoples. Moreover, it is significant that such movements articulate their claims within a broader discourse of the familiar demands of modernity (land, democracy and citizenship). They are thus transformative and radical *within* rather than in opposition to modernity and development (see Bebbington and Bebbington 2001).

Social movements are examples of subaltern agency within development. They are 'fostering resistance to capital as a nexus of social relations on a global scale', not through specious ideas about the nation as with the early postcolonial movements, but through far more complex and uncertain ideas of local sovereignty. They work around a fluid notion of politics where outcomes are not guaranteed and, as such, they 'are a genuine and hopeful alternative to the contemporary totalitarianism surrounding us again' (Patel and McMichael 2004: 251).

## The problem of 'speaking for'

While some feminist and participatory approaches and some social movements come close to the ideals of postcolonialism in attempting to create spaces in which subaltern agency can be activated, they are still sometimes problematic in their attempts to allow the subaltern to speak. Thus, they can still be subject to postcolonial critiques that reveal the problems in attempts to give subalterns in the South a voice. Drawing on Spivak's writings, Kapoor (2004: 636–8), for example, explores three attempts within development to allow the subaltern to speak: PRA, ecofeminism and social movements. The three scholars he examines (Chambers, Shiva and Escobar) are taken seriously in development because of their systematic critiques and their political vision in positing alternatives to mainstream approaches. However, their work reveals how decolonization in both North and South has largely failed and thus remains an unfinished project. Kapoor's critique is worth outlining in detail:

## Participatory Rural Appraisal

PRA, as outlined by Chambers (1994), has become fashionable in both governmental and non-governmental development circles. It covers a range of approaches and methods designed to enable local rural and urban people to express, enhance, share and analyze their knowledge of life and conditions, and to participate in planning and action on the ground. It is critical of top-down approaches developed by external experts and instead aims to valorize local knowledge and empower local people so that they can determine the development agenda. Outsiders thus become facilitators of a participatory process, in which *'the poor, weak, vulnerable and exploited . . . come first'* (Chambers 1997: 11, emphasis in original).

The problem with PRA is that it rests on the idea that the subaltern voice can be rendered transparent so that the external expert can hear and represent subaltern desires and interests. However, Chambers ignores the fact that all knowledge framing (PRA included) produces power relations. For example, PRA takes care to include women in public space, but ignores the fact that in some parts of the South they can feel intimidated (and be intimidated) when speaking in public, especially on issues such as sex, rape and violence, which are often considered taboo. Attempts in parts of South Africa to create participatory local democracy, for example, have often failed to include women because local cultures forbid them to speak in public (McEwan 2003, 2005). In societies where age brings status, young people are similarly marginalized. Thus, even if subalterns speak, they very often perform roles that are expected of them by their own communities, the facilitator or the funding agency. Other important power relations that affect participation, such as those generated by socio-economic inequality, are also ignored in PRA. This is a crucial issue when it comes to differences in land tenure and access to resources. Even when the subaltern does speak, her voice is relayed, filtered, reinterpreted, appropriated and even hijacked by intervening institutional structures. Some attempts have been made to circumvent this by using direct testimonials, but there is no avoiding who edits and translates the stories, how they are presented, for what purpose and so on (a good example of the dilemmas created by attempts to let the subaltern speak is Townsend *et al.* 1995). The subaltern voice is always mediated, even within participatory approaches.

## Ecofeminism

Vandana Shiva's *Staying Alive* (1989) is one of the first feminist critiques of development and mainstream science, as well as seeking to validate grassroots environmental movements in the South. However, as Kapoor (2004) argues, Shiva essentializes subaltern women by depicting them as closer to nature and thus as the natural guardians of the environment. This move effectively reinforces gender roles and stereotypes. Shiva also champions Chipko, the forestry movement in the North Indian Himalayas, as the feminist and ecological alternative to mainstream development and science. However, she ignores the problematic patriarchal political structures of Chipko (significant male leadership despite women's overwhelming participation) in favour of romanticizing the movement. The heterogeneity of ecofeminist struggles goes unrecognized and instead there is a tendency to fix both the 'nature' of Nature and women. The subaltern woman is eulogized, giving the impression that subalterns are transparent to themselves and immune to struggle or failure. In short, Shiva not only 'speaks for' the subaltern, but attempts to produce an authentic and heroic subaltern. She is an example of what Spivak calls the 'native informant'. Although the native informant might provide a greater insight into local developmental concerns than a Northern 'expert', she is still removed from the experiences of local people by virtue of privilege in terms of relative wealth, access to education, caste and so on. The native informant, therefore, still exists in a problematic power relationship with the subaltern and, in the process of 'speaking for' the subaltern effectively negates her agency.

## Social movements

Escobar's work on social movements has been criticized for romanticization of the 'local' and for assuming that social movements are always necessarily benign. In fact, many are internally and externally sexist, racist, homophobic, xenophobic and undemocratic (e.g. right-wing Hindu movements in India, left- and right-wing guerrilla movements in Columbia). According to Kapoor (2004), Escobar also misses the diversity of social movements and the fact that, far from being uniformly anti-development, many are struggles *for* development (e.g. based on demanding access to health, education, irrigation, land rights, technology, etc.). Like Shiva, therefore, Escobar both essentializes and romanticizes social movements in the South. The heterogeneity of social movements goes unrecognized and instead there is a tendency to fix their political orientation. The subaltern subject is eulogized, again giving the impression that subalterns are immune to struggle or failure. In short, Escobar not only 'speaks for' the subaltern but, like Shiva, attempts to produce an authentic and heroic subaltern subject.

As Kapoor (2004) argues, in providing profound critiques of development, both Shiva and Escobar are required to produce a hyperbolic construction of the subaltern – the Other as heroine or hero – which reflects not the voice of the subaltern but the desire of the intellectual to be benevolent and progressive. Ultimately, as Spivak would argue, this produces another form of silencing of the subaltern. These attempts are progressive in attempting to enable the subaltern to speak or to listen to the subaltern, but they can all too easily do the opposite. This needs to be borne in mind when advocating social movements and other radical alternatives as postcolonial in practice or as postcolonial alternatives to mainstream development.

## Contesting development from the South

The increasing economic power and influence of countries in the global South suggests that it can no longer be assumed that it is the responsibility of the western academy to make space for subalterns to be heard. This implicitly assumes western agency and, as Raghuram *et al.* (2014) argue, posits a western centre of knowledge, which is no longer entirely relevant when the conversations are happening elsewhere. This is certainly true of development. Global economic power is shifting inexorably towards Asia – to China and India in particular – such that scholars now refer to the twenty-first century as the 'Asian Century' (Roy 2014). There is also similar optimism about the prospect of economic development in African countries among African scholars and Afro-futurists (Mbembe 2016; see Chapter 8). The Asian powers already have the potential to drive development in different directions. As discussed, China is attempting to raise its poorest out poverty and find paths to more economically sustainable development this through its five-year plans. India, through its own five-year plans, is pursuing market-oriented inclusive growth – what Roy (2010) has termed 'bottom billion capitalism' – in attempt to leverage economic growth for the benefit of the poorest people.

The success of these reworkings of modernization of promoting inclusive growth through integration of the poor into market rule remains to be seen, and postcolonial scholars remain cautious (see Roy 2014). But these shifts suggest that at the level of the state, concerns with agency are no longer an issue: the BRICS are pursuing their own agendas irrespective of models and programmes devised in the WB. The question of how the poor in these countries can exert agency in their own development is likely to remain important but, as discussed previously, examples have emerged across the world of First Nations, Aboriginal and indigenous peoples challenging dominant understandings of development and attempting to forge alternative trajectories. In some cases, this has been scaled up to the level of national development strategies. The most well-known examples are Ecuador and Bolivia's experiments with 'buen vivir', a 'local, decolonial, Indigenous alternative to long-term Western hegemony over defining what development is' (Ranta 2016: 426; see also Radcliffe 2015).

Western worldviews have been reduced to a notion of endless development based on the separation of society and nature, emphasis on the rights of the individual, and capitalism oriented around materialism and consumption. In contrast, buen vivir is an alternative to the development-centred approach, based on the belief that true well-being (or 'living well') is possibly only as part of a community, which extends to nature, plants, animals and the Earth. Buen vivir began to develop in the early 2000s in Latin America for several reasons. First, it is a response to development projects implemented by governments and the IFIs that promised increased wealth, but instead depleted natural resources and increased poverty levels. Second, the increased visibility and voice of indigenous peoples provides an alternative discourse about ways of living that does not include modern development and capitalism. Third, climate change and the rapid destruction of the Amazon in the name of development require a higher value to be placed on the importance of respecting and protecting nature.

Buen vivir was incorporated first into the Ecuadorian constitution in 2008, followed by the Bolivian constitution in 2009. In Ecuador, buen vivir has a solid legal framework, whereas in Bolivia it sits alongside third generation human rights such as social and gender equality and social justice. It takes its legitimacy from the fact that indigenous peoples lived sustainably for centuries. The governments of both countries are struggling to reconcile the principles of buen vivir with their current economic and political environment, and the need to balance social change with the financial needs of the country and the demands for development from large corporations. These tensions coalesce around the extraction of natural resources, such a lithium and oil, which have continued in both countries. Ecuador, for example, remains capitalist and heavily dependent on agricultural and energy exports. However, the nationalization of oil and electricity in Bolivia, and the renegotiation of contracts with foreign corporations, has enabled the state to redistribute land to indigenous people and some wealth and higher wages to workers. This saw the numbers living in poverty fall from 60 to 50 per cent of the population and those in extreme poverty from 38 to 25 per cent between 2005 and 2010 (Rist 2014). Buen vivir acknowledges the diversity of informal sector economic activities and the role of caring and reproductive labour, and it 'envisions a form of solidarity economy in which collective well-being, redistribution, use values and human needs prevail' (Radcliffe 2015: 259). However, it is difficult to implement in full, not least because of the entrenched social inequalities (not all Bolivians and Ecuadorians share the same vision of buen vivir) and failure to tackle colonial legacies (which, as Radcliffe points out, are also gendered). However, the critique of development within buen vivir at least posits a realizable alternative to western models of modernity, exposes the limits of western development, and centralizes a different ethos of human-environment relations, from which different ways of living can at least be imagined. The goal of development is to establish buen vivir and 'the construction of buen vivir enables a new vision of human and social development' (Walsh 2010: 9). In Ecuador, the state does not envision a universal welfare state, but a form of 'development justice committed to dealing with discrimination, labour insecurity, uneven development, and income gulfs' (Radcliffe 2015: 260). The anti-state discourses of neoliberalism are reversed and the state assumes responsibility for redistributing wealth and guarantee in rights to people and nature.

While western and indigenous ontologies are profoundly different, especially in how they understand the relationship between society and nature, some of the principles of buen vivir are being taken up in western cultures, for example: through carbon taxation and environmental accounting that aim to reflect the costs of products and processes that harm the environment; by those seeking to adopt zero waste lifestyles and the increased popularity of austere living; by those advocating sustainable de-growth, slowing down consumption and reducing energy use; by those advocating smaller, local production. Critiques of Northern-based institutions that emerge from these indigenous perspectives are also finding increasing resonance in developed countries. For example, Bolivian President Evo Morales has denounced global initiatives, such as the 2009 UN Copenhagen Climate Change Summit, for failing to address the true nature of climate change in poorer countries.

Even in the face of new forms of dispossession, subaltern peoples are exerting their agency and resistance to posit development alternatives and their stories are increasingly being heard. One example is captured in Kurt Langbein's 2015 film: *Land Grabbing – The Movie*. The film gives voice to those involved in and affected by global land grabs: farmers, local and international activists, politicians, investors, financiers and speculators. It tells the story of Cambodia, where state and agribusiness use violence to dispossess and displace rural villagers from their farmland to make space for massive mechanized sugar plantations. Primitive accumulation and proletarianization have created conditions in which villagers have become landless peasants and wage labourers on land they once owned. The film traces similar developments in Ethiopia, where vegetables produced by exploited wage workers in huge greenhouses are now being eaten by Dubai's super-rich. As discussed in Chapter 4, this is happening while Ethiopians live on the very edge of starvation. The film also explores the complicity of the European Union, whose 'Everything but Arms' trading initiative has lowered trade barriers and facilitated land grabbing. This is a process not confined to the global South, but is also seen in countries like Romania. However, the film also focuses on examples of resistance against land grabbing and struggles to reoccupy land. One example is the adoption of small-scale agro-ecological farming practices for food sovereignty in Ethiopia's Tigray highlands, which provide an alternative to 'food security' based on large-scale industrial agriculture. The film charts numerous examples of similarly successful local political struggles for indigenous rights and responsible agriculture, which, if scaled up to the level of the EU and IFIs, might provide solutions for more sustainable rural development.

## Shared imperatives and the ethic of care

Participatory, feminist and post-development approaches have raised significant questions for scholars and practitioners in development.

These relate to asking by what right and on whose authority does one claim to speak on behalf of others? On whose terms is space created in which they are allowed to speak? Is this merely an attempt to incorporate and subsume 'other' voices into Northern canons? Such questions form the core of postcolonial approaches to rethinking development theory and practice. As we have seen, it is no longer feasible to represent formerly colonized peoples as passive, helpless victims. The voices of people in the South are now being heard in some quarters, and their ideas are increasingly being incorporated into grassroots development policies and, in a small number of cases, also at the level of national governments. Through the shifting terrain of power in international governance, they may yet begin to be heard in international development agencies and financial institutions. Where the idea of agency has crept into development policy it has done so in problematic ways. We might consider, for example, the ways in which neoliberalism constructs agency as responsibility – the poor themselves, rather than the state, are responsible for their own survival. This is evident, for example, in antipoverty policies in the UK, USA and Germany. Thus, in advocating agency within development, it is essential that this is not conflated with the worst aspects of neoliberalism in the North, in which the blame for poverty and the responsibility for survival is shifted onto the shoulders of the impoverished.

Participatory and feminist approaches to subaltern agency in development cohere to some extent with postcolonial imperatives of allowing the subaltern to speak, be heard and act. These approaches promise to inform development policies and practice that allow individuals and communities in the South to determine their own development priorities and identify appropriate responses. Despite this, however, the fact that development, geopolitics and security are so often intertwined at macro-levels means that it is unlikely that Northern governments and the IFIs will wish to relinquish their hold on determining trajectories of development in the South (see Duffield 2007). Development studies, meanwhile, has changed a great deal in terms of its acknowledgement and recognition of agency in the global South. One aspect that has not changed is the normative concern with emancipation from inequality and poverty. The way in which this concern is articulated, however, is beginning to move away from patronizing, ethnocentric representations, rooted in colonial discourses, of people in the South as passive recipients or incapable victims. A concern with the excluded means that development studies is ideally placed to respond to postcolonial critiques that make visible processes by which subalterns become and remain excluded, even within well-meaning development projects. Tackling the question of what precisely excluded peoples should be included in is still tricky for development studies and requires a concerted effort to encompass subaltern voices in defining the meanings, processes and outcomes of development in particular localities.

The moral imperative underpinning development studies has also not changed. This moral imperative is focused on a responsibility for distant others that should not be lost in arguments about acknowledging the agency of people in the South. It would be an easy step, for example, from an argument, on the one hand, that posits the agency of the subaltern and resists neocolonial representations of passivity to, on the other hand, an argument that allows the North to abrogate its moral responsibilities to distant others. Avoiding such a dangerous move requires instead that moral responsibility be reformulated in ways that do not replicate neocolonial power relations. Preventing the starvation of millions of 'others' or justifying the welfare rights of distant peoples remains a moral responsibility for those who are privileged in a profoundly unequal world (Smith 2002). One's position of advantage in an increasingly uneven development surface is a matter of good fortune rather than something that is earned or deserved: 'the needs and rights of strangers could easily – and but for the "accident" of birth – be the needs and rights of ourselves' (Corbridge 1993b: 464). The challenge is to enact responsibility in ways that also erode rather than deepen existing unequal power relations (discussed further in Chapter 7). Creating spaces in which subaltern voices might be heard and subaltern agency valued should be part of a broader ethic of care that shapes development studies.

To this end, it is worth repeating the point made in Chapter 3 that development is not simply a form of intervention in the global South, but a response by governments to the problem of 'surplus people' left out of the telos of capitalism (Silvey and Rankin 2011). Just as the economies of the global North are converging towards those of the South, so new forms of protest and resistance are also converging (e.g. the waves of popular protests against austerity measures in Europe since 2010). This also opens the possibility for progressive and experimental development projects that have emerged out of these protests in the global South to also travel northwards. The push for society-wide basic income grants currently being experimented with in some European countries are inspired by similar, long-established programmes in the global South, such Brazil's Bolsa Familia – an enormous cash transfer programme initiated in 2003 (Comaroff and Comaroff 2012) – or even China's *dibao*. In January 2017, Finland became the first country in Europe to pay its unemployed citizens an unconditional monthly sum, and this is being watched closely by other European countries. Ontario in Canada also began trialing a similar programme in 2017. As more work in advanced economies is displaced by automation and artificial intelligence, universal basic income schemes are garnering interest across the political spectrum: on the left because the schemes are potentially redistributive of wealth and protect displaced workers in an increasingly post-work economy; on the right because they simplify welfare spending and, once people are freed up from the need to earn money to live, they have time for self-improvement and providing help in their communities, relieving the state of its obligations.

Despite this increasing convergence between global North and South, international development is still often framed by a responsibility to distant others, grounded in the fact that, as discussed at the beginning of this chapter, our lives are intertwined with those of distant strangers, through flows of capital and commodities, modern communications technologies, shared and hybrid cultures and so on (Corbridge 1993b, 1998). The desire of many postcolonial intellectuals is to reveal injustice and oppression. The desire of many development scholars is to find ways to tackle injustice and oppression on the ground. Both articulate an ethic of care and both share a concern with subaltern agency and voice. Inequitable structures often make it very difficult for subaltern voices to be heard and for subaltern peoples to assert themselves as active participants in development processes. However, development studies has the potential not only to make visible the position of the marginalized but, through its commitment to an ethic of care, to disrupt the power relations of speaking for distanced others and to create space for a more equitable development practice. To do so, it might embrace postcolonial approaches that enable it to resist the establishment of sovereignty over the minds and bodies of distant others.

## Summary

- Widespread international resistance and increasing economic power of some developing countries raises important questions about the need to democratize development institutions, the dissemination of development, policy and intervention and the need to create space in which subaltern voices can be heard and subaltern agency be incorporated into development theory and practice.

- Many arenas exist in which the monopolization of knowledge about development by IFIs, Northern NGOs and Northern 'experts' is being contested. Within academic circles, and at the level of grassroots development, it is recognized that other ways of knowing are possible, with indigenous sources seen as crucial to development thinking and practice.

- These alternatives focus on democracy, abolition of debt, freedom from poverty and violence, gender equality and rights and highlight the importance of self-determination. This coincides with postcolonial imperatives and shares links to earlier anti-colonial movements (Chapter 2).

- Postcolonialism challenges the notion of a single path to development and demands acknowledgement of a diversity of perspectives and priorities. Key questions are: Who voices the development concern? What power relations are played out? How do participants' identities and structural roles in local and global societies shape their priorities and which voices are excluded as a result?

- Postcolonialism attempts to overcome inequality by creating spaces for the voices and agency of people in the South. However, poverty and lack of technology make this difficult.

- It is no longer feasible to represent the peoples of the South as passive, helpless victims; evidence is growing that agency in the global South may be more effective in tackling developmental problems than traditional 'top-down' international development. Further efforts are required to ensure that development is truly opened up to the presence and significance of subaltern voices, knowledge and agency.

- Through an engagement with postcolonialism and a shared ethic of care, development studies has the potential not only to make visible the marginalized, but to disrupt the power relations of speaking for distanced others, and to create space for a more equitable development practice that acknowledges and values agency in the South.

## Discussion questions

1 Explain how postcolonialism challenges temporal and spatial distancing and how this relates to moral responsibility for distant strangers.

2  In what ways and with what consequences do contemporary discourses of partnership echo colonial discourses of trusteeship?
3  Explain how dominant knowledge closes off the possibility of alternatives and what effects this has had on development studies.
4  Why is it important to acknowledge the agency of peoples and governments in the South and how might this change the nature of development studies?
5  What are the similarities and differences between postcolonial, feminist and participatory approaches regarding subaltern agency?

## Further reading

Baaz, M.E. (2005) *The Paternalism of Partnership. A Postcolonial Reading of Identity in Development Aid* London, Zed Books. A critique of discourses of partnership through a case study of Scandinavian donor agencies in Tanzania.

Cooke, B. and Kothari, U. (eds) (2001) *Participation: The New Tyranny* London, Zed Books. An edited collection that explores how radical approaches to development have been appropriated by various development agencies, very often with depoliticizing and negative effects.

Corbridge, S. (1998) 'Development ethics: distance, difference, plausibility' *Ethics, Place and Environment* 1, pp. 35–53. An important paper that relates debates in political philosophy about responsibility to distant others/strangers to development studies.

Hickey, S. and Mohan, G. (eds) (2004) *Participation: From Tyranny to Transformation? Exploring New Approaches to Participation in Development* London, Zed Books. An edited collection that moves the agenda on from Cooke and Kothari's critique.

Kapoor, I. (2004) 'Hyper-self-reflexive development? Spivak on representing the Third World "Other"' *Third World Quarterly* 25, 4, pp. 627–47. Essay on the politics of representation and a useful deployment of Spivak in a critique of attempts to 'let subalterns speak' in feminist and participatory approaches to development.

Walsh, C. (2010) 'Development as *Buen Vivir*: institutional arrangements and decolonial entanglements' *Development* 53, 1, pp. 15–21. Excellent discussion of an alternative notion of development based on indigenous knowledge.

## Useful websites

www.actionaid.org ActionAid International.

www.actionaid.org.uk/323/reflect.html Website detailing ActionAid's Reflect approach linking empowerment to adult learning.

www.nepad.org/ NEPAD website, with information on its origins, strategies and programmes, country reports and action plans.

www.china.org.cn/english/features/guideline/156529.htm Contains details of China's development guidelines in the 11th 5 Year Plan (2006–2010) as well as historical data on previous plans.

www.wombles.org.uk/zapatista/ News and information on the struggle of the Zapatista communities in Mexico, the extension of their campaign into the global North and developments in the 'Other Campaign'.

www.ourplanet.com/imgversn/82/shiva.html An account of Vandana Shiva's politics regarding ecological destruction in India and beyond.

www.mstbrazil.org/?q=about Website of the Movimento dos Trabalhadores Rurais Sem Terra (Brazilian Landless Workers' Movement).

www.odi.org.uk/ Website of Britain's leading independent think tank on international development and humanitarian issues.

# 7 Towards a postcolonial development agenda

## Introduction

Having traced the mutual criticisms and intersections between postcolonialism and development studies in preceding chapters, this chapter sets out an agenda for a postcolonial development studies, both in theory and in practice. Previous chapters highlighted the significance of acknowledging location and the partial nature of all knowledge in producing postcolonial development knowledge, and the value of engaging with regional scholarship to overcome the parochialism of many Northern-based and Northern-dominated disciplines. Transforming the production and circulation of knowledge and developing a more cosmopolitan scholarship is thus of critical importance. Effecting such transformations is vital in allowing the theoretical advancements of postcolonialism to reshape development practice. This chapter outlines some of the strategies that might enable the production of postcolonial knowledge.

The challenge of transformation – what might be thought of as decolonizing knowledge – is important because, while cultural difference and identity tend not to be a feature of many books on development, generations of students in the North have been socialized into specific modes of thinking about the 'Third World' or the 'less developed world'. In response, as we have seen, postcolonial approaches to development have encouraged critical reflection on:

- the binaries that shape relations between North and South (First World/Third World; 'more' and 'less' developed; developed/ underdeveloped; advanced/backward; non-problem/problem; knowing-subject/needy-object; industrialized/non-industrialized) (Figures 7.1 and 7.2);

**Figure 7.1**   *Lakeside poverty in Dhaka, Bangladesh*

Source: Brian Cook 2007


**Figure 7.2**   *Development and modernity on the same lake in Dhaka, Bangladesh*

Source: Brian Cook 2007

- linear notions of time and historicist understandings of modernity that see the South as 'back there' in terms of progress and the North as the originator of development;
- the vast array of difference and complexity in what we call 'the Third World' and the dangers of generalization;
- the problems inherent in perpetuating images of people in the South as helpless victims and of the endless recycling of disaster discourses and alarming images of starvation and desperation;
- taking responsibility for producing the patronizing, ethnocentric, negative attitudes that societies learn to have about other parts of the world;
- the importance of acknowledging how peoples of the South are rich in economies and cultures, a point that is often neglected in development theory;
- the possibilities of theories of development and modernity emerging in the South;
- the ways in which postcolonialism offers new ways of thinking about development – of what it is, what it does and what it might do. It also allows us to reconsider the practice of doing development.

Reconsidering the practice of development is of critical importance. Within development studies literature there has been growing recognition of the need to understand the 'multiple spaces of development knowledges and the variety of centres from which these knowledges are produced' (Power *et al.* 2006: 232). This involves reflecting more critically and clearly about our position as researchers and seeking a more 'cosmopolitan' theorization of development. A postcolonial lens on development makes this possible through politically engaged scholarship concerned with the cultural, political and material aftermath of various forms of colonialism (Radcliffe 2005). Postcolonial critiques destabilize some key assumptions within development studies, highlighting the need to 'provincialize' (Chakrabarty 2000; Robinson 2003a) and decolonize the discourses and practices of development. This involves shifting away from taking Europe, or the North more broadly, as the theoretical and normative reference point for theorizing about countries in the South. It also involves creating possibilities for novel epistemological and methodological approaches and for transforming both research and teaching.

## The politics and ethics of postcolonialism

As we have seen, broadly speaking postcolonial perspectives can be said to be anti-colonial (Ashcroft *et al.* 1995; Radcliffe 1999). However, the politics of postcolonialism often diverges sharply from other perspectives and its radicalism rejects established agendas and accustomed ways of seeing. Postcolonial critical agendas in different places are shaped by the different nature, form and timing of colonialism and anti-colonial resistance, different levels of social division and new forms of neocolonial domination and transnational connections. As Nash (2002: 227) argues, these differences 'work against postcolonialism becoming a set of impressive theoretical tools that are never challenged by the particular, complex, messy material of social relations in different places'. They also work against the positing of a singular postcolonial politics or set of political strategies. Therefore, in raising questions about the politics of postcolonialism and what postcolonial development studies would look like, it is important to acknowledge that these are always positioned within different and interconnected colonial and post-colonial contexts and legacies.

One of the challenges for postcolonialism is connecting its powerful discursive insights to material consequences, and this inevitably involves thinking about what the political might mean for postcolonial analysis in development studies. It could be argued that political imperatives have driven postcolonialism from its beginnings because of its anti-colonial stance. However, as Quayson (2000) argues, there are also tensions between an activist engagement with the 'real' (material) world and a more distanced participation through analyses of texts, images and discourses. The pretext for postcolonial criticism – the desire to speak to western paradigms of knowledge in the voice of otherness, to show how the constitution of western subjectivity depends on interactions with subjected others, to destabilize centres and peripheries (Bhabha 1994; Ong 1999; Stoler 1995) – is that it is an ethical enterprise, pressing its claims in ways that other theories, such as postmodernism and poststructuralism, do not. Paradoxically, however, the idea of a postcolonial politics is also problematized by a constant reluctance to take radical ethical standpoints. Thus, postcolonial theory and criticism is riven by a contradiction that has attracted much criticism:

> . . . social referents in the postcolonial world call for urgent and clear solutions, but because speaking positions in a postmodernist world

are thought to be always already immanently contaminated by being part of a compromised world, postcolonial critics often resort to a sophisticated form of rhetoric whose main aim seems to be to rivet attention permanently on the warps and loops of discourse.

(Quayson 2000: 8)

It is possible, of course, to speak and to indicate an existential tentativeness in whatever has been spoken (Katz 1995; Storper 2001), and this shapes the questions that remain over what the ultimate objectives of a responsible postcolonialism should be. What, for instance, is the use of a discursive analysis of the language of development when this does not address the economic and social disjunctures produced in countries in the South by the imposition of structural adjustment policies or the enabling of neoliberal incursions on land and livelihoods? What is the use of undermining discourses of power when 'we never encounter any specific scenario of injustice, domination, or actual resistance from which we may gather intimations of the passage through the postcolonial ordeal' (San Juan 1998: 2)? What do academic postcolonial studies contribute to the *experience* of postcolonialism in the contemporary world? These questions inform much of what follows.

It is difficult, if not impossible, to separate postcolonial discourse from an ethical project, even though the means by which its ethical ends are to be achieved remain a highly contentious issue. What precisely postcolonialism does in development studies and what it is meant to do is the subject of intense debate. Consequently, it is perhaps helpful to think of postcolonialism as an 'ethico-politics' of becoming (Ferguson 1998), a 'process of postcolonializing' (Quayson 2000: 9) or an 'anticipatory discourse' (Childs and Williams 1997: 7), recognizing a condition that does not yet exist, but working nevertheless to bring it about. Postcolonial development studies has the potential to provide a careful grounding of the specificities of the local and to embed phenomena in a variety of social, cultural, historical and political contexts through which 'a transfigured and better future might be brought into view' (Quayson and Goldberg 2002: xiii). Postcolonialism is not just a viable way of interpreting events and phenomena that pertain directly to postcolonial contexts, but a means by which to understand a world thoroughly shaped at various interconnecting levels by 'the inheritance of the colonial aftermath' (Gandhi, 1998: x). Thus the 'process of postcolonializing' should refer to the critical process by which to connect modern-day phenomena to their explicit and implicit relations to this heritage.

Drawing on these insights, we might more carefully consider what postcolonialism might mean for development studies. Ashcroft (2001: 19) argues that 'a theory which may more faithfully engage the actual practice of post-colonial subjects . . . is a poetics and politics of *transformation*'. On the one hand, a poetics of transformation is concerned with the ways in which writers and readers contribute constitutively to meaning, the ways in which formerly colonized societies appropriate dominant discourses, and how they interpolate their voices and their concerns into dominant systems of textual production and distribution. Transformation recognizes that power is central to cultural life and resists by adapting and redirecting discursive power, creating new forms of cultural production. Examples of these new forms of cultural production include the anti-colonial political and cultural movements of the twentieth century (Chapter 2). Within development they might include postcolonial fiction, photography and film (Chapter 4), as well as alternative theories about development, such as post-development (Chapter 3) or different notions of what development means (Chapter 6). On the other hand, a politics of transformation works constantly within existing discursive and institutional formations to change them. Through taking hold of writing itself, in political discourse or political structures, in educational discourse and institutions, conceptions of places, peoples, development and even economics are transformed. This is where the agency of people in the South in the creation of development knowledge (Chapters 5 and 6) is significant. Ultimately, a poetics and politics of transformation can bring about 'transformation of the disciplinary field' (Ashcroft 2001: 19).

A transformed, postcolonial development studies provides a route to a more reflexive understanding of the complexities of postcoloniality. Rather than 'sort out postcolonialism once and for all' (Nash 2002: 228), and devise theoretical or political frameworks that are all-encompassing, settled and complete, it might be more productive to keep the notion of a postcolonial politics within development studies as provisional and constantly under review, able to respond to different spatialities of the postcolonial, but constantly in question. What is apparent in this is the continuing centrality of culture to a transformative postcolonial development studies and the recognition that underlying all economic, political and social resistance is the struggle over representation, which occurs in language, writing and other forms of cultural production. The potential of postcolonialism is to discriminate between the continuing reality of imperial power and subaltern peoples, and to resist the submergence of the neocolonial subaltern within global power relations. Whereas globalism erases differences between people on the basis that they are all consumers, postcolonialism works to reveal the gaps between peoples that remain. It reveals that people belong to a society as well as an economy, and that society is still controlled by a cultural dominance established by imperialism (Ashcroft 2001) and often perpetuated through economic and geopolitical relations. Transformation of representation is crucial because such practices are situated in a material world often with direct material implications.

Central to this is an understanding of the importance of place-based and local/global machinations of postcoloniality. Critical reflections on the geographical dynamics of postcoloniality also need to herald a clear and coherent ethical and political position on the present. With this in mind, we might draw on the philosophical and historical referents to the importance of place, the local, the grounded and the performative in colonialism and postcoloniality. Chapter 2, for example, explored the geographies of postcolonial politics and how these differed in through time in different places, despite often being connected through anti-colonialism and anti-racism. Similarly, diverse places in both North and South are postcolonial and often deeply intertwined, but experiences of postcoloniality often differ markedly within and between places. Chakrabarty's (2001, 2002) elevation of the notion of 'dwelling' in modernity and the colonial past to an ethico-political principle is instructive. Modernity always unfolds within specific cultures or civilizations and different starting points of the transition to modernity lead to different outcomes (Gaonkar 2001). Chakrabarty is concerned with recognizing the struggle that many elements of society have with the communal and spatial dislocation in the modern, or what he calls 'the struggle to be *at home* in globalized capitalism' (2001: 145).

Similarly, the writings of Derrida (2000, 2001) and Spivak (2002) on hospitality and cosmopolitanism in relation to transnationalism and the lived experiences of migrants and asylum seekers are also significant. They raise profound questions about the ethical and political relationships between people who experience modernity, globalization and development in very different ways. These issues are important globally and not least in the South, where transnational flows of people are increasingly significant. We might also consider the unethical neglect and destruction through imperialism of the colonial Other's locational attachments (see, for example, Mehta 1999) and how this informs contemporary geopolitics and geo-economics. Exploring the ways in which individual and collective identities are expressed, lived and performed in specific locales can also foreground the agency of indigenous peoples in contrast to the erasures of dominant cultures and romantic representations of victimry, tragedy and nostalgia (Vizenor, 2001). Concerns such as these are beginning to inform critical postcolonial development studies, casting light on what a politically and ethically informed understanding of postcoloniality might look like.

## Postcolonializing development studies

Critical and progressive approaches to development and postcolonialism already share much common ground in both principle and practice. Postcolonial approaches are part of broader, progressive epistemologies and methodologies that seek to valorize and mobilize politically around cultures, identities, rights, knowledge and worldviews of marginalized or subaltern groups. Examples of these epistemologies include the localization and locality-based anti-globalization agendas mentioned in previous chapters (e.g. Escobar 2001; Escobar *et al.* 2002), and empowerment-oriented participatory approaches to development (Cooke and Kothari 2001; Golooba-Mutebi 2005; Mohan and Stokke 2000; Simon, *et al.* 2003). As concerns have arisen around development as an imperialist or postcolonial project, development practice has increasingly turned towards participatory approaches and the discourse of the local in 'an effort to recapture the emancipatory potential of development' (McKinnon 2006: 23). Significantly, ideas about cultural diversity and liberty have also been mainstreamed within development. For example, the 2004 UNDP Human Development Report *Cultural Liberty in Today's Diverse World* argued that cultural flexibility and appropriateness are essential to progressive and participatory approaches to development. Of course, there are issues about who decides on appropriateness and the kinds of flexibility, but the recognition that individuals hold and utilize a complex multiplicity of identities according to specific circumstances is a step forward. Adding culture to development does not in and of itself postcolonialize development. However, it is encouraging that some of the central concerns of postcolonialism are beginning to feed into mainstream development thinking.

The challenge remains to link the focus on local identities, practices and agendas to broader and multi-scale campaigns for progressive and radical change that are 'substantively postcolonial and critically (post) developmental' (Simon 2006: 17). Simon proposes four broad ways forward:

1   The establishment of new, and the extension of existing, North–South alliances and partnerships for progressive research, information sharing, dissemination of good practice and joint political action among and across different constituencies (progressive politicians, agency staff, NGOs, scholars and activists). Escobar's (2004: 210) 'meshworks' of anti-globalization movements and Routledge's (2003) 'convergence spaces' of grassroots anti-globalization social movements are examples of such alliances.
2   Strengthen alternative and progressive trading structures, such as Fair Trade and Ethical Trade Initiative alliances of companies, trade unions and NGOs (Figures 7.3 and 7.4). Such campaigns

**Figure 7.3**  *Fair trade raisin production, Eksteenskuil, Northern Cape, South Africa*

Source: Author

**Figure 7.4** *Ethical trade: Marks & Spencer flowers in the packhouse, Flower Valley, Western Cape, South Africa*

Source: David Bek 2007

are making a real difference to conditions of workers in the South (see, for example, Bek *et al.* 2007; McEwan and Bek 2006). They currently represent a small proportion of global trade, but the force of anti-globalization and postcolonial criticism is beginning to have some effect, with consumers in the North turning increasingly to more ethically produced goods (Box 7.1). The challenge is how to expand and scale up these initiatives to the point that they are mainstreamed.

## Box 7.1 Ethical consumer concerns: anti-slavery campaigns in the twenty-first century

In 2001, following newspaper reports, British chocolate firms admitted they could not guarantee their cocoa products were exempt from child slavery in Africa:

Much of the cocoa produced in west Africa is grown and harvested with the help of child labour, and much of that is done in slave conditions . . . [In May 2001] authorities in the west African state of Benin said a ship, believed to be the one at the centre of an international hunt for a cargo of suspected child slaves, was thought to be nearing the west African country. The vessel was carrying about 250 children believed to have been 'bought' from their parents and re-sold into slavery. The ship was short of food and water and it is feared the children may have been thrown overboard somewhere along the coast of west Africa. In October [2000], chocolate firms including Cadbury, Nestlé and Mars launched an investigation into claims their products are tainted by slavery. The companies were forced to act following a damning television documentary which claimed that in the Ivory Coast, the world's largest cocoa producer, 90 per cent of plantations use slaves. The workers are unpaid, beaten if they try to escape and sold at markets for a pittance. According to the programme, the Ivory Coast's massive share of the world cocoa market means up to 40 per cent of the chocolate eaten in Britain may be tainted with slavery. Cadbury, Nestlé and Mars, which dominate the UK's chocolate market, all buy Ivory Coast cocoa. They say they have seen no evidence of child exploitation and have accused the documentary makers of exaggerating. However, a spokesman for Mars, which has one million cocoa farms dotted around the country, said he could not rule out child slavery in Africa. He said: 'You simply can't give guarantees. It's almost impossible to police when you're dealing with a country five times the size of the UK'. Nestlé also conceded: 'I don't think anybody could guarantee that they don't buy from countries using child labour. We have not seen any evidence of child slavery, but we are concerned'.

Source: 'The Modern Face of Child Slavery'

http://intranet.csreurope.org/news/csr/one-entry?entry_id=114226, accessed 02/08/07

In response to strong public reaction in the UK, one of its largest supermarket chains, the Co-op, announced in November 2002 that it intended to switch its entire own-brand bars to Fairtrade chocolate in its 2,400 stores, bringing fairly traded cocoa firmly into the mainstream market:

This move sends an important message to manufacturers and retailers of chocolate that using ethically produced cocoa is commercially possible. This will help to protect children and adults in West Africa from forced labour in this sector and guarantee producers a fair price for their product. Importantly, the Co-op is

calling on chocolate manufacturers to make at least one product in
their range carry the Fairtrade mark, and for retailers to follow their
example with their own-label block chocolate. *'This is a significant
move. The best way consumers can be confident that the produce
they use is free from exploited labour is by buying products that
carry a fair trade label,'* Anti-Slavery International Deputy Director
David Ould said.

Source: 'The Co-op switches to Fairtrade chocolate in move to fight slavery' 26 November 2002,
www.antislavery.org/homepage/news/coop261102.htm, accessed 02/08/07

3   Develop the potential for new analyses of agendas and roles
    of key institutions, assessing not only structures and specific
    practices and devices, but also impacts (e.g. Bell 2002; Pritchett
    and Woolcock 2004). This would 'provide evidence for both
    intellectual and political action to effect change in the institutional
    architecture, conventions and practices of global governance'
    (Simon 2006: 18). It would also counter tendencies to essentialize
    the North, notably institutional knowledge in the North, and the
    exclusive and oppressive nature of institutional power, showing
    greater sensitivity to the historical record and the importance of
    causation, context and chronology (Bell 2002: 508). For example,
    Lewis *et al.* (2003) demonstrate the importance of cultural
    differences within and between development organizations, which
    has received little consideration in development studies. Drawing
    on their work in Bangladesh, Burkina Faso and Peru, they
    demonstrate that where development projects involve multiple
    organizations (such as donors, government agencies, NGOs and
    grassroots groups) an analysis of cultures both within and between
    organizational actors can help explain important aspects of project
    performance. Organizational culture is constantly produced within
    projects, sometimes tending towards integration, often towards
    fragmentation. The latter indicates the range of cultures within
    development organizations, which is an important reason why
    some projects fail and why ideas stated in project documents
    (e.g. empowerment) are often not realized (see also Li 2007).
    Morag Bell's analysis of the Carnegie Inquiries (funded by the
    philanthropic US Carnegie Corporation in the early 1980s) is
    another example, interpreting postcolonialism not only as critique,

but as practice that is sometimes taken up to radical effect by erstwhile liberal institutions (Box 7.2).

Bell's analysis of the Second Carnegie Inquiry raises some key points:

● Uncertainty and reflexivity are often apparent within Northern institutions, highlighting the need for closer investigation of the extent to which confidence and authority drive their initiatives. For example, internal dissent within the WB became apparent when Joseph Stiglitz was sacked (see Chapter 5), but this is often ignored in critical accounts of its policies and activities.
● There is a need to focus on specific practices and devices sponsored by Northern institutions (e.g. the use of inquiries, and within these, the use of photographs), rather than solely on institutional structures.
● There is a need to move beyond the characterization of institutions or devices as orthodox or unorthodox, liberal or radical, since the boundaries between these are frequently blurred. The Carnegie Inquiry was run by a liberal charity with an established South African university (University of Cape Town) and thus appeared to be both orthodox and conformist, but the reality was far more complex. Analyzing the postcolonial aspects of the Inquiry suggests that the Carnegie Corporation was aware of the limits of its credibility. Thus, 'the public reputation of Northern institutions may be enhanced by adopting a decentred approach and by accepting the fluidity of their power' (Bell 2002: 521). At the same time, the Inquiry demonstrated that the identity of Northern institutions can be manipulated by groups in the South to facilitate their own transgressive practices.
● Cultural devices like inquiries can work to the mutual benefit of groups in the North and South, revealing their ambivalence. This ambivalence is also often forgotten in critiques of Northern institutions.

## Box 7.2  Postcolonial analyses of Northern institutions: the Carnegie Inquiry

The second Carnegie Inquiry into poverty in South Africa (1982–1984) was based on co-operation across the American and African continents. The Carnegie Corporation saw itself as an intermediary and a vehicle for influence rather than as the central agent of control. It was concerned with a need for 'sensitivity', 'humility' and deep 'skepticism' of the relevance of any American experience to understanding conditions in South Africa (Carnegie Corporation 1980, in Bell 2002: 513). Thus, the impatience of black South Africans with attempts to 'discover' the poverty that they experienced every day was incorporated in determining the Inquiry's focus on producing an action plan to attempt to tackle it. The Inquiry was also co-designed by South African architects, ensuring its hybridity – both as an endeavour to reveal the realities of poverty to a wider world and as a means of producing something socially useful in South Africa. It had to tackle suspicions within apartheid South Africa about the involvement of US money and questions about why it was interfering in South Africa, rather than highlighting and resolving problems of poverty within black communities in the US. The Inquiry addressed this problem by emphasizing its philanthropic credentials and continuities with the first Inquiry that had shed light on white poverty amongst Afrikaners in the 1930s.

The Inquiry drew upon the experiences of South African community leaders, social workers, NGO representatives, academics and activists who all had close experience of poverty. The report used a thousand photographs to paint a picture of the realities of poverty and the everyday lives of black people. A quarter of the papers at the Inquiry's 1984 Conference were produced by black authors, which went some way to countering accusations that the Inquiry was authored by white liberals. The final product was the revelation of the extent of racialized poverty in South Africa, and a fundamental challenge to outward projections by the South African government that it was a 'First World' country when most South Africans were living in impoverished 'Third World' conditions. It successfully raised awareness of poverty among the white population at a time when anti-apartheid rioting was a common occurrence in South Africa's streets, and it also raised international awareness of poverty under apartheid.

The findings of the Inquiry were shown in the mid-1980s at exhibitions around the world and were notable in challenging Eurocentric visions of Africa, particularly through the Inquiry's use of documentary photography. The focus of these on both poverty and blacks as the victims of apartheid (which itself was a revelation to many in the North who had previously refused to accept this as reality), and on portraying how people survive, resist and cope was a radical departure from usual depictions. Following the 1984 Conference at which the findings were first reported, the concept of learning from the South was enshrined in the Corporation's funding activities. It avoided the sensationalist reporting of South Africa that was common in the Northern press at the time and instead rendered meaningful the history and contemporary struggles of a specific country and its people.

Source: adapted from Bell 2002

4   Exploring the potential for integrating political ecology and multi-scale or multi-local sustainable livelihoods analyses as progressive approaches. These recognize the spatial and politico-economic interdependencies and interactive landscapes of the contemporary world, but they avoid the implicit privileging of the local above the non-local that characterizes

some of these approaches (e.g. post-development). Such approaches could be a practical way of combining post-development and postcolonialism (see, for example, Schroeder's (1999) research on agro-forestry and gender politics in the Gambia, and Peet and Watts' (1996) edited collection on liberation ecologies and social movements).

These four elements and associated examples are suggestive of how radical approaches in development studies might engage productively with postcolonialism to produce a progressive agenda for research and action. Putting this into practice also requires a consideration of the implications of postcolonialism for research methods.

## Postcolonial method?

The convergence between the concerns of postcolonialism with subaltern histories, voices and identities and progressive approaches to development has produced what some describe as a 'disabling angst' (Simon 2006: 15) about issues of representation and reflexivity, especially amongst Northern scholars in respect of 'others' in the South. Some Northern scholars have disengaged entirely, but this leaves open the problem of privileging and legitimizing diverse and often conflicting Southern voices and knowledge with little basis for discriminating between them. Moreover, cultural resources can be exploited by community members for their own processes of self-development and empowerment at the expense of others – gender, class, ethnic and age-related power relations are thus of significance (see Watson 2003). They can also be used by outsiders to impose externally defined goals and priorities that do not accord with local aspirations and priorities, perpetuating oppression, exploitation and what might be thought of as a form of neocolonialism (Simon 2006:16). It is important not to forget, then, that culture and power are intertwined, but that there has also been a long tradition of engaging with debates about ethics in research by scholars in the North.

Postcolonial approaches are by no means the first to attempt to foreground and problematize power relations in research. There exists a long tradition of problematizing power relations (from interpersonal to global) in development research (see, for example, Corbridge and Mawdsley 2003; Eade and Rowlands 2003; Hickey and Mohan 2004; Robson and Willis 1997), and of attempting to address these power relations directly (Chambers 2005; Mohan 1999; Nagar and Ali 2003; Tuhiwai Smith 2012). In addition, feminist methods have much to offer postcolonial approaches (see, for example, Butz and Besio 2004; Sharp 2005). However, there is still concern that development research does not fully address the problems of those who are researched (Dreze 2002), and that widespread discussion of methodological issues has not fundamentally altered practices of 'doing' development (Raghuram and Madge 2006: 273).

## Theory as practice

One problem is the disjuncture between the theoretical concerns of researchers in the North and those in the South. Scholars in the North are sometimes accused of using social theory that does not travel (Raju 2002) or translate easily in other contexts. This creates distance between the theoretical concerns of scholars in different parts of the world, which is rendered more problematic by the fact that there is still little attempt in Northern academies to engage with modes of theorizing in Southern contexts. As Raghuram and Madge (2006) argue, a postcolonial method requires holding in tension the mutual constitution of the North and South and the importance of each for the other's theorizations, while also acknowledging and accepting that the South is not entirely constituted by the North. This is essentially a challenge to Eurocentric thinking, or what Chakrabarty (2002) has referred to as a requirement to 'provincialize' Europe (see Chapter 2). A postcolonial approach is thus essential in decentring dominant cultures and their place in development knowledge. Chakrabarty suggests that within social theory, the fundamental categories that shape thinking are devised in Europe; all other contexts are matters of empirical research that flesh out an essentially European theoretical skeleton. Thus, western knowledge and culture are saturated with informal developmentalism – a 'first in the West, and then elsewhere' structure of global time (Chakrabarty 2002: 6) in which cultural objects from the South can be understood only with reference to the categories of European cultural history. Europe is the origin and the South is a pale or partial reflection, to be seen ultimately as coming late, lagging behind, and lacking in originality. Chakrabarty's aim is to decentre Europe so that the former colonies might also be considered originators and Europe becomes one among many locales of knowledge generation and cultural production. In addition,

Said's (1993) notion of contrapunctality demands an understanding that societies on either side of the imperial divide now live deeply imbricated lives that cannot be understood without reference to each other. Drawing on Said, Mufti (2005: 476) calls for a 'global comparativism' to counter the tendency towards hierarchical understandings of knowledge systems and formations that have been in existence from colonial times.

These ideas are of critical importance to development research. The question, then, is how we might create a meaningful academic discourse within development studies, where research can open up and create new spaces of dialogue, crossover and exchange? Central to this is reflecting critically on the questions we ask and how we theorize, and how both are embedded in our academic and institutional contexts (Raghuram and Madge 2006). In other words, we need to think about:

- why we are doing research in the South in the first place;
- how we come to and produce our questions;
- how we analyze and represent our findings based on our subject positionings;
- recognizing that theorization is also a part of method;
- recognizing our multiple investments – personal, institutional and geopolitical – and how they frame possibilities for change;
- asking who gains from the research and why.

The intention should be to avoid what Sidaway (2000: 606) terms Eurocentric 'world-picturing'. According to Raghuram and Madge (2006: 270), this involves:

- complex and potentially contested processes that require recognition of the specificities of historical and spatial production of inequalities in particular places;
- a thoroughgoing dialogue with those who inhabit these places, that takes into account the conceptual landscape of those with whom we engage;
- the willingness to challenge existing hierarchies;
- a commitment to take up issues raised by those who are being researched (see Peake and Trotz 2002; Townsend *et al.* 1995);
- a desire to participate in emancipatory politics;
- a commitment to collaborative research (e.g. Monk *et al.* 2003; Nagar and Ali 2003).

Commitment to dialogue and collaborative research probably comes closest to meeting the ethical concerns of postcolonialism, but this requires a radical opening up of the research process. Without this, there is a danger of appropriating the experiences and knowledge of peoples in the South without radically altering the power relations that structure knowledge production. As Briggs and Sharp (2004: 664) put it, 'the experiences of the marginalized are used in the West, but without opening the *process* to their knowledges, theories and explanations'.

Opening up the process of research presents a number of challenges. According to Raghuram and Madge (2006: 271), these include:

1   Reflecting on the politics behind the process of producing research questions and considering whether these can arise out of dialogue with the research subjects.
2   Modifying the exclusionary practice of theory to address the materialities of the postcolonial world.
3   Thinking through and acknowledging how the identities we bring to the 'field' are linked to our investments in the broader geopolitical context of the neo-liberal Northern academy.
4   Developing an agenda for scholarly action to enable the development of a postcolonial method.

There is thus a need to think about a broader set of issues, including the moral imperatives underpinning development research, the ethics of research, techniques, ways of writing and so on.

## Moral imperatives and ethics in development research

Postcolonial critiques raise profound issues concerning the ethics of research and knowledge production more broadly. The responsibility of researchers towards their 'subjects' has long been recognized in development studies. For many years, most researchers have been bound by formal codes of ethics (for example, those outlined by the American Anthropological Association or the British Sociological Association). These codes attempt to articulate concern through a regulatory framework that outlines the ethical obligations of researchers towards the peoples, places and materials they study as well as to the people with whom they work. These concerns also resonate with participatory approaches in both community development and social research and have stimulated further questions about the power relations between researcher and researched, and the moral imperatives underpinning development research in the field. For example, Chambers (1983, 1997) and other advocates of participatory approaches have argued that the primary purpose of the development professional is to help the 'targets' of development interventions to achieve a better quality of life, emphasizing the importance of working '*of* the community *for* the community' (Crespo *et al.* 2002: 63).

These moral imperatives have brought about a shift from seeing the development professional as the expert with the specialized knowledge that will help the underachieving South 'catch up' with the North. Instead, the development professional has a responsibility to act as a facilitator in helping communities achieve their own vision of a better future. An ethic of 'putting the people first' in development discourses parallels a sense of duty to the researched in feminist and participatory approaches, with priority given to the needs, desires and perspectives of people in the South. For feminist scholars this means privileging the local and abandoning claims to be able to represent an objective truth; for postcolonial scholars it means avoiding speaking on behalf of others, since this very act silences those others and negates their agency. McKinnon argues that embracing these ethics has created a mythical figure of the 'pro-local', 'good' professional subject and her research in Thailand suggests that this is precisely the subjectivity adopted by professionals working there. However, the conundrum remains of 'how to live up to the desire to be a good, participatory, pro-local professional, although participation so often seems to fail' (McKinnon 2006: 28).

As we have seen, participatory methods often look like they have become the new orthodoxy but are often only a superficial engagement with the rhetoric of participation and fail to bring about meaningful empowerment (Cooke and Kothari 2001; Mohan 2001). Despite this, McKinnon suggests that a rhetoric of pro-localism and participation has been instrumental in creating conditions in which local voices can be heard, often for the first time, by those in positions of power – representatives of the state and decision-makers in international organizations like the UN and WB. In addition, characterizing development as always inescapably imperialist is problematic and could have detrimental effects:

> If we are to take seriously the contribution these professionals hope
> to make, then the postcolonial critique, and its expression through
> a pro-local disposition and a participatory development approach,

> must be harnessed in a re-imagining of development professionalism.
> This has to be one that recognizes the long lineage of altruistic aims
> and emancipatory ideals that have shaped a contemporary vision of
> the good development professional; one that, at the same time, takes
> seriously postcolonial critiques of development and their implications.
> This has to come about whilst supporting the aspiration to a project of
> social justice and emancipation, fully attentive to the complex politics
> of development.
>
> (McKinnon 2006: 32)

The promise of postcolonialism within development is its focusing of
critical attention on spatial difference and interconnection, while also
foregrounding the materialities and lived experiences of colonialism
and its legacies. Drawing on the lessons of participatory approaches,
postcolonial development approaches can combine discursive and
textual strategies with material concerns through an understanding
of local scale analysis that reveal localized resistances and
re-appropriations. Hearing, speaking and writing tactics are significant
within this.

## Postcolonial practice: hearing, speaking and writing tactics

Feminists have long demonstrated how dominant narratives might
be disrupted through writing strategies (this is sometimes known as
*écriture féminine*, or gendered women's writing, after the French
poststructuralist feminists who pioneered this tactic). Where this
overlaps with postcolonialism it can have particularly powerful effects.
One example is Trinh's essay, 'Difference: "A Special Third World
Women's Issue"' (1989). Trinh adopts numerous strategies to disrupt
the reader's attempts to place her authorial voice. The rhythms of the
narrative are poetic and playful, disrupting usual academic narrative
styles. The text is polyvocal, littered with quotations, references to
women's writings and fragments of poetry. Trinh shifts from using the
authoritative 'I' to a less authoritative 'i' or hybrid 'I/i'; she also shifts
from first personal narrative to a more removed, third person voice.
The essay proposes a shift away from binary notions of difference (for
example, between 'Third World women' and 'First World women', or
between North and South), towards a hybrid notion of difference in
which identities and subjectivities are not fixed but fluid and through
which connections can be identified as differences are recognized and
acknowledged. While difficult to incorporate into development studies,
these discursive strategies problematize ways of writing that privilege

the authorial voice and silence the voice and agency of those being represented.

Postcolonial approaches can employ tactics through which to 'hear' alternative knowledge or voices of resistance, to reveal historical and contemporary agency (Barnett 1998b; McEwan 1998), or to analyze different forms of resistance that reveal the lived experiences of people otherwise silenced by hegemonic relationships of power. One such tactic is to focus on textual production from the South. This includes autobiography and testimonials (*testimonio*) by people marginalized by impoverishment. These provide a rich site for postcolonial analysis because they demonstrate the ways in which individual lives are affected by a global system of capital. One example of the power of testimony is Domitila Barrios de Chungara's *Si me permiten hablar* (*Let me speak!*) (1978, transcribed and edited by Brazilian journalist Moema Viezzer). Domitila was a Bolivian indigenous woman, daughter and wife of tin miners and political activist. In her testimonial, she recounts her personal life in the Bolivian tin mines. She draws parallels between her suffering and powerlessness at home and as a worker in the mines. She recounts her experiences of exploitation by both the mine owners and the patriarchal system in Bolivia, and tells of hardships and abuse that are part of everyday life in the mining towns. However, she is not simply a victim, but a long-time activist who fights for the well-being of Bolivian women, advocating education and political action as the basis for social change. Her role as one of the leaders of the Housewives' Committee that confronted the government and campaigned for better working and living conditions in her local community resulted in persecution, jail, torture and relocation to try to silence her protest. Her testimonial was her means of resisting this silencing. It was first voiced at the 1975 International Women's Year Tribunal held in Mexico. Domitila agreed to tell her story to make public the struggles and the suffering of the people in Bolivian mining communities. In the introduction to the printed version she writes:

> I don't just want to tell a personal story. I want to talk about my people. I want to testify about all the experience we've acquired during so many years of struggle in Bolivia, and contribute a little grain of sand, with the hope that our experience may serve in some way for the new generation, for the new people.
>
> (Barrios de Chungara 1978: 15)

On publication and translation, Domitila's testimony made her internationally famous (see also Menchu 1984). Her story resonated with both the struggles of Andean women in the face of gendered oppression and international subaltern struggles against social injustices, human rights violations and political corruption that went hand in hand with capitalism. It was thus a powerful critique of the exploitation of indigenous peoples by international capital and a feminist challenge to local patriarchies by advocating a new role for women within Andean communities. Her testimony affirms the value of women's oral narratives in revealing injustices, but it also highlights the differences between western feminism and feminist struggles in the South. The latter, as Domitila argues, fight not against men but alongside them, to work together towards a common objective – a community based on respect of all its members regardless of their gender or their role in society (Barrios de Chungara 1978: 199).

In addition to such narrative forms, as we saw in Chapter 4, Sylvester (2006) and other scholars (e.g. Eshun and Madge 2012; Madge and Eshun 2016) advocate the use of poems and creative writing to access, understand and learn from the lived experiences and subjectivities of people in the South. Similarly, Townsend *et al.* (1995) attempt to produce a polyvocal text in which the voices and agency of rural Mexican and Colombian women can be articulated. Postcolonial strategies can work *for* and *with* poor people so 'that the law of genre will no longer dominate the representation and expression' of people from different parts of the world (Kaplan 1998: 215). They allow for the appropriation of the dominant language for the purpose of 're-inscribing place to produce a regional, or localized, worldview', and thus disrupt 'one of modernity's most pervasive effects – the emptying out of local space by colonialism and neo-imperialism' (Ashcroft 2001: 30–1).

The role that academics can play in creating spaces for the articulation of voices of resistance through textual production is important, since very few marginalized people are able to make their voices heard within the global economy of publishing. Artist and academic Shelley Sacks's work on social sculpture is particularly inspiring in this regard. Her exhibition, *Exchange Values: Images of Invisible Lives*, is an installation of stitched dried banana skins, each corresponding with an oral testimony by the farmer who grew them in the Windward Islands (Cook *et al.* 2000). It attempts to use art to connect the largely voiceless people at one end of the commodity chain (in this case the Caribbean), whose labour remains invisible, to consumers at the other end, using the product itself (bananas) to make this connection. This is a radical critique of the effects of 'free trade' with the potential to empower people at both ends of the commodity chain: the producers by giving them a voice through which to engage with consumers and facilitating connections that the functioning of the global economy often mitigates against; and the consumers who are made aware that the choices they make, over something as simple as a bunch of bananas, can have a direct bearing on the lives of people elsewhere. The project blends art, discourse and materiality to connect disparate places and peoples, breaking down boundaries between core and periphery, former imperial metropole and former colonial hinterland. It suggests ways in which postcolonial theory might be translated into methodology, transgressing boundaries, creating connections and alternatives, facilitating participation and empowerment and giving voice to the previously voiceless. It is also suggestive of the possibilities of using other visual methodologies in radical and productive ways.

A different example comes from Geraldine Pratt's (2012) research on Filipina migrant workers in Canada. Pratt's original research revealed the trauma and racism that Filipina nannies experience, many having left their own children behind to earn money for their families by looking after the children of wealthy middle-class Canadians. One of Pratt's Filipina participants questioned the useful of the research and asked if she was telling her story only for the research findings to gather dust on a shelf. This led to collaboration with theatre artists, who used the interview transcripts to write a testimonial play. The play was shown first in Canada and shone a spotlight on the experiences of the nannies and the lack of empathy and understanding of their Canadian employers. It was then shown in the Philippines, where the government has long encouraged emigrant labour and remittances as part of its national development plan, glossing over the realities of working overseas. As Pratt argues, the play created a performative space of politics and ensured that the voices and experiences of the participants were communicated in an embodied, performative and challenging way.

The constraints of the western academy and academic publishing often impose constraints on collaborative writing: collaborative research is often difficult to translate into collaborative writing when the requirement is a journal article in English or a single-authored PhD thesis. However, collaborative writing or co-production of research outputs is one of the most ethical ways of avoiding some of the pitfalls and violence of representation. Commitment by the western researcher to co-author outputs in the language of their collaborators, in non-academic or popular outlets in the research site (e.g. magazines, blogs, newspaper articles), through different media (e.g. participatory video), or through outlets in which collaborators or research participants would like to disseminate their knowledge, is essential in ensuring that participatory research does not become entirely extractive at the point of dissemination.

## Participatory video as postcolonial method

Video is not a neutral medium, 'having distinct class overtones in terms of ownership' and being 'redolent with ideologies about technology' (Cook and Crang 1995: 70). However, it can form the basis of participatory techniques in development studies that bring to the fore the agency and knowledge of subaltern peoples. It allows participants to produce video diaries and their own filmic representations of their world, as well as developing the skills to critique other people's representations. There is a long tradition of using participatory video in community development practice, particularly in the South. Its application usually involves a scriptless video process, directed by a group of local people, moving forward in iterative cycles of shooting–reviewing. This process aims at creating video narratives that communicate what those who participate in the process really want to communicate, in a way they think is appropriate (Kindon 2003). Participatory video is perceived to be an effective way of reaching and including the most powerless people, traditionally lacking agency in community development, as a means of increasing equitable outcomes (Kindon 2003). Lara Bezzina (2017), for example, uses it to highlight the significance of opening up spaces in which disabled people's voices in the global South can be heard and to ensure these voices are heard by development practitioners and INGOs funding interventions (Figure 7.5). In contrast to disabled people's activism in the global North, her participatory research in Burkina Faso revealed that disabled people here desire economic independence over and above equipment or accessible spaces because they see this as an essential element in challenging the predominant

Figure 7.5  *Participatory video preparation, Disabled People's Organization in Burkina Faso*

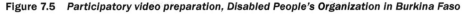

Source: Lara Bezzina (2017)

perception of disabled people as a burden. Communicating this to INGOs is important in ensuring the specific developmental needs of disabled people in global South countries are met. Participatory video is often used in processes of public consultation and mobilization, policy dialogue and to communicate the outcomes of participatory development processes within and between communities or to funding agencies.

One example of effective participatory video was that produced by the Masithembane Housing Association (MHA), South Africa. MHA was located in Khayelitsha, one of South Africa's largest townships and one of the most impoverished in the Western Cape. It was established in Site B, an area dominated by informal settlements. It was a grassroots organization founded by women and used the People's Housing Process as a framework based on the assumption that people and communities can deliver better and more cost-effective housing than the government. The women of MHA worked with a national NGO (People's Housing Partnership Trust), a local NGO (Development Action Group) and an experienced developer in the area (Marnol) and organized to build 220 houses in under a year. This first phase was highly successful and MHA proceeded with further developments to house more families living in shacks. A written evaluation of the project was completed and a video produced by the women and the People's Housing Partnership Trust to document its remarkable success and the stories of families involved. The video was screened for the first time at the community hall in Khayelitsha on 25 March 2001 and copies were handed over to the community.

The video served two purposes. First, it shared the achievements of MHA and the knowledge and experience gained through the People's Housing Process with other communities in South Africa. The aim was to inspire poor people in other areas to build their own homes, to share knowledge on how to form partnerships with provincial governments, NGOs and local developers and to mobilize and employ in local grassroots development many of the existing skills in poor communities that often remain untapped. Second, it recorded the agency of women who have been most marginalized by poverty and gender relations in homes, communities and society at large. It demonstrated to other community members and NGO and local government agents that poor women have agency in decision-making and development, and have considerable abilities given the opportunity to lead local development projects. It also served to underline the nature of women's achievements; seeing themselves on film as role models for others enhanced their sense of self-esteem. For women who have long been oppressed by colonial and apartheid power relations, and persistent unequal gender relations, this self-esteem was essential in providing the foundation for further human and community development.

Participatory video is regarded as an effective tool for participatory research (Frost and Jones 1998); it also allows for translation of postcolonial imperatives into practice. By producing, watching and discussing the video with local people, the researcher can more quickly comprehend their complex perceptions and discourses. This is also a dialogic process that allows communities to analyze their own realities critically and has the potential to create more equitable relationships between researcher and research participants (Kindon 2003). Participatory video often helps the latter to become 'so familiar with the technology of image production that they become image makers themselves' (Banks 2001: 122). One example of this is the work of Eric Michaels, who facilitated Australian Aborigines' use of video in ways that enabled him to understand differences between Aboriginal and European uses of visual representation and information, but that also empowered his informants to respond in a global media context in which they might be seen as marginal or 'invisible' (Pink 2001: 593).

Sara Kindon (2003) has also used participatory video successfully with a group of Maori people from the North Island of Aotearoa/New Zealand, exploring relationships between place, identity and 'social cohesion' in communities in the Rangitikei district. She built deeper relations of trust with the community members with whom she worked and destabilized the usual researcher–researched relationship in a more equitable, participatory research process. This, Kindon argues, allowed other transformations to occur. In this case, participatory video facilitated the sharing of technical and 'local' knowledge between the research team and participants. Using video to record the research process of negotiations and decision-making, as well as the production of locally embedded audio-visual texts, enabled an ongoing or iterative reflection upon the knowledge produced. They thus provided important visual information that, when combined with field notes and interview transcripts, has the potential to create new knowledge and critiques. Since the project was participatory from the beginning, many problems of access and confidentiality that often limit the research process were circumvented, with community members able to set the terms of the research process through negotiation at the beginning of the project. Moreover, transformation occurred in the research process with many of the ideas for the use of video coming from community members rather than the researchers. Many of the community participants also indicated the transformative potential of participatory video at the personal level by expressing their feelings of empowerment, agency and self-worth (Kindon 2003).

There is, of course, a caveat in the use of video – it is time-consuming and relies on building trust with participants. Researchers, therefore, need to be aware of the commitment required when contemplating using participatory video and the importance of maintaining ongoing research relationships once the 'fieldwork' has finished (Kindon 2003). Video is also comparatively expensive but, as equipment costs have fallen, is becoming more accessible. Aboriginal and Inuit communities, for example, are using video in trying to move from being images in the culture of hegemonic groups to producing their own images of themselves and to digitize their own histories. Engaging with such peoples and methods allows the researcher to gain access to various processes whereby numerous communities are seeking to redefine themselves in the modern world (Cook and Crang 1995). Leading scholars are engaging with tribal communities and indigenous peoples in the creation of powerful, postcolonial online archives (see Povinelli 2011b). The groundbreaking Mukurtu content management system, for example, is built on the premise of indigenous curation or co-curation. It lets indigenous people control which materials are archived and made visible, and under what conditions. As Senier (2014) argues, indigenous communities are rapidly creating their own digital spaces that take a range of forms from impromptu practices like sharing and annotating historic photographs on Facebook to sophisticated apps for language revitalization.

The use of testimonials, novels, art, images, films and photography, as well as digital technologies, is potentially useful to development researchers in the context of postcolonial critiques and the urgency of moving away from purely economistic analyses (Prabhu 2003). These resources provide a means by which to understand the lives and agency of subalterns in all their complexity. Such an understanding opens possibilities for reading development that can take into account multiple spaces, especially those inhabited and generated by marginalized groups, and unconventional types of action, with a new understanding of 'production'. This raises the possibility of a more complete, nuanced understanding of subaltern peoples in the context of development, which engages with their creative resistances to various hegemonic forces. As Appadurai (2004) argues, there is a need to strengthen the capacity of the poor to exercise 'voice' – treating voice as a cultural capacity – because it is not just simply a means of inculcating democratic norms, but of engaging in social, political and economic issues that work best in their cultural worlds.

These examples suggest ways in which development researchers might respond to the challenges of postcolonialism, employing diverse methodological procedures for producing a postcolonializing discourse and creating postcolonial archives. As Quayson (2000: 21) argues, postcolonializing is meant to 'suggest creative ways of viewing a variety of cultural, political and social realities both in the West and elsewhere via a postcolonial prism of interpretation'. Sometimes this involves: a rigorous analysis of existing theories; a careful analysis of conditions governing particular subject positions in the modern world; an interdisciplinary reading of the cultural and the political; or finding creative ways beyond the dominant modalities of analyzing particular social issues. In this way, procedures censured as 'facile textualist thought' that contrive to block 'the appeal to any kind of real-world knowledge and experience' (Norris 1993: 182) can be avoided, and the 'politics of the symbolic order' do not displace but complement the theory and practice of politics (Parry 2002: 67). The case of postcolonial digital archives is one example through which hegemonic forms of knowledge might be challenges. They counter the often violent, usually unethical and often exclusionary appropriation of indigenous artifacts and knowledge in (neo)colonial archives in museums, universities and antiquarian societies, and challenge the fabrications associated with them. Development studies would be better informed if it engaged with these archives and present-day development needs would be better understood through engaging with indigenous knowledge, artworks still in current use and living oral traditions.

## The role of academics and how to have ethical encounters

These examples also relate to Spivak's (1993) interrogation of the role of academics in a global context in which her model of a continuing politico-intellectual global activism is useful. Throughout her work, Spivak alludes to the significance of unlearning privilege as loss (see Chapter 2). As we saw in Chapter 6, privileges may have prevented us from gaining access to Other knowledge, not simply information we have not yet received, but the knowledge that we are not equipped to understand by reason of our social positions. Spivak's 'unlearning' of privilege involves working hard to gain knowledge of others who occupy those spaces most closed to our privileged view and attempting to speak to those others in a way that they might take us seriously and be able to answer back. This is especially important in postcolonializing development studies. Despite the problems of the inevitable partiality of the privileged academic view, recognizing the effectiveness of knowledge 'creates an important role for research as an activity of producing and transforming discourses, creating new subject positions and imaginative possibilities that can animate political projects and desires' (Gibson-Graham 2002: 105).

Although Spivak does not systematically spell out what an ethical practice in relation to the subaltern might be, her work is suggestive of how an ethical encounter might be possible. These suggestions are worth bearing in mind for those working in the field of development – as practitioners, as Northern academics, as native informants, as students – who, by virtue of their privileged positions and identities are always necessarily positioned in relationships of power vis-à-vis subaltern subjects. Spivak's strategy can be summarized as follows (adapted from Kapoor 2004: 639–44):

1. Critique from within

Although Spivak argues that we are inevitably positioned within power relations, this should not preclude changing and transforming discourses from within – 'transforming conditions of impossibility into possibility' (1988b: 201). In a sense, we can use Spivak to caution the likes of post-development theorists (e.g. Escobar) against throwing the baby out with the bathwater by being uncompromisingly anti-development. Post-development theorists claim that development is all-encompassing, dominant and oppressive, but if this were the case then none of the following would be able to exist:

- a post-development critic who could claim to be outside of this dominant discourse, and those who do thus ignore their complicity;
- a post-development critic who could represent the subaltern as pure and unentangled in the dominant discourse, which amounts to essentialism and romanticism;
- a post-development critic who could posit a utopian alternative to development, since where would this alternative come from if not from the dominant discourse of development itself, and how would it avoid creating its own disciplining and power structures.

Spivak's alternative is to be vigilant about one's practice and she intimates that it is possible to work within dominant discourses like development and to engage in a persistent critique of hegemonic representations. As Kapoor (2004: 640–1) argues, 'Development may indeed have become a shady business, but this does not mean one cannot retrieve from within it an ethico-political orientation to the Third World and the subaltern'.

2. Acknowledging complicity

One cannot stand outside of discourses, and personal and institutional desires and interests are always written into one's representations.

Acknowledging this is the basis for practising vigilance – in other words to be hyper-self-reflexive about our complicities. At the very least, this should reduce the risk of personal arrogance and 'geoinstitutional imperialism' (Kapoor 2004: 641) and create the conditions for a nonhierarchical encounter with the South.

3.   Unlearning privilege as loss

One cannot do fieldwork without first doing homework. The starting point is to recognize the limits of knowledge – that privilege might mean that parts of the world and of people's lives are not immediately knowable. This recognition challenges the taken-for-granted notion that dates back to the Enlightenment that the West holds the intellectual keys to understanding the world. It also reverses the gaze from 'over there' to 'in here', since it forces one to look inwards to acknowledge one's prejudices and learned habits, as well as the fact that these prevent one from seeing what might be hidden from one's view.

4.   Learning to learn from below

For Spivak, 'learning from below' (which is considered a politically correct way of doing development) is insufficient. Serious and meaningful learning from the subaltern first requires a more profound step – learning to learn, or the on-the-ground version of unlearning privilege as loss. Kapoor (2004: 641–2) summarizes it thus:

● I must clear the way for both me and the subaltern before I can learn from her/him.
● It is suspending my belief that I am indispensable, better or culturally superior.
● It is refraining from thinking that the Third World is 'in trouble' and that I have the solutions.
● It is resisting the temptation of projecting myself or my world onto the Other.
● It is resisting assuming that concepts such as 'nation', 'democracy' or 'participation' are natural, good or uncontestable and to impose them unproblematically in the field is to forget that they were written elsewhere in the social formations of the West.
● It is remembering to unlearn – stopping oneself from always wanting to correct, teach, theorize, develop, colonize, appropriate, use, record, inscribe, enlighten, speak and write.

By following these steps, it is possible to create openness to the subaltern in which the subaltern becomes not the object but the subject of development. It also lays the foundations for a two-way conversation, in which knowledge can flow from the South to the North and not always in the other direction, and for non-exploitative and mutual learning. This is not an evasive or delaying gesture, as some critics have accused Spivak, but a crucial step in being able to encounter the South in an ethical manner and without which learning from below continues to be benevolent and self-consolidating.

5.  Working without guarantees

It is important to remember that the subaltern is heterogeneous; coming to terms with the Other's difference means reckoning with the impossibility of knowing it, accepting that it exceeds one's understanding or expectations. In development, then, one ought to accept the 'nonspeakingness' of the subaltern and her/his refusal to answer or submit to the gaze and questioning of the outsider. Silences (as opposed to silenc*ing*) need to be recognized as a form of resistance and agency. These can take the form of reticence, equivocations, lies, secrets, refusal to be named or labelled and so on. Working without guarantees means being aware of the blind spots of one's power and representational systems, and either accepting failure or learning to see it as success. For development professionals it means being open to the limits of knowledge systems and of the profession: 'enabling the subaltern while working ourselves out of jobs' (Kapoor 2004: 644).

Scholars working with postcolonial and decolonial theories need to be mindful of their location in a political economy knowledge. This becomes problematic when these theories are circulated in western academic spaces in ways that fail to challenge the colonialist routes and continuing inequalities of those spaces. Silvia Rivera Cusicanqui (2012; see also Noxolo 2017) describes how theories and concepts that originate with indigenous scholars are appropriated, repackaged and disconnected from the political struggles and experiences in which they were formed and become instrumental in the academic careers of already privileged scholars in the global North. Western scholars embracing indigenous and subaltern knowledge may be well meaning, but unless they are actively working to decolonize the spaces of the academy and working in ethical ways with theories and ideas from elsewhere, they remain part of a system that perpetuates inequality and exploitation.

## Becoming authors of postcolonial development knowledge

Becoming authors of postcolonial knowledge within development studies requires careful consideration of numerous issues, some of which relate to the politics of speaking and writing and others to questions about the kinds of topics we choose to explore, the methods we employ and the ethical relationships through which knowledge production takes place. Postcolonializing development studies is a political project that demands engagement with debates and practices from 'the margins' and, in so doing, works against divisive geopolitics of knowledge. No longer is development knowledge thought of as putatively Northern, but instead it is re-conceptualized as both local and global, situated and networked, relational and complex, mutually constituted and sometimes conflictual. Such an approach opens the possibility of 'theorizing back' from the South and also of transversal knowledge production. In other words, through creatively crossing and redrawing borders that mark significant politicized differences, differences and commonalities can be acknowledged and the possibility of learning from the experiences of others becomes possible.

Such an approach also allows simple assertions that sometimes inflect critical development studies about Northern power, especially institutional power, to be broken down. Instead agency and resistance in the South is brought to the fore. An example of this is Donald Moore's (2000) sophisticated ethnographic study of the cultural politics of resistance by a rural Zimbabwean community to successive attempts to impose modernization in the name of development by colonial and postcolonial states. Moore's focus on agency and resistance reveals the insights that a postcolonial and critical analysis of development can yield. Topics previously neglected by development studies also provide fertile ground for producing postcolonial development knowledge. One example is the biopolitics of development (Chapter 5). Biopolitics began under colonialism, is perpetuated in development, and has not been uprooted by Third Worldism, anti-colonialism or postcolonialism. Postcolonial perspectives on development could highlight the ways in which biopolitics continue to regulate and control the bodies of people in the South, very often to their detriment.

Two examples of this are health care and human trafficking (Patel and McMichael 2004: 240–1). The current global trajectory of privatization of services has affected access to health services in countries in the

South. It also heightens the policing of bodies through payment systems, which demand an accounting system at the level of the individual body, a prior history of health and access to cash (and hence paid labour) for the patient. It invokes a system of state monitoring, evaluation by capital and control of individuals, individuating bodies as repositories of unequal and delimited (market) rights. This is also a transnational phenomenon increasingly under the scrutiny of supranational organizations such as the WTO through its treaty on General Agreement on Trade in Services (GATS). A postcolonial approach would make visible and critique this policing of bodies and work towards producing more emancipatory models of health care delivery. Human trafficking is an example where biopolitics does not need a state, since capital can recognize and regulate the body through criminal activity. It is estimated that 16 million people have been trafficked for forced labour (ILO 2017), and 71 per cent of trafficking victims around the world are women and girls. After drug smuggling and gunrunning, human trafficking is the third largest illegal trade, earning profits of roughly $150 billion a year for traffickers (ILO 2014). Child trafficking already dwarfs the transatlantic slave trade at its peak by a magnitude of ten. Destinations include farming, restaurant labour, domestic servitude, fishing, mail-order brides, market stalls, shop work and the sex trade. The *New Internationalist* (July/August 2001: 18) refers to this as twenty-first-century slavery. This is not simply a political-economic issue, but also a biopolitical one.

Finally, a comparative approach, rather than a problematic area studies approach, demonstrates the interconnectedness between histories, experiences and struggles in diverse parts of the world. Rather than focusing on 'out there', this requires understanding both the historical and experiential specificities/differences and connectedness of people's lives. It allows for research into social and economic processes and specific issues such as alternative economies, environmental justice, sex work and community development, highlighting points of contact and connection as well as disjuncture, and the foregrounding of issues of domination, agency and resistance.

## Postcolonialism, field research and working in the South

Having done our 'homework' on hearing, speaking and writing tactics, it thus becomes possible to consider further ethical issues raised by the practicalities of working in the field. There are many possibilities of learning in the field of development between development scholars and practitioners working in both North and South (McFarlane 2006). As we have seen, this requires a conception of learning that is critically reflexive of the power relations between different groups and that does not simply fall back into familiar, historical patterns of colonization or into contemporary tendencies of aid-based conditionality. Instead, this conception of learning between different contexts should help pluralize the production of knowledge and lead to a more globally informed – what might be termed postcolonial – social science (Spivak 2003a).

Many people who embark on development work do so out of a sense of solidarity with the people with whom they work in the South. Their actions and intentions are often couched within a discourse of solidarity. However, the unequal living conditions that the development worker encounters in the field stand in stark contrast to this discourse. As Baaz (2005: 85) argues, for Europeans, development work and fieldwork more generally within the South involves both a 'discovery of whiteness as a marker of difference' and a 'discovery of wealth and privilege'. This might seem strange, given that most development workers are aware of global inequalities and a desire to combat these inequalities is often what drives their ambition. However, recounting the experiences of Scandinavian development workers in Tanzania, Baaz (*ibid.*) writes:

> [T]he process of relocation increases this awareness . . ., it involves a certain 'rediscovery' of wealth and privilege, in the sense that it is one thing to experience global inequalities mediated through the media and quite another to encounter them in everyday life, at work, and even within the home (through domestic workers, whom most development workers eventually come to employ). But most importantly, relocation seems to involve the rediscovery of privilege and inequality even though the discourse of solidarity, by emphasizing that development workers should 'relate to colleagues at their level', tends to mask the inequalities.

The rediscovery of wealth and privilege creates both feelings of estrangement and guilt, and hence the tendency of development workers to seek refuge in ex-patriot communities rather than to be reminded of inequalities.

Interactions by researchers with and representations of people in the global South are inevitably loaded. They are determined by one's favourable historical and geographical position, one's material and cultural advantages that have been shaped by imperialism and capitalism, and one's identity as either privileged Northerner or 'native informant'. Spivak suggests that when the investigator, either knowingly or subconsciously, disavows his/her complicity, or pretends that the research 'is doing no harm', s/he is liable to speak for the subaltern, 'justifying power and domination, naturalizing Western superiority, essentialising ethnicity, or asserting ethnocultural and class identity, all in the name of the subaltern' (Kapoor 2004: 631). Put another way:

> Though the speaker might be trying to materially improve the situation of some lesser-privileged group, the effect of her discourse is to reinforce racist, imperialist conceptions and perhaps also to further silence the lesser privileged group's own ability to speak and be heard.
> (Alcoff 1991:26)

An example of a well-meaning but ultimately problematic attempt to engage with the opinions of people in the South was the WB's *Voices of the Poor*. In the late 1990s, the WB conducted an analysis of 60,000 poor men and women in 47 countries around the world. Seventy-eight Participatory Poverty Assessments (PPAs) were conducted, which the WB argues were based on 'discussions' with poor people. Although there are many problems with how the WB drew on these 'voices of the poor' to legitimate its discourses about the opportunities of markets (Narayan 2000), the study engaged in an ambitious attempt to understand people's own perceptions and experiences of poverty across many diverse cultural contexts. However, there is a need to problematize how issues of subjectivity and definitions of knowledge and poverty are articulated in this and similar surveys, and the extent to which history and culture are considered important to understanding poverty in its spatial variation.

The report demonstrated that fieldwork and conducting research in developing countries was difficult for the WB team and emotionally challenging for many of those employed to conduct the interviews. Some required counselling for the trauma they experienced because of what they had seen or heard (Narayan 2000). Fieldwork also allows us to explore locally grounded issues and globally informed praxis and to transcend and reinterpret the relationship between researcher and researched, and the gulfs and separations between the different worlds of global development. However, fieldwork is also deeply problematic.

Studies such as the WB's *Voices of the Poor* survey claim to use 'participatory' approaches and methodologies and assume that they are successful in creating a space in which the marginalized can speak. Yet their very representation of these people as 'poor' (and very few of them are named in the report) is an erasure of their voices.

Fieldwork is embodied and racialized (Abbott 2006; Raghuram and Madge 2006). As discussed, within development studies various forms of participatory research are seen as parallelling a postcolonial epistemology, because they claim to co-produce knowledge, minimize power differentials and capitalize on cultural hybridity (Mohan 2001). However, such ideals often remain aspirational since they are constantly frustrated by racial, gender and class hierarchies. Despite this, more field-based studies are now beginning to position and reflect upon the researcher's subjectivity and to reveal the spatiality and temporality of power relations in the research process. As Power *et al.* (2006: 233) argue, this kind of work can disturb and unsettle what are usually taken as the central concerns of development studies and, at the very least, demand that we think more critically about what development means and how 'we' envision and practise it.

While acknowledging its problematic nature, it is important not to be deterred from fieldwork by trenchant criticisms. As Madge and Raghuram (2006: 275) argue, fieldwork brings us most directly into contact with the everyday lives and material realities of others and through a democratic impulse has considerable potential for developing a dialogic version of research whose outcomes are also accountable. According to Briggs and Sharp (2004: 673), 'rather than abandon fieldwork, it is perhaps now more than ever necessary to de-centralise Western centrism'. To do fieldwork ethically, therefore, it is important for Northern academics to decentre themselves: geographically, linguistically and culturally. Participatory methods are a step towards this in creating meaningful connections with researchers in the South, ensuring that fieldwork is collaborative and engaging with local researchers at each stage of the research process, from project formulation and design to publication.

## Students and field study

One area of difficulty for development academics in the North is the issue of student field trips. Field study visits are seen as one means of bridging the distance between North and South and helping to disrupt some of the stereotypical assumptions that relatively affluent students in the North may consciously or subconsciously hold about lived experiences in the South. However, the problem lies in the fact that fieldwork is often seen by some academics in the North as a 'normalized, legitimate, apolitical practice' (Abbott 2006: 327). It is incumbent, therefore, on teachers to encourage ethical awareness and promote postcolonial field practices. Across several disciplines, teacher-led overseas field study has taken on board the critiques associated with feminist, development, cultural and environmental studies to question issues such as cultural representations (see Mains 2004), environmentalism (see Binns 1992) and voyeuristic development tourism. There is greater critical reflection than previously on the impacts on local people, especially in the South, of groups of relatively affluent Northern university students continually arriving in groups, often on tourist package trips, to carry out their field studies (Robson 2002). Despite this progress, however, Dina Abbott (2006: 328) cautions that much of this literature on long-haul field study 'continues to side-step the awkward questions raised by race, gender, poverty and power', with uncomfortable echoes of colonial period expeditions and 'discovery'.

Proponents of fieldwork argue that it provides an opportunity for cultural exchange, rather than a one-way learning process. It can be viewed as positive for all parties and a means of improving cross-cultural understanding. For example, Robson (2002) discusses the possibilities for European students to reciprocate African dancing displays with displays of western dancing styles. However, as Abbott (2006: 329) argues, this assumption of 'cultural exchange' often fails to question the meaning of this in a context where 'the two groups (the local community and the students/tutors) are interacting in historically racialized spaces and places in the midst of conditions of material poverty'. It could be argued that such activities are haunted by and even replicate past racialized practices.

There is a need, therefore, to locate fieldwork within a context of historical and racialized power relationships to understand the standpoint of those we 'study' and on whom we impose ourselves. Ethical approaches to fieldwork might learn from feminist critiques, which have argued that subjective interpretations are central to understanding societal and power relations. Using the example of the Gambia, Abbott points out that white student groups are often indistinguishable to local communities from tourist groups, since they are usually part of the same tourist package. Thus, tactics for 'fobbing off' (Abbott 2006: 335) tourists with stories they wish to hear are also applied to field groups. Certain 'authentic' local customs are performed without any deeper explanation, because the reward is much needed revenue, which enables them to survive the non-tourist season. Community codes of conduct remain hidden from visitors' and students' eyes. Gift giving by visitors, for example, might be well meaning but may insult communities whose social norms of hospitality include welcoming visitors with gifts. Gambians are distressed that children have become skilled in begging, since this is a source of family shame. These invisible realities are rarely considered if fieldwork is regarded as an apolitical 'cultural exchange'.

The question of fieldwork is one that still poses unresolved ethical difficulties in development studies. Some academics believe it is unethical to take affluent Northern students to countries in the South unless they can speak the language of the country they are visiting. This is one way of militating against the objectification of peoples and places, but it does not completely resolve the ethical dilemmas. Increasingly, too, ethical issues are also being raised about another type of experience students in the North may have of the South – the gap year. This is often perceived as a more ethical engagement by affluent students with peoples and communities in the South since it is based on volunteerism. However, it can often perpetuate stereotypes and ethnocentric assumptions about the needy, passive South and the dynamic, capable, generous North. In addition, what was once seen as a form of charity has now become big business in ways that are often deeply unethical (Box 7.3). There are no easy solutions, therefore, to the dilemmas raised by fieldwork and, since many students have some experience of the South, either as tourists or as gap year visitors, fieldwork can be a means of challenging perceptions, or at least raising awareness of the profound ethical questions that being in the field provoke. At the very least, students should be asked to consider a range of issues and questions, such as:

- How are you positioned in power relations (historical and contemporary) between North and South?
- How are you aware of your relative wealth?

- Why are you doing research over 'there' (e.g. Africa, Latin America, South East Asia) rather than 'here' (Britain, Europe, USA, Japan, Australasia)?
- Who benefits from your field visit and how?
- How can doing fieldwork blur boundaries between 'here' and 'there'?
- In what ways is fieldwork a learning process?
- What does fieldwork teach us about ourselves?
- How might fieldwork foster knowledge and understanding of other people's cultures?
- Fieldwork can be used to claim an authoritative voice to represent the other; What methodologies can be adopted that challenge the process of 'unequal exchange' (working with and for and giving back to local peoples)?
- How might fieldwork be used to establish connections with people and other cultures?
- Would you consider working in the international development sector (EU, UN, DfID, USAID) or with NGOs and civil society organizations?
- Would you reflect on power relations, ethical issues and development theory the next time you are a visitor to a country in the global South?

---

## Box 7.3  The great gap year debate (from *The Times*, 14 August 2007)

*'Gap year students told to forget aid projects'*
*(by Rosemary Bennett)*

One of Britain's leading charities has warned students not to take part in gap year aid projects overseas which cost thousands of pounds and do nothing to help developing countries. Voluntary Service Overseas (VSO) said that gap year volunteering . . . has spawned a new industry in which students pay thousands of pounds for pre-packaged schemes to teach English or help to build wells in developing countries with little evidence that it benefits local communities. It said that 'voluntourism' was often badly planned and spurious projects were springing up across Africa, Asia and Latin America to satisfy the demands of the students rather than the needs of locals. Young people would be better off simply travelling the world and enjoying themselves, it added. Judith Brodie, the director of VSO UK, said: 'While there are many good gap year providers, we are increasingly concerned about the number of badly planned and supported schemes that are spurious – ultimately benefiting no one apart from the travel companies that organise them'.

VSO is drawing up a code of good practice to help gap year students to find genuine voluntary work abroad. The charity cited the case of a volunteer teacher in Africa, who was surprised to be shunned by her fellow teachers, then discovered that her placement had led to a colleague being made redundant. In another case, a volunteer in Mexico who thought that she would be working on a rural conservation project, spent six months behind a desk in an office inputting data onto spreadsheets. Another volunteer was asked to survey an endangered coral reef in the Indian Ocean and discovered that it had been surveyed countless times before by previous volunteers.

Taking a gap year used to be the preserve of only the wealthiest students, but it is now big business. Up to 200,000 people do it every year, including 130,000 school-leavers. The average gap year traveller spends £4,800, and numerous companies have sprung up to get a slice of the market by offering pre-packaged trips to projects for just two weeks at a time. Gapyear.com, one of the biggest players, is offering places on dozens of voluntary projects, including work on a South African horse safari for £2,400 or two months observing coral and marine life in Borneo for £1,895. Another firm, i-to-i, is offering work with orphans in Argentina for £1,095. In most cases the price does not cover the flight, but in-country travel, accommodation and an orientation session on arrival is included. Ms Brodie urged students to go backpacking instead. 'Young people want to make a difference, but they would be better off travelling and experiencing different cultures, rather than wasting time on projects that have no impact . . .', she said.

Tom Griffiths, founder of gapyear.com, defended his business. 'Some companies raise the expectations of students to unrealistic levels and make them think they will change the world. When they get there they discover they are only small players in the project and feel disappointed', . . . Raleigh International backed VSO's call for caution. 'Students should be very careful about the voluntary work they choose', a spokeswoman said.

Questions to consider having read this article:

- Why do development aid, charity organizations and gap year travel companies take young people overseas?
- Is on-the-ground experience useful in combating the superficiality of much media reporting about the South?
- What are the problems with student expectations of 'wanting to help', 'wanting to make a difference' and 'wanting to save the world', and is the gap year experience useful in challenging these expectations?
- Is the learning experience of the volunteer as important as what they might contribute in the South?
- Who benefits from the gap year activities?
- Do volunteers become longer-term advocates rather than just 'voluntourists'?
- What would be the most important consideration in choosing an organization through which to organize a gap year in the South: the location in which it is working or having a record of well-structured development work in specific locales?
- Is making a business out of volunteering ethically problematic?

## Postcolonial pedagogies

### Globalizing the curriculum

Postcolonialism alerts us to the significance of the dissemination of cross-cultural knowledge through pedagogical strategies to 'internationalize' the curriculum. It shares with feminism a commitment to foregrounding the politics of knowledge in bridging the 'local' and the 'global' in development. How we teach is as significant as what we teach. And the micro-spaces of the academy are intertwined with, rather than separate from, the macro-spaces of the global economy (see Spivak 1988, 1990, 1993). Crossing cultural and experiential borders allows us to position historical narratives of experience in relation to each other and to theorize relationally, which determines how and what we learn. Drawing on Mohanty's (2002: 518–24) feminist approach, three pedagogical models can be proposed for internationalizing development studies curricula, each based on different politics of knowledge. Each is grounded in specific conceptions of the local and the global, of agency and of national identity and presents different ways of crossing borders and building bridges. The first two are problematic; the third posits a comparative model rooted in notions of solidarity and is the most useful and productive pedagogical strategy for cross-cultural development studies.

1. Development-scholar-as-tourist model (or colonial discourse model)

This involves making brief forays into cultures in the South and addressing these from a Eurocentric development studies gaze. People of the South are represented as global victims. The effects of such stories are that students and teachers are left with a clear sense of the difference between the local ('here', self, western, the North) and the global ('there', other, non-western, the South). As Mohanty (2002: 519) argues, 'The strategy leaves power relations and hierarchies untouched since ideas about center and margin are reproduced along Eurocentric lines'. We might add accounts of Indonesian workers in Nike factories, but such examples stand for the totality of lived experiences in such cultures. They fail to position workers in the reality of their everyday lives, and instead emphasize the difference from the western norm in stereotypical terms. This strategy produces the monolithic 'Third World' that postcolonialism seeks to disrupt.

## 2. Development-scholar-as-explorer model

This strategy is common in area studies where the foreign 'other' is the object and subject of knowledge. Here the local and the global are both defined as non-Euro-American; international implies outside of 'here'. This strategy can solidify notions of difference in an 'us-and-them' dichotomy but, unlike the tourist, the explorer can provide a deeper, contextual understanding of development issues in tightly defined geographical and cultural spaces. However, as Mohanty (2002: 520) suggests, 'unless these discrete spaces are taught in relation to one another, the story told is usually a cultural relativist one, meaning that differences between cultures are discrete and relative with no real connection of common basis for evaluation'. While this is sometimes seen as a culturally sensitive way of internationalizing the curriculum, questions of power, agency, justice and common criteria for critique are silenced. Courses on Africa, Latin America or South East Asia can be sophisticated and complex studies, but they are viewed entirely as separate from western intellectual projects. This can be seen in UK Geography, where economic geographers are almost exclusively concerned with advanced economic contexts, while those working on economic issues in the South are referred to as 'development geographers' or 'area studies specialists' (Jones 2000). This division of academic knowledge tends to preclude learning from the South. Similarly, US women's studies courses on Latin America or the Third World are not part of the intellectual project of US race and ethnic studies. In both examples, the UK/US is not seen as part of 'area studies'; rather area studies is about places 'out there'. The clear problems with this strategy are that it reinforces a tendency to universalize from the (unacknowledged, British/European/American) parochial, while globalization, as an ideological, economic and political phenomenon, actively brings the world under connected and interdependent discursive and material regimes.

## 3. Comparative, solidarity-based model

This strategy is based on the premise that the local and the global exist simultaneously and constitute each other. It foregrounds the material, conceptual and temporal links and the relationships between places. It assumes a comparative focus – differences and commonalities exist in mutual relation and tension in all places. As Mohanty (2002: 521) argues, 'relations of mutuality, co-responsibility, and common interests' are emphasized. A comparative course would demonstrate the interconnectedness between histories, experiences and struggles in diverse parts of the world; the story of development in the North being historically shaped by its relationship with the South through imperialism, slavery and now globalization is but one example. Students can move away from the 'add and stir' and the relativist 'separate/different but equal' to a 'co-implication/solidarity' model (Mohanty 2002: 522). This requires understanding the historical and experiential specificities and differences of people's lives and the historical and experiential connections between people in different national, ethnic and cultural communities. It thus suggests organizing curricula around social and economic processes and histories of various communities in substantive areas (geopolitics, alternative economies, environmental justice, sex work, citizenship and human rights) and looking for points of contact and connection as well as disjunctures. This also allows for the foregrounding of issues of domination, agency and resistance. Moreover, agency and resistance can be formulated across the borders of nation and culture, providing the basis for an ethics and politics of solidarity. Such pedagogies can thus theorize experience, agency and justice through a more cross-cultural lens, rather than a Eurocentric or culturally pluralist one. In practice, a postcolonial educational framework aims to avoid a model based on compassion and seamless development. Instead, it outlines an approach that attempts to go beyond ethnocentrism, essentialism, reversed racism and Orientalism. Table 7.1 illustrates this difference in general terms.

## Decolonizing the university

An alternative to the struggle to decolonize traditional spaces of education through, for example, indigenizing research, teaching and governance in traditional educational establishments, is to create different educational establishments in which indigenous or subaltern ways of knowing form the basis of scholarly and pedagogical practices. One example is the creation of intercultural universities – 'rooted in and connected with the political and social struggles of marginalized groups' (Cupples and Glynn 2014: 58) – across Latin America. In contrast to the universalizing and globalizing ambitions of most universities, these institutions emphasize knowledge and learning as dialogue between universal discourses and local knowledge. The universities include URACCAN (University of the Autonomous

*Table 7.1*   *'Compassion' versus postcolonial educational framework*

| | A compassion/seamless progress framework | A postcolonial framework |
|---|---|---|
| *Problem* | Poverty, helplessness. | Inequality, injustice. |
| *Nature of the problem* | Lack of 'development', education, resources, skills, culture, technology, etc. | Complex structures, systems, assumptions, power relations and attitudes that create and maintain exploitation and enforced disempowerment and tend to eliminate difference. |
| *Justification for positions of privilege (in the North and South).* | 'Development', 'history', education, harder work, better organization, better use of resources, technology. | Benefit from and control over unjust and violent systems and structures. |
| *Basis for caring* | Common humanity/being good/sharing and caring. Responsibility *for* the other (or *to teach* the other). | Justice/complicity in harm. Responsibility *towards* the other (or to *learn with* the other) – accountability. |
| *Grounds for acting* | Humanitarian/moral (based on normative principles for thought and action). | Political/ethical (based on normative principles for relationships). |
| *Understanding of interdependence* | We are all equally interconnected, we all want the same thing, we can all do the same thing. | Asymmetrical globalization, unequal power relations, Northern and Southern elites imposing own assumptions as universal. |
| *What needs to change?* | Structures, institutions and individuals that are a barrier to development. | Structures, (belief) systems, institutions, cultures, individuals, relationships. |
| *Why?* | So that everyone achieves development, harmony, tolerance and equality. | So that injustices are addressed, more equal grounds for dialogue and power are created. |
| *What individuals can do* | Support campaigns to change structures, donate time, expertise and resources. | Analyze own position/context, participate in changing structures, assumptions, identities, attitudes and power relations in their contexts. |
| *Basic principle for change* | Universalism (non-negotiable vision of how everyone should live, what everyone should want or should be). | Reflexivity, dialogue, contingency and an ethical relation to difference. |

*(Continued)*

*Table 7.1 Continued*

|  | A compassion/seamless progress framework | A postcolonial framework |
|---|---|---|
| *Goal of global citizenship education* | Empower individuals to act (or become active citizens) according to what has been defined for them as a good life or ideal world. | Empower individuals: to reflect critically on the legacies and processes of their cultures and contexts, to imagine different futures and to take responsibility for their decisions and actions. |
| *Strategies for the global dimension in education* | Raising awareness of global issues and promoting campaigns. | Promoting engagement with global issues and perspectives and an ethical relationship to difference, addressing complexity and power relations. |
| *Potential benefits of the approach* | Greater awareness of some of the problems, support for campaigns, greater motivation to help/do something, feel good factor. | Independent/critical thinking and more informed, responsible and ethical action. |
| *Potential problems of the approach* | Feeling of self-importance or self-righteousness and/ or cultural supremacy, reinforcement of colonial assumptions and relations, reinforcement of privilege, partial alienation, uncritical action. | Guilt, internal conflict and paralysis, critical disengagement, feeling of helplessness. |

*Source*: Andreotti 2007: 8–10

Regions of the Nicaraguan Caribbean Coast) and Bluefields Indian Caribbean University, which were created in Nicaragua in the 1990s. Others have since been established across the continent: for example, Universidad Autónoma Indígena Intercultural in Colombia, Centro Amazônico de Formação Indígena in Brazil, Universidad 'Amawtay Wasi' in Ecuador and Universidad Intercultural de Chiapas in Mexico. These universities are open to non-indigenous students on the principle that white and mestizo people ought to respect and learn from indigenous and other marginalized ways of knowing. As Cupples and Glynn (*ibid.*) point out:

> practising interculturality is often hardest for dominant groups, as
> black and indigenous populations are used negotiating hegemonic

cultures and knowledges while holding on to their own, whereas white privilege allows the dominant to ignore subordinated ways of knowing and being.

These decolonizing spaces of higher education could be sites from which alternative ideas about development might emerge. URUCCAN's vision statement, for example, expresses a desire to be 'a leading community intercultural university that accompanies the indigenous peoples and mestizo and Afro-descended communities of the region in development processes which promote citizenship' (in Cupples and Glynn 2014: 61). Its students are 'outspoken, engaged, articulate, and willing to share their ideas about globalization, development and culture' (*ibid.*: 62). The students are clearly confident in a space in which they feel supported and in which their ways of knowing and being are both validated and legitimated. The university is connected into local communities through diverse forms of engagement and is focused on responding to local needs and realities, but is also connected to a strong international network, including both mainstream and intercultural universities elsewhere. As Cupples and Glynn (2014: 64) argue, the existence of intercultural universities disrupts (neo)colonial assumptions about indigenous and black peoples, challenging ideas that they are incapable of scholarly activity or that their knowledge stopped developing after contact with Europeans. By contrast, indigenous knowledge is dynamic, contested and heterogeneous and, as with European knowledge, will 'continue to develop, hybridise and function as sites of productive disagreement' (*ibid.*). They continue to engage with western knowledge and do not discount the development value of Enlightenment thought, but they create a space in which scholars can reflect on 'different ways of knowing and on the strengths and pitfalls of different models of development' (*ibid.*: 65). Such decolonized spaces of learning are one possible model for those advocating decolonizing the curriculum in other parts of the world (e.g. the Fees Must Fall campaign and broader debates about decolonizing education that have emerged out of the Rhodes Must Fall campaign in South Africa – see Chapter 1).

## Ethics and hope in postcolonial pedagogies

As we have seen throughout this book, postcolonialism has the potential to create more ethical development thinking and practice. 'Not infected (as such) by the know-all history of development studies, but just as embroiled in thinking about the West, it is freer to criticize colonialism and creeds of progress' (Sylvester 1999: 717). However, it is important that postcolonialism goes beyond merely criticizing; it needs to challenge students' perceptions and offer more in the way of a hopeful politics. This begins by challenging stereotypical images:

> To the millions of people who live in Africa, China and India such . . . [images] are a constant reminder of what they are not. Whatever else they may be is banished into oblivion by a universalizing metric that rank orders peoples by the average cash value of the nation's market basket of consumption. They are pitied, 'wretched of the earth', living in the periphery of the world system. Young American (and European) graduates are told that Africans and Indians live in less developed countries. Young as they are, they know that they 'rank' higher than millions of those 'other' people from underdeveloped countries.
>
> (Yapa 2002: 43)

In addition, if we constantly critique but pose no alternatives we are at risk of discouraging affective attachment to the world; it is essential that hope is not extinguished in critique. For teachers, this issue is particularly acute, since teaching generates affect, which might include empathy or detachment. There is a delicate balance to be struck, for example, in persuading students of African development that they should read Jeffrey Sachs (2005) with a profoundly critical eye (Chapter 4). Some students like Sachs because he appeals to their sense of moral indignation about African poverty and to their desire to do something about it. Sachs makes them believe that, like him, they *can* do something. Sachs can appeal to an innate sense of community-mindedness or appease feelings of guilt about relative privilege. Exposing *The End of Poverty* as less scholarly research, and more like a nineteenth century travelogue/missionary report infected with 1950s modernization theory, risks disaffecting our students. Therefore, teachers have a responsibility to offer an alternative that retains this underlying sense of a need to engage but in less problematic ways.

Sachs is perhaps an easy target – his Africa is replete with disease and starvation, at the mercy of a cruel international community, with an appalling climate, permanently and uniformly suffering, only able to get to the next stage of development with the help of the West (egged on by Sachs himself, Bono and Bob Geldof), and specifically through a strong dose of Northern medicine in the form of neoliberalism. Postcolonial critiques can allow us to deconstruct this perpetuation of

'Afro-pessimism' (Ahluwalia 2001), where Africa's future is seen as bleak, meaningless and already a tragedy, in which African peoples have neither the political will nor the capacity to deal with their own problems. It is essential that this image of Africa's future (and that of the South in general) as meaningless is contested (see Chapter 8). As we have seen, 'Africa' and 'the Third World' is an invention or a western signifier. This means of signification is constantly being reconstructed by Northern institutions that exert their power through 'knowing' what is best for the South – namely development and modernity. In terms of Africa, 'it is through their amassing of statistics and surveillance that an underdeveloped, primordial, traditional and war-ravaged [continent] is (re)produced' (Ahluwalia 2001: 133). Postcolonial critiques reveal how specific ideologies and normative expectations have 'flowed from a colonial discourse into a development discourse' (Kothari 1996: 5). They also challenge these negative representations by telling different stories about the South.

It is important to find hope-filled counter-discourses. Visual imagery provides one example (McEwan 2006b). In contemporary development, photographs are used in a variety of ways (see Chapter 4). Photo-essays have been used by NGOs to record successes in community development. For example, the Philippines Homeless People's Federation photo-essay (VMSFDI 2001) describes its success in bringing together low-income community organizations that have formed housing savings groups. It records the strong emphasis on community-managed savings schemes, on community-to-community exchanges (so that members can learn from each other's experiences and formulate solutions for specific problems), and on the importance of negotiating with governments with clear, carefully costed proposals that demonstrate what communities can do for themselves. Similarly, Espelund *et al.* (2003) use photography deliberately to unsettle stereotypical images of contemporary Africa in their book *Reality Bites*, shifting the focus from stories that tell of war, hunger and crisis to more positive representations. Photographs are used alongside narrative accounts to tell alternative stories about parts of Africa that have hit the headlines, focusing on hopeful stories of ordinary people (usually women) who have made a positive difference in their communities and whose abilities to survive are more impressive than the achievements made by people in more privileged parts of the world. The photographs tell alternative stories of women reverting to traditional farming during severe droughts in Malawi, and not just surviving, but creating a business out of cassava production. They tell of people in different countries risking their lives to make peace with murderers, of a woman's success in the corporate world of male-dominated Morocco, and of shack owners in South Africa selling the township experience to tourists. In the preface to *Reality Bites*, Jan Pronk writes:

> . . . the bits of reality displayed in this book are proof of an enduring resilience of African people. Resilience in times of hunger and drought, resilience when deadly diseases manifest themselves, resilience when the world economy turns against any glimpse of prosperity, resilience in times of war, resilience when leaders fail, rulers govern badly or despots suppress denizens. If people in Africa are resilient, others should stay away from gloom and doom. Those who consider themselves partners in an effort to develop, allies in a struggle for survival, or friends on the road toward peace should never despair in the sight of decline. African people . . . pick themselves up, wipe the dust from their clothes and continue on their way. Their friends should do the same. This book shows us why despair would be unfair. Hope is justified, because people in Africa do not give up. Supporting them is legitimate, because they are determined to face their destiny, whether or not support will reach them. The stories in this book are more than reality bits. They are reality jewels, shining, worthy of exhibition. The narratives deserve to be told, remembered, retold and relived.

The potency of both examples lies in telling a more hope-filled story about the Philippines and Africa and of the nature and prospects for human development there.

It is also possible to contrast books like *The End of Poverty* with more radical accounts of development without generating disaffection. James Ferguson's *Global Shadows* (2006) is a radical critique of the kinds of prejudice that are perpetuated about Africa. Ferguson does not offer any easy solutions to the challenges that face contemporary African countries. Rather, he reveals a picture of diverse economies, social orders and forms of governance in which a bewildering range of complex transformations are taking place. In juxtaposing Ferguson's critique of portrayals of Africa – as a land in permanent crisis, a place of failure and seemingly insurmountable problems, and as a moral challenge to the international community – with those very same representations in Sachs, students can develop a more sophisticated understanding of Africa's place in the world. Such juxtaposition enables them to develop a more ethical stance towards Africa that does not involve them wanting to 'save' Africa along age-old ethnocentric and paternalistic lines that in reality have done nothing at all to 'save' Africa. Rather, it convinces them of the need to learn about and from Africa in all its complexity. It is important that developing critiques of accounts like *The End of Poverty* do not lead to complete disenchantment, disengagement and detachment. *Global Shadows* is one example of a rather different approach, informed by postcolonialism, that can both critique and foster different ways of relating to and with Africa, specifically, and the South more broadly.

*Global Shadows* is not necessarily full of hope, and it realizes the impossibility of replacing 'Afro-pessimism' with a sunny 'Afro-optimism' given the low (in some cases declining) rates of economic growth, increasing inequality and marginalization. However, it encourages a more politically and ethically informed engagement with Africa by painting a rather different and richly sophisticated picture. It shares similarities with Cindi Katz's (2004) *Growing Up Global*, in which she explores what global processes, specifically economic restructuring, mean in different spaces and places and the impacts this has on young people and their families. Despite presenting a radical critique of development and globalization, Katz examines the ways in which young people in Harlem, New York and Howa, Sudan and their communities cope. She juxtaposes the despair created by revanchism – 'the vengeful social, cultural and political-economic policies and practices of ruling groups and nations' (2004: 241) that is responsible for devastating local livelihoods – with more hopeful stories. Specifically, she explores people's involvement in 'resilience, reworking and resistance' that ensures their lives continue and that positive opportunities are created for children. Both books encourage a 'semblant solidarity' (Bhabha 2001, cited in Ferguson 2006: 22), that encompasses likeness and difference and that can recognize the significance of, and connect with, new forms of politics that are emergent in Africa and in communities similarly affected by the detrimental outcomes of globalization in North and South. As Ferguson (2006: 23) argues, these forms of politics are not captured in the old frameworks for nationalism and development. Instead, they include the emergence of a politics of international civil society, claims of transnational moral accountability, claims to an imagined supranational authority that is can recognize rights that are denied by nation states, and attempts to assert transnational responsibility directly through migration. Students of development will need to nurture new habits of thinking if they are to grasp the 'true originality and importance' (Ferguson 2006: 23) of these emergent politics.

Taking seriously experiences of the global in the South requires 'a discussion of social relations of membership, responsibility, and inequality on a truly planetary scale' (Ferguson 2006: 23). These new habits of thinking are implied in moves towards decolonizing development studies. At the very least, students in the North should be able to recognize their complicity in development in its broadest sense, through consumption of goods and images, through the involvement of their governments in the affairs of other countries, and through the interconnectedness between the micro-politics of the spaces of the academy and the macro-politics of the global economy. 'Out there' is always connected to 'in here', implying both an ethical responsibility to think critically about those connections, how they work and to what effect, and a politics of solidarity rather than of difference (but a solidarity that is always transient and contingent, and the terms of which ought to be determined by the people we work in global South contexts). There is a danger of postcolonialism becoming yet another colonizing discourse and practice directed from Northern centres of global power, 'one which plays to the political rhetoric of inclusion within the academy while adding further layers of delusion and amnesia to the material exclusion beyond it' (Raghuram and Madge 2006: 284). Foregrounding postcolonialism as both politics and ethics guards against this, since it takes seriously the need to develop these new habits of thinking when seeking to understand the place in the world of peoples, communities and countries in the South. However, as Cupples and Glynn (2014) argue, we must exercise caution that it is not privileged, critical first world scholars who are transformed and empowered, while indigenous peoples continue to be denied their land rights or black people continue to be disadvantaged by structural racism and inequalities.

As we saw in Chapter 2, decolonizing is necessarily unsettling in both senses of the word: it aims to reverse white colonial settlement and it is profoundly uncomfortable to those who continue to occupy positions of relative power and advantage. Postcolonial pedagogies can be unsettling and deeply uncomfortable for those present in the classroom. However, as Tuck and Yang (2012: 19) argue, such practices must not 'stand in for the more uncomfortable task of relinquishing stolen land' and addressing the profound injustices that continue to affect subaltern peoples globally. A postcolonial approach to development reveals that 'the era is over when development expertise could be thought to flow unidirectionally from global North to global South' (Cupples and Glynn 2014: 68) so that the latter was seen to be a put on what was assumed to be a singular, linear and inevitable development trajectory defined by Europe. The new era requires that development scholars listen and learn so that diverse ways of knowing the world can be part of the solution to the planetary problems caused by uneven development and environmental destruction in the name of development.

## Summary

- Postcolonialism implies a political and ethical project, but there are debates about the means through which its political and ethical ends might be met.
- A postcolonial approach to development might be thought of as a process of postcolonializing – a poetics and politics of transformation that effects change within the disciplinary field.
- This involves producing a progressive and ethical agenda for research and action.
- A postcolonial method requires that we question why we are doing research in the South, how we come to and produce our questions, how we analyze and represent our findings based on our subject positionings, who gains from the research and why.
- A postcolonial approach also requires recognition that theorization is also practice, that our multiple investments – personal, institutional and geopolitical – frame possibilities for change.
- Postcolonial approaches imply the need for an ethical engagement with the South through the development of radical hearing, speaking and writing tactics and through postcolonial pedagogies.

## Discussion questions

1  Explain the meaning of 'postcolonializing', and why it is necessary in development studies.
2  Is it possible to produce a postcolonial method and what would this involve?
3  What implications does postcolonialism have for field research, gap years and other forms of engagement through travel in the South?

4    Why is 'hope' important in development studies, and are there any problems in countering pessimism about the South through hope?
5    What might an ethical engagement with peoples and places in the global South involve?

## Further reading

Cupples, J. and Glynn, K. (2014) 'Indigenizing and decolonizing higher education on Nicaragua's Atlantic Coast' *Singapore Journal of Tropical Geography* 35, pp. 56–71. A discussion of the ways in which spaces of Higher Education are being actively decolonized in Latin America and why this matters.

Ferguson, J. (2006) *Global Shadows. Africa in the Neoliberal World Order* Durham, NC and London, Duke University Press. A collection of essays that develops a radical critique, informed by postcolonialism, of prejudices about Africa and is richly suggestive of a more ethical way for scholars in the North to engage with Africa.

Mohanty, C.T. (2002) ' "Under Western Eyes" revisited: feminist solidarity through anticapitalist struggles' *Signs* 28, 2, 499–535. This essay draws on feminist and postcolonial approaches to posit a new politics of solidarity between North and South.

Raghuram, P. and Madge, C. (2006) 'Towards a method for postcolonial development geography? Possibilities and challenges' *Singapore Journal of Tropical Geography* 27, 3, pp. 270–88. An insightful essay that problematizes development research and proposes ways in which a postcolonial method might be effected.

Sachs, J. (2005) *The End of Poverty* London, Penguin. An important book deserving serious and critical review, and one that students should read with a critical eye. It is revealing of the profoundly unethical drive in dominant economic ideologies to blame global poverty on the impoverished and then to posit aid as the solution to deprivation, not through a sense of social justice, but through benevolent imperialism. Not an agenda for postcolonializing development studies.

Simon, D. (2006) 'Separated by common ground? Bringing (post)development and (post)colonialism together' *The Geographical Journal* 172, 1, pp. 10–21. A useful essay that lays out an agenda for bringing postcolonialism into radical development studies.

# Useful websites

www.indigenousgeography.net/ethics.shtm 'Research Ethics: A Source Guide to Conducting Research with Indigenous Peoples': this free, online resource attempts to keep abreast of scholarship, protocols, and other documents relating to the conduct of research in indigenous settings and is hosted by the Indigenous Geography website.

www.mukurtuarchive.org/ Mukurtu is an open-source Content Management System with the distinction of being developed alongside indigenous communities themselves. Mukurtu is mobile-ready, and the development ethos is to be responsive to the needs of its core base of users. Its key tenet is to facilitate the management, sharing and exchange of digital heritage to empower and connect communities.

www.exchange-values.org/ Link to Shelley Sacks' social sculpture project, *Exchange Values: Images of Invisible Lives*, containing details of the project, images of the sculptures and the testimonials of banana growers in the Caribbean.

www.insightshare.org/ Website of the UK/France-based organization pioneering the use of participatory video as a tool for empowering individuals and communities.

documents.worldbank.org/curated/en/131441468779067441/Voices-of-the-poor-can-anyone-hear-us Link to the World Bank's Voices of the Poor archive.

www.ioufoundation.org/component/content/article/46-iou-news/347-intercultural-open-university-foundationuniversidad-aztecacentral-university-of-nicaragua-consortium-a-global-trend Intercultural Open University Foundation/Universidad Azteca/Central University of Nicaragua Consortium with information about the ethos of intercultural education. (Spanish speakers can also go to www.uraccan.edu.ni/ on URACCAN – The University of the Autonomous Regions of the Nicaraguan Caribbean Coast, an intercultural university community for indigenous peoples and ethnic communities.)

# 8 Beyond development and decolonizing life in the 'Anthropocene'?

## Introduction

In the ten years since the publication of the first edition of this book, debates about human existence and planetary life have changed considerably. There is a growing realization in the global North that capitalism and the conditions of the 'Anthropocene' – the naming of a new epoch in which human beings are considered a geologic agent and whose impact on planetary life will be visible in the geological strata of the earth – no longer create the conditions for life. While this represents a profound unsettling of the philosophical foundations underpinning ideas about development, arguably for millions of people around the world it is nothing new since colonized peoples have experienced this reality since the advent of capitalism. What is new, however, is that the global North is now facing the challenges of environmental and social destruction wrought by capitalist development. Of concern is the relationship between climate change and carbon-fuelled development, and the dilemmas of tackling this without exacerbating already existing global inequalities. The Anthropocene thus presents considerable challenges for theorizing development. It even raises the question of whether the idea of development is defunct as a means of framing the challenges facing and the aspirations of countries in the global South. It also presents considerable challenges for postcolonial theory. Before we consider these challenges, it is worth reflecting on what Spivak might call the epistemic violence done by naming contemporary globalism as 'Anthropocene'.

## The Eurocentrism of the Anthropocene

The Anthropocene – the age of humans – is part of the naming practices of the West. It implies that humans are a single type, a species, and an essential human nature has created the current crisis.

When approached from a postcolonial perspective, there are obvious problems with this. As Klein (2016: 12) argues, 'systems that certain humans created, and other humans powerfully resisted' – capitalism, colonialism, patriarchy – are 'let off the hook'. Elevating the frame of analysis to the level of the human species erases the racialized history of extractive colonialism that has given rise to this form of globalism. For this reason, Donna Haraway (2016) rejects the term Anthropocene, preferring the terms Capitalocene and Plantationocene to signal that the current and future state of the planet is a direct consequence of 500 years of capitalism, central to which has been the global transportation of people and plants and forced labour systems (see also Tsing 2015). As Sidaway *et al.* (2014: 5) argue, Anthropocene is thoroughly Eurocentric in that it 'is measured and dated in science whose origins are in, and which expresses the power of, a Western *weltanschauung* (world-outlook)'. Scientists are currently debating the origins of the Anthropocene, but any origin point – 1492 and the conquest of the Americas; 1610 and the Colombian Exchange (the transfer of human populations, plants, animals, cultures, technology and ideas between the Americas and Europe that was triggered by colonialism and extraction); the 1800s and the origins of the fossil and plantation economies that were founded on slavery and drove European industrialization; the 1940s and nuclear testing, first on Japanese cities and subsequently on Aboriginal lands in Australia, Pacific atolls and native American land in the Mojave desert – is marked by racial inequalities. The erasure of these histories in the naming of the present and future as Anthropocene could be considered another act of epistemic violence (see Yusoff, forthcoming). Talk of the Anthropocene also erases the existence of human systems that have organized life differently:

> systems that insist that humans must think seven generations in the future; must be not only good citizens but also good ancestors; must take no more than they need and give back to the land in order to protect and augment styles of regeneration. These systems existed and still exist, but they are erased every time we say that the climate crisis is a crisis of 'human nature' and that we are living in the 'age of man'.
> (Klein 2016: 12)

Such erasure forecloses possibilities that these different human systems may also inform alternative futures. And, importantly, the Anthropocene should not prompt us to abandon the fundamental concerns of social science and the terrain of postcolonialism: the theorization of culture and power. Instead, an 'increasing recognition of the potency of social relations of power to transform the very conditions of human existence should justify a more profound engagement with social and cultural theory' (Malm and Hornborg 2014: 63).

Inequalities and dispossession have underpinned the advent of the Anthropocene. The environmental destruction that is currently taking place has been caused by approximately 25 per cent of the planet's population – many millions are not party to the fossil economy, for example – and yet its effects are felt most keenly by those least responsible: indigenous peoples in the Arctic whose territories are literally melting away, or small island states in the Indian and Pacific oceans whose very existence is threatened by rising sea levels, for example. The great majority of the potential victims of climate change are in Asia and, as Ghosh (2016) argues, the threats faced here are stark. In India and Bangladesh 125 million people, and around 10 per cent of the population of Vietnam, could be displaced by rising sea levels. As much as 24 per cent of India's arable land is at risk of desertification. Forty-seven per cent of the world's population depend on the waters that flow out of the Himalayan ice sheets: if the glaciers continue to shrink at current rates, the most populous parts of Asia will face catastrophic water shortages within the next 10 to 20 years. Ironically, European imperialism has likely been an agent of delay in the onset of the climate crisis by retarding the development of Asia. For example, Ghosh points out that Burma had been extracting oil at Yenang-yaung for centuries before the British began seizing territory in the 1850s and there is no reason to suppose that Burma could not have navigated the twentieth century global petroleum economy had it been free to do so. Instead, the British conquered the whole of Burma by 1885, established the Rangoon Oil Company in Glasgow (absorbed into BP in 2000) and ensured that oil wealth flowed out the country. The surge in global carbon emissions since the 1980s is a result of the rapid industrialization of China and India (now the world's the first and fourth largest polluters, respectively, alongside the US and the EU), with the further irony that it is Asian populations that are particularly vulnerable to its effects. As with development more broadly, responses to climate change are dominated by a search for technocratic solutions rather than addressing inequalities and dispossession, which are rooted deeply in colonial history and inhere powerfully in the present. And one of the uglier features of environmentalism is the insistence by the haves that in order to 'save the planet' the have-nots should be denied the patterns of life (electricity, refrigeration, transportation) which the former take for granted. As Ghosh argues, the injustices of colonialism are perpetuated by claims that the poor should make sacrifices so that the rich can continue to enjoy the fruits of their wealth.

New forms of Orientalism accompany the reverberations of climate change. In the Middle East and North Africa, for example, violence is caused by both extraction of fossil fuels and the burning of fossil fuels, and the people displaced by this violence are dehumanized in western political and popular discourses. This is especially so if they make it alive across the Sahara Desert and the Mediterranean Sea to arrive on European shores or at Europe's borders in desperate need of refuge. Weizman and Sheikh (2015) trace this violence across an aridity line that stretches from the Horn of Africa, runs through Somalia, Ethiopia and West through Eritrea, Sudan, Northern Nigeria and Mali, and then east across the coast of North Africa to Gaza, Syria, Iraq, Iran, Afghanistan and into Pakistan. The line connects places marked by drought, water scarcity, scorching temperatures and military conflict, and in which climate change is a profoundly colonial project. These places are also hotspots for western drone strikes, which goes largely uncommented upon precisely because the people living here are discursively dehumanized. As Klein (2016) argues, bombs following oil and drones following drought fill boats with refugees who are dehumanized (see Box 8.1). The needs of others for security is cast as a threat to 'our' own security and in response walls are being built – literally in Israel, the USA and at the margins of Europe, and symbolically within Britain with its retreat from the EU, from principles of free movement of people and towards ever tougher and inhumane immigration policies. Black and brown lives are valued so little that thousands can be lost at sea – for example, according to a report to the UN (Fargues 2017), at least 34,000 people died or went missing attempting to cross the Mediterranean from North Africa and the Middle East between 2000 and 2017 – or incarcerated in prisons on islands – for example, Nauru, off the coast of Australia – which themselves are at risk of slipping beneath the rising seas. As Stoler (2016: 337) argues, 'Displaced peoples have become the "toxic" refuse of our contemporary world' and terms like 'environmental refugees' deflect 'from the historical and political conditions that have produced these effects'. The time is ripe for a sustained postcolonial critique of the Anthropocene, but this may also require a rethinking of postcolonial theory.

## Postcolonialism and the Anthropocene

Postcolonialism is concerned with racial power and the myriad ways in which groups of people have been, and continue to be, designated outside the category of human. The Anthropocene calls the category of human into question – it is no longer understood as internal to itself (and in many non-western cultures it never was), but is a reflection of the totality of the Earth system. As Baldwin (2017) argues, Anthropocene requires that dehumanization is historicized at the very moment in which the human becomes post-human, but attempts to theorize the Anthropocene have mostly ignored race. This is problematic because one of the prevalent themes in debates about the Anthropocene is human survival and contained within survivability is the biopolitical question of which bodies will be designated as best suited for survival. These are fundamentally racial questions, which raise questions about:

> whether postcolonial theory is capable of deciphering the politics of otherness contained within them, and, additionally, whether postcolonial theory is able to contribute toward a progressive politics capable of responding to the Anthropocene crisis alongside its resurgent contemporary fascisms.
>
> (Baldwin 2017: 294)

Critical geographers and environmental historians have long documented environmental racism and injustices and their relationship with (neo)colonialism – for example, the grossly unequal distribution of pollution, waste disposal and biowaste among impoverished peoples everywhere; the egregious practices of multinational corporations, conglomerates and governments that destroy ecologies, resources and livelihoods. However, little of this work has made its way into postcolonial scholarship (Stoler 2016). And despite growing scholarly concern with the inequalities and dispossession underpinning the current ecological crisis, postcolonial theory has struggled to engage critically with the Anthropocene debate. In part, this relates to the human-centred focus of postcolonial analysis and its concern with issues such as identity, cultural hybridity, political heterogeneity and epistemology. In contrast, theorizing the Anthropocene requires an account of how human beings are entangled in ontological aspects of wider relational and ecological processes. In part, this relates also to the concern of postcolonial theory to reveal the Other in text by revealing the structures of representation that draw meaning through references to a colonial past. In response, Stoler (2016) calls for a focus on how the 'ruins of empire' endure in the present (in Palestine, Australia and with respect to Native American lands, for example) and how people live with and in these ruins. In contrast, Baldwin (2017) argues that engaging with the Anthropocene requires that postcolonial methodologies shift their temporal gaze towards the future to better come to terms with the subaltern in formation or the figure of the yet-to-come (for example, the racialized figure of the climate change migrant – see Box 8.1). Finally, the disconnection between the Anthropocene and postcolonialism also in part relates to the requirement to theorize at the planetary rather than the human scale, which also brings the humanism at the heart of much postcolonial theory into question.

Dipesh Chakrabarty has written some of the most thoughtful reflections on the problems with species thinking in debates about anthropogenic climate change and the challenges this poses for humanist philosophies. He argues that the Anthropocene is the moment when the human becomes fully manifest in earth history and, paradoxically, the moment in which we lose our ability to comprehend this effect (2012: 14). Its primary historical significance lies in the way its proponents give finality to the blurring of Nature and the human, even while we know this binary to be an artefact of European imperial power. Chakrabarty argues for:

> the need to view the human simultaneously on contradictory registers: as a geophysical force and as a political agent, as a bearer of rights and as author of actions; subject to both the stochastic forces of nature (being itself one such force collectively) and open to the contingency of individual human experience.
>
> (Chakrabarty 2012: 15)

This marks a significant shift in postcolonial thinking about the human and has potential to connect in significant ways to the decolonial movements that are challenging the Eurocentrism of the Anglo-American academy. It also has some links with Paul Gilroy's (2015) call for a reinvigorated left humanism that draws inspiration from the writings of Césaire, James, Fanon, Senghor and Du Bois (see Chapter 2). Gilroy's argument is that what he calls a 'posthumanist humanism' ought to engage with Black scholarship and its critique of racialism located in anti-racist struggles. However, both Gilroy's and Chakrabarty's attachment to humanism means that they cannot escape endorsing it as a necessary project, with all the attendant risks of shoring up Eurocentric notions of beneficent rescuers in wealthier countries and darker skinned victims in poorer countries needing to be saved (see Box 8.1).

---

## Box 8.1 The climate migrant, humanism and challenges for postcolonialism (adapted from Baldwin 2017)

The figure of the climate migrant designates a particular kind of racialized Other within the context of climate change, but there are challenges for postcolonial theory and methodology in addressing otherness and climate migration. These are twofold:

First, while postcolonial methodology is predicated on distinguishing difference in the dialectical idiom of *different from*, the form of difference pertinent to Anthropocene, climate change and migration is the *yet-to-come*. Postcolonialism relies on tracing the colonial past in the present and appears inadequate to the task of building an analytics and politics of otherness specific to the future-conditional grammar of climate change. However, to have relevance in debates about climate change, postcolonial methodology needs to better attend to the ways in which the yet-to-come configures colonial imaginaries, both past and present. The yet-to-come can be promissory; recognition of an open, malleable future inspires the ethics and politics of invention and, while it can never be known in advance, it can inaugurate new forms of categorization, domestication and colonization. One example concerns the figure of the climate migrant, which does not exist in international law. Werz and Conley (2012: 12), writing for the US-based Center for American Progress, argue that a potential hotspot for climate migrants is the 'arc of tension', a space said to encompass Nigeria, Niger, Algeria and Morocco in which the combined effects of climate change, Islamic fundamentalism and increasing northward migration from Nigeria to Tunisia threaten Europe's southern border. This is a racial-militarist fantasy space, nowhere in which can be found an actual climate migrant. However, by the logic of racial fantasy, everyone in this space is a *potential climate migrant*, living on the verge of ungovernable excess and thus a threat to the security of Europe. As such the 'arc of the tension' is an ontogenetic space – it is a zone of potential instability in which the source of that instability, the potential climate change migrant, is the subaltern in formation. A revitalized postcolonial theory is thus required to address the ways in which the *yet-to-come* is itself colonized in debates about Anthropocene and climate change, and also to the ways in which the subaltern becomes central to its colonization.

Second, the figure of the climate migrant is deployed as a biopolitical tactic of humanist and thus racial renewal in response to the crisis of climate change. The danger in rethinking humanism to take account of the human as both political agent and geophysical force, as Chakrabarty (2012) suggests, is that this rethought humanism locates itself in the saviourism of climate refugees or climate migrants. Locating humanism in the spontaneous act of rescue is laudable, as is recalibrating what it means to be human amid the tragedies unfolding around the world. However, as with much of colonial and development discourse, saviourism is a newly articulated climate racism aimed at colonizing the future. It seeks to shore up a waning humanism at the present moment in which modern humanism is being called into question. Colonizing the future in this way amounts to a biopolitical tactic of racial subordination that re-inscribes a naturalized hierarchy onto planetary population survival. Since climate change designates a kind of ungovernable future – a future of chaos or of catastrophe – the figure of the climate migrant is invented precisely in order to render the ungovernable governable. Moreover, epistemological violence occurs when people are labelled climate migrants or climate refugees because their humanity is cast into a political non-space.

For humanism to be an imperative of progressive political thought, postcolonialism needs to expose the invisible racism at stake in the discursive phenomenon of climate change migration and climate refugeeism. The risk in not doing so is that the discourse on climate change and migration constructs climate change as a problem of global population management, which is deeply problematic given the rightward shift in contemporary politics in Europe and North America. A progressive politics of climate change might rethink climate change and migration not as a problem to be solved or as an object of state management, but as a racial relation to be historicized. By exposing the ideological structure of this relation and tracing the kind of political work it does in political, economic and cultural life, postcolonialism might more fully appreciate the material conditions obscured by this relation, develop an account of the way in which capital is adapting to the crisis of the Anthropocene through a new and emerging spatial fix, and develop a progressive and pluralist humanism in response to these emerging conditions.

According to Chakrabarty, humanity is universal and climate change represents a planetary threat for all humanity: 'Unlike in the crises of capitalism, there are no lifeboats here for the rich and the privileged (witness the drought in Australia or recent fires in the wealthy neighbourhoods of California)' (2009: 221). Chakrabarty (2012: 14) has more recently acknowledged that the crisis of climate change will be routed through 'anthropological differences', but as Malm and Hornborg (2014: 63) argue, it is important to understand the realities of differentiated vulnerability on all scales of human society. They point to the differential impact of Hurricane Katrina in black and white neighbourhoods of New Orleans, or of Hurricane Sandy in Haiti and Manhattan, or sea level rise in Bangladesh and the Netherlands. They suggest that, 'For the foreseeable future – indeed, as long as there are human societies on Earth – there *will* be lifeboats for the rich and privileged'. Indeed, members of the world's uber-rich business elite – notably Elon Musk of SpaceX and Jeff Bezos of Blue Origin –

who have profited most out of the global capitalism that threatens ecological devastation are already investing in making human life multiplanetary, with plans to colonize Mars. We might speculate about how many places will be available on those lifeboats if they ever transpire, who will manage to secure a berth and through what means. Meanwhile, less privileged and poorer people whose lives have been constrained by colonialism and racism are left to try to maintain livelihoods in already marginal areas (Figure 8.1)

Gayatri Spivak has responded incisively to these critiques of universalizing narratives of the Anthropocene and planetary change. She argues that planetary ideas risk re-inscribing the universalism of globalism by problematically assuming an 'undivided "natural" space rather than a differentiated political space' (Spivak 2003b: 72). She calls on us to 'imagine ourselves as planetary accidents rather than global agents, planetary creatures rather than global entities' so that 'alterity remains underived from us' (Spivak 2012: 339). Spivak's

**Figure 8.1**    *Farmers on the Orange River (South Africa) after a second 'once in 100 year' flood event within two months, February 2011*

(Source: Author)

notion of planetarity is a strategy for learning planetary difference. It challenges us to decolonize knowledge about the world by inviting us to know it from outside of the categories of western thought (Krishnaswamy and Hawley 2008). It calls for a move away from reading non-western epistemologies, ontologies and subjectivities from the vantagepoint of western knowledge and through the orientalizing canon. It demands that difference is known in ways that do not prescribe otherness in one's own terms and thus 'offers a topos for learning the worldings that liberal multiculturalism and the cosmopolitan imagination may not be equipped to recognize' (Jazeel 2011: 88). It creates the possibilities for a 'multiplicity of critical, decolonizing perspectives against and beyond Eurocentered modernity, from the various epistemic locations of the colonized people of the world' (Grosfoguel 2012: 97). This is important because progressive ideas need to emerge from the critical thinkers in dialogue across cultural difference.

The invitation opened by Spivak's notion of planetarity works against the hopeless scenarios for future international relations painted by some about the Anthropocene. Pessimists paint a bleak picture in which Anthropocene politics are reduced to management of the post-apocalyptic present: the governance of polluted oceans, flooded cities and desertified landscapes in which survival is all we can hope for. In contrast, planetarity and the collapse of the modernist universe create unique possibilities to decolonize international relations, to reflect again on who counts as human, to become attuned to the needs of nonhumans, and to engage with and learn from non-western indigenous cosmologies. In turn this might reinvigorate debates about development by renegotiating and re-politicizing ideas concerning security, participation or well-being, and to establish new forms of political cooperation. However, this is unlikely to happen while debates remain decidedly Eurocentric in focus and outlook.

Indigenous scholars and race theorists have raised serious concerns about the Euro-western academy's current approach to human-environmental relationships, which they argue erases race, colonialism and slavery and poses difficult challenges for theorizing development. Posthumanism and the ontological turn have been challenged as Eurocentric because they erase non-European ontologies, referring instead to a foundational ontological split between nature and culture as if it is universal (Sundberg 2014). As Todd (2015: 245–6) argues, scholars critical of this Eurocentrism re-centre the *locus* of thought, offering a reconfiguration of understandings of human-environmental relations towards praxis that acknowledges the critical importance of land, bodies, movement, race, colonialism and sexuality. For example, Sundberg urges scholars to enact the 'pluriverse' (a world in which many worlds fit) as a decolonial tool, in her case drawing on Zapatista principles of 'walking the world into being'. For Sundberg (*ibid.*: 39), walking and movement bring a decolonizing methodology to fruition because, 'As we humans move, work, play, and narrate with a multiplicity of beings in place, we enact historically contingent and radically distinct worlds/ontologies'. Planetary thinking, therefore, acknowledges and embraces difference, in contrast to global thinking, which assumes universal planetary sameness.

## Development and sustainability in the Anthropocene?

The question of what 'Anthropocene' means for development and whether development can truly be decolonized is one of utmost importance. It seems that we are at a cusp in which development as modernization still appears to be happening, but the faith in this as progress is diminishing rapidly, not only because of the environmental catastrophe it heralds, but also because of the bleak prognosis for the human condition. Hardly anywhere in the world is untouched by the political economy that was built from post-war development. Much of what is happening in the global South resembles modernization of the 1960s: vast infrastructure projects built in poorer countries by richer ones in return for resource extraction and market access, but this time with the major players being China, India, the Gulf countries and so on. We might ask, however, to what extent do people in poorer countries still have faith in the promises of modernization? That everyone will benefit? That the future is known? That development will provide jobs, stable employment and wages, and other benefits?

One of the ironies of the contemporary moment is that while 'the promises of development still beckon' (Tsing 2015: 3), people's lives are ever more precarious. At a global scale, value chain capitalism that realises primitive accumulation and jobless economic growth generates this precarity. In 2018, on current projections by Oxfam, 2.5 individuals own the same net wealth as 3.5 billion people, or half the world's population. The extreme wealth of the few is built on the impoverishment and precarity of the many. Irregular livelihoods are now the norm everywhere and, as Comaroff and Comaroff (2012) have argued, in this regard Euro-America is evolving toward Africa. Development in terms of improving human lives is no longer happening in large parts of the global North. Importantly, this is a consequence of political choice rather than an effect of a presumed transcendent and implacable capitalism. For example, according to Danny Dorling, since 2011, under successful Conservative-led governments, life expectancy in the UK has flatlined for the first time since 1841 and is lower than in all other affluent countries apart from the US: 'The most plausible explanation would blame the politics of austerity, which has had an excessive impact on the poor and the elderly . . . The stalling of life expectancy was the result of political choice' (Dorling, 2017: 19). This is not the progress that was assumed to accompany post-war development. Development as progress is in question because of the reality of diminishing livelihoods in parts of the global North and because of the growing rejection of it as an idea in much of the global South, where decolonial and postcolonial critique represent 'an irrefutable indictment of the developmentalist fallacy' (Escobar 2007: 198). As Rist (2014) argues, the effects of the global financial crisis in 2008 have meant that countries in the global North have lost interest in development and *growth* is now what mobilizes hope (as well as despair among those concerned about its environmental implications). Growth has also largely replaced 'development' in countries like China, Brazil and India. In North Africa, political dimensions are taking precedence over socio-economic concerns. In much of sub-Saharan Africa, development appears to be at breaking point, with land being leased or sold off to governments or private investors and with several countries still beset by political troubles that exact a huge toll on lives and livelihoods. The ongoing conflict in the Democratic Republic of Congo (also involving Rwanda, Uganda and Burundi) has 'reduced 'development' to a matter of humanitarian intervention' (*ibid.* 246), which is compounded by the effects of diseases such as AIDS and malaria.

Global genealogies of persistent neo-colonial inequality, exclusion and violence mean that postcolonial critiques of development are as relevant as ever, but global ecological crisis suggests they need to respond to demands to broaden who and what constitutes Others (human and non-human). As this book has demonstrated, postcolonial critique has exposed the continuing legacies of colonialism and imperialism with the contemporary neoliberal practices and theories of development. However, while development is no longer thought of as Euro-American modernity, it perhaps is not entirely defunct and is, instead, actively being rethought in both the global North and global South in new and challenging ways by postcolonial, decolonial and indigenous scholars and activists. Both China and (until recently) Brazil have used their economic growth in quite diverse ways to attempt to alleviate poverty and drive forms of state developmentalism that differ markedly to the market orthodoxies on the global North. Inclusive growth, as a decisive departure from global neoliberalism, is intended to mark different Asian pathways of development, with the Asian Development Bank prioritizing poverty and inequality as the main risk that Asian leaders will need to manage (Roy 2016). Radical ecological democracy in India and buen vivir in Bolivia and Ecuador (see Chapter 6) offer radical alternatives to market orthodoxy and, in the case of the latter, have already delivered real improvements in people's lives.

There are still profound questions to be asked about how improvements in lives and livelihoods are understood and the claims about development that are made in the light of these improvements. For example, poverty cannot be reduced to accountancy (reducing the number of people living on below $2 per day) because it is experienced or constructed within a social relation. China has seen an enormous reduction in people living below the $2-dollar-a-day marker of extreme poverty, but as Rist (2014) asks, does a Chinese wage-earner whose pay as risen by the equivalent of a few dollars per day feel less poor when they compare their situation to the escalating wealth and conspicuous consumption of the middle and upper classes that is still far beyond their reach? In addition, rises in monetary income are usually linked with economic growth, and thus to environmental damage. Therefore, we need to ask do Chinese urban-dwellers feel better off when their lives are being shortened by air and water pollution? Do Brazilians feel better off when, according to locals, the bay of Rio de Janeiro is 'beginning to look like a toilet and a rubbish dump' and is poisoning fishing resources (Rist 2014: 249). Ensuring the SDGs are met by 2030 is, of course, desirable, but there is likely to be a price to be paid, which is not acknowledged in the Rio+20 global 'sustainable development' agenda (see Chapter 6) because it refuses to acknowledge that the pursuit of economic growth is irreconcilable with protecting the environment. India's twelfth Five-year Plan is framed around the theme of 'Faster, More Inclusive and Sustainable Growth' with seemingly little acknowledgement of the contradictions in this. How people experience growth will not necessarily bear any relation to their understanding of the betterment of life. In so many contexts in which people struggle to survive, it is difficult to see how anyone can still dream of development. However, as we saw in Chapter 6, people still do dream of better lives and, for this reason, alternative approaches to development are continuing to emerge. The idea of degrowth is one such alternative.

In the context of the Anthropocene and global climate change, managed degrowth, economic recession or 'de-development' in some places appear to be the only courses towards shifting human impact on the earth towards human responsibility for the earth (Head 2016). Degrowth activism is not necessarily anti-development, although it shares terrain with post-development agendas, including its origins in the global South: Latouche (2009) traces it back to Benin scholar Tévoédjrè who in 1978 advocated rethinking the economy based on anti-consumption and collective wealth and building on a changed value system. Degrowth is seen by many contemporary advocates as a progressive alternative path to advancing human development that does not rest on economic growth (see Box 8.2). Yet degrowth among the global middle and upper classes may also generate negative short-term impacts on people struggling to survive in the global South and some scholars caution that it could entrench a neo-colonial global agenda. As Dhawan (2013: 139) argues, 'though we might be facing the same storm, we are not all in the same boat'. It is thus imperative that debates about degrowth do not themselves become neo-colonial by assuming normativity and are instead attentive to Southern perspectives on degrowth (Dengler and Seebacher forthcoming). As Spivak (2008) argues, degrowth requires the rearranging of desires at both ends of postcoloniality, that is in both global North and global South.

## Box 8.2  Degrowth as a radical alternative to sustainable development

While the concept of sustainable development rests on a notion of false consensus (Hornborg 2009), the idea of 'socially sustainable degrowth' (Schneider *et al.* 2010) – or simply degrowth – has emerged in the twenty-first century as a proposal for radical change. The movement emerged among diverse activists and theorists with diverse concerns: reduced availability of energy resources, especially oil; global warming, pollution and threats to biodiversity; environmental destruction and the threatened extinction of the flora and fauna upon which humans depend; the rise of negative societal side effects of over-consumption, such as unsustainable development, poorer health and poverty; and the neocolonialism underpinning the ever-expanding use of resources by global North countries to satisfy lifestyles that consume more food and energy, and produce greater waste, at the expense of poorer countries in the global South. Degrowth ideas coalesce around anti-consumerist, anti-capitalist and ecological economic ideas to propose political, economic and cultural programmes for radical change, based on reducing both production and consumption.

Advocates of degrowth do not aspire for it to be adopted as a common goal by the UN, OECD or other international organizations. Rather, in reaction to the foreclosure by neoliberal capitalism of the possibilities of alternatives, degrowth is an attempt to re-politicize the debate about socio-ecological transformation and to critique the current development hegemony, especially the post-Rio+20 notions of 'green growth' or 'green economy'. In addition, to taking hold in alternative ecological macroeconomics, it is also based in social movement activism – anti-car and anti-consumption activists, cyclist and pedestrian rights campaigners, supporters of organic agriculture, critics of urban sprawl and promoters of solar energy and local currencies (Demaria *et al.* 2013). Degrowth thinkers and activists advocate for the contraction of economies by reducing production and consumption, arguing that over-consumption is the principal cause of long-term environmental destruction and social inequalities. An important element of degrowth theory is that reducing consumption should not require a decrease in well-being. Instead, its proponents argue that happiness and well-being can be maximised through non-consumptive means, such as the sharing work, reducing consumption, and devoting more time to creativity (art and music), family, nature, culture and community.

Degrowth is opposed to the current ideas and policies concerning sustainable development. While the concern for sustainability does not contradict degrowth, sustainable development is rooted in mainstream development ideas that aim to increase capitalist growth and consumption. Degrowth theorists, therefore, view sustainable development as an oxymoron (Latouche 2004) because any development based on growth in a finite and environmentally stressed world is considered inherently unsustainable. Critics of degrowth argue that a slowing of economic growth would result in increased unemployment, increase poverty and decrease income per capita, with particularly severe consequences for the global South. However, degrowth advocates argue that the key to avoiding this is in transforming the economic system, abandoning the global economy and relocalizing economies. This, they argue, would allow people of the global South to become more self-sufficient at the same time as halting the overconsumption and extraction of resources from the South by the North.

## Decolonizing development and global South futurism

Responding to the profound challenges posed by the irreconcilability between development and environmental sustainability requires development scholars everywhere to decolonize their imaginaries. Discourses of development are being moved away from a human-centric focus to consider the relationship between sustainability and development, with questions of racial inequality and agency bound up with questions of environmental destruction. Plural democratic imaginaries in which humans and non-humans actors are shaping new ideas about responsible and progressive planetary futures and are far removed from a foundational narrative of economic-growth-equals-modernity-equals-development. These ideas create new ethical perspectives on nature, life and the planet that privilege 'subaltern knowledges of the natural' and articulate 'in unique ways the questions of diversity, difference and inter-culturality – with nature . . . occupying a role as actor and agent' (Escobar 2007: 198).

The concern of postcolonial scholars to provincialize Europe (in its widest sense as 'the West') remains pertinent considering the urgency to imagine and enact responsible and progressive planetary futures, and the maintenance of colonial relationships in global geopolitics and economics. However, in recent years, this concern may have become less urgent precisely because the energies driving alternatives are no longer necessarily located in Euro-America. Western countries appear to be provincializing themselves while African, Asian and other developing countries are advancing economically, socially and politically. As Mbembe (2013) argues, the fixation in Europe and the USA with the question of immigration is jeopardizing the creation of more dynamic relations with Africa and Asia. Both Europe and the US are building fortresses around themselves and the only other response to a receding American empire is more militarism. As we have seen, as Europe and the US are withdrawing, other countries such as China, India, Turkey and Brazil are increasingly playing a role in the unfolding geopolitical and economic reconfiguration. They are also the places in which solutions are being sought to the irreconcilability of economic growth and environmental sustainability. As Europe closes its borders, countries in the global South may begin to open their borders and play a more influential role in global affairs. In this context of declining western influence, the racist stereotypes that have been the concern of postcolonial critics (see Chapter 4) perhaps also become less relevant: as Mbembe wryly observes 'we should leave it Europeans to deal with their own stupidities because we have more urgent tasks and projects to attend to . . . Europe will have to deal with its own mental illnesses, racism being the first of these' (*ibid.*).

Despite the pessimism concerning environmental crisis, global South scholars like Mbembe are optimistic about the greater role that African, Asian and Latin American countries will play in global affairs. Mbembe argues that the African continent has three attributes, rooted in African philosophies such as Ubuntu, which are creative and increasingly significant in a contemporary world of rapid change and profound uncertainty. The first is multiplicity – the profusion of cultures, knowledge, world-views and philosophies that colonialism set out to erase (e.g. through the imposition of individuality and monotheistic Christianity), but which Mbembe considers a resource for remaking the continent. The second is circulation and mobility – the idea of the African historical cultural experience being one in which almost everything was on the move. In contrast to the Hegelian notion of Africa as a closed continent, Mbembe argues 'it was always a continent that was on the move' (*ibid.*) and he suggests that this concept of movement can be mobilized in creative ways. The third is composition – African lives are compositional and relational, for example in the ways in which the economy is lived on an everyday basis or in the ways in which people relate to one another. Mbembe's point is that Africa can make a creative contribution to the world of ideas and praxis that can be to the benefit of the world at precisely the moment of western withdrawal. This, he argues, has implications for all manner of things in an age of ecological crisis: 'theories of exchange, theories of democracy, theories of human rights, and the rights of other species' (*ibid.*). Moreover, he argues that 'the very future of our planet is being played out in Africa' (Mbembe 2015) in the responses to the challenges posed by ecological crises, climate change, refugees and so forth.

These trends lend weight to the belief of Afro-futurists that the twenty-first century will be an African century and that, along with China, what happens in Africa will have planetary significance. Similar arguments are being made by theorists and commentators in the countries of Latin America and the resurgent economies of Asia. As Ananya Roy (2016: 318) argues, crisis and uncertainty pervades the West and is 'foreclosing the future', while in Asia 'ambition and aspiration promise a better future'. Her point is not that the future is better in Asia, but that discourses of a prosperous future here do political and even economic work. India's Five-Year Plan may be fraught with contradictions, but it is significant that inclusive growth offers a different politics of futurity to neoliberal development and it is being framed as a shared, democratic aspiration among India's population. The promise of inclusive growth may have limits (and degrowth theorists would take issue with it), but the fact that Asia, and the global South more generally, are asserting themselves as the laboratories of future development is important (Roy 2016: 219). In addition, because economic crisis in Europe and North America is 'increasingly framed in the language of structural adjustment – austerity, a lost decade of growth, informal work' (*ibid.*) – the South can be thought about as prefiguring the futures of the North (see also Comaroff and Comaroff 2012). It is no longer the case that the global South needs to 'catch-up' with the global North, rather that coalescence is being driven by global economic and environmental processes. Therefore, if it is to avoid becoming provincialized, anachronistic and irrelevant, development studies needs to acknowledge and engage with these new realities and the innovative solutions that are being sought in the global South.

Both development studies and postcolonial theory face challenges in shifting away from ideas of development and decolonization as solely human concerns towards post-human strategies for survival in what Tsing (2015) refers to as the 'ruins of capitalism'. Post-human and more-than-human thinking is at the forefront of debates that sound the death-knell of development-as-progress, generating an ethos of life in the ruins that human beings have made and will continue to make, and engaging at a planetary scale for thinking about presents and futures attuned to ecological and non-human demands. Tsing (2015), for example, explores companionship with non-human assemblages, the possibilities of latent commons in the ruins of capitalism and the emergence of ways of surviving at the very edges of capitalism. Her starting point is the rare and valuable matsutaki mushrooms that thrive in the ruins of pine forest plantations, which she argues assemble human and non-human ecologies that in turn create alternative means of surviving. She uses the idea of cohabitation to ask how capitalism might look without assuming progress and how ecologies might support multi-species life without assumptions of progressive human mastery. She also attempts to create space for politics, arguing that 'We will need a politics with the strength of diverse and shifting coalitions – and not just for humans', and an 'alternative politics of more-than-human entanglements' (2015: 135).

The idea of the commons – what we hold together and what holds us together – has been core to the ideas of critical development scholars. This is because the destruction of the commons by neoliberalism is seen as a central problematic in primitive accumulation, generating uneven and unequal development. Reclaiming the commons is often at the heart of alternative forms of development advocated by social movements and emerging in state policy in places like Bolivia. The possibilities of the formation of new commons is a source of optimism in Tsing's writings. Postcolonial theory can play an important part in decolonizing the commons – not just for subaltern human beings, but also for subaltern non-humans – and in reconstructing the commons in post-humanist terms. Watson (2014: 93) argues that 'subalternist cosmopolitics' are important for the creation of 'cohabitable worlds and corollary forms of local governance with actors that don't share the cognitive, bodily, and metaphysical forms of human being'. He argues that the idea that the current political system 'constructed to represent human interests and constituencies' might be able to 'know and care for the ontologically disparate worlds and experiences of other-than-human beings may amount to the present era's predominant ideological delusion' (*ibid.*). In other words, our political systems cannot act ecologically because, as artefacts of previous eras in which western notions of modernity and progress were considered universal, they were not designed to do so. The more precarious present age thus requires more heterogeneous and experimental forms of governance in which one group of humans should not exercise sovereignty over other groups of humans and humans as a whole 'should not exercise sovereignty over the entirety to the Earth as we know (and don't know) it' (*ibid.*), as has happened throughout the history of development. Again, pluriversal thinking – the idea of many worlds within one – and the idea of a co-inhabited cosmos are critical to producing more just and liveable worlds. Rather than dwelling on doom-laden scenarios of the inevitability of catastrophe or apocalypse, pluriversal thinking generates the kind of optimism that both western scholars, like Haraway and Tsing, and First Nations, Aboriginal and indigenous scholars are cultivating by exploring 'alternative politics of more-than-human entanglements' (Tsing 2015: 135). The hope is that this might create the possibilities for new commons in the ruins of capitalism from which humans and non-humans can survive. At the very least, it fosters a postcolonial ethos of care:

> Care means taking representation slowly. Care means attending to the gaps, the ignorances produced by our knowledges, the subaltern holes and knots in the fabric (Chakrabarty 2000: 106). Care means hospitality. Cosmopolitan care might also mean anarchism, as an activist reconstruction of the commons in a finite world with no recourse to the holy trinity of God, Nature and Neoliberalism.
>
> (Watson 2014: 93)

Care has often been at the discursive heart of development, but as we have seen it has often been deployed in support of illiberal, violent, patronizing or paternalistic forms of intervention. In contrast, cosmopolitan care may revivify development as a more ethical and decolonized endeavour. Cosmopolitan care means engaging with and learning from experiments in the global South that are rooted in subaltern ontologies, attempting to script alternative models of development and to open new trajectories. The anti-state discourses of neoliberalism are being reversed and in some countries the state has assumed responsibility for redistributing wealth and guaranteeing rights to people and nature. From buen vivir in the Andes, to Gross National Happiness in Bhutan, from Ubuntu in southern Africa to Swaraj in India, and beyond, there are multiple experiments in redefining and realizing development. These movements are not limited to the global South. As we have seen, in the global North, there is a growing recognition – manifested movements such as degrowth, but also in reclaiming commons, Transition Towns, steady-state economics and permaculture – of the need to transition to a post-materialist, post-developmental paradigm. Postcolonial approaches are important because, in their focus on racial politics rooted in colonial and neo-colonial socio-economic relations, they offer alternatives to the technocratic approaches to dealing with current challenges of addressing climate change, sustainability, food security and so forth. These challenges cannot be addressed without also acknowledging continuing dispossession, uneven economic development and social inequality that emerged in the colonial past. As Klein (2016: 11) argues, Edward Said 'may not have had time for tree-huggers, but tree-huggers must urgently make time for Said – and for a great many other anti-imperialist, postcolonial thinkers – because without that knowledge, there is no way to understand how we ended up in this dangerous place, or to grasp the transformations required to get us out'.

The ethos that underpins much of postcolonialism, decoloniality and critical post-humanism will remain relevant in responding to the current ecological crisis and the challenges it poses for development. As Mbembe (2016) argues, this ethos is concerned with reopening the future of the planet to all who inhabit it and learning how to share it again – 'amongst but also between its human and non-human inhabitants, between the multiple species that populate our planet'. Any possibility of surviving the current ecological crisis rests on a growing awareness of our precariousness as a species in the face of ecological threats, decoupling the idea of development from the pursuit of economic growth and recoupling it with a postcolonial ethos of sharing. The planetary turn is still unfolding, but it signals a conceptual shift that realizes one aim of postcolonialism in that Europe and the West – and western ideas about development – will be thoroughly provincialized. As Mbembe (2016) argues, 'older senses of time and space based on linear notions of development and progress are being replaced by newer senses of time and of futures founded on more open narrative models'. To have any relevance in understanding these shifts and their possible outcomes, development studies needs to remain attuned to the fact that China and other Asian, African and Latin American countries will undoubtedly play significant and possibly quite divergent roles in shaping these new futures.

## Summary

- The concept of the Anthropocene has been criticized as Eurocentric and part of the naming practices of the West. Its roots in colonialism, and the inequalities and dispossession that are a feature of the Anthropocene are often overlooked.
- The Anthropocene poses challenges for postcolonial theory in deciphering the politics of otherness and contributing toward a progressive politics capable of responding to the ecological crisis alongside its resurgent contemporary fascisms. This requires a shift in postcolonial thinking about the human, planetary theorizing and pluriversal approaches that are attentive to difference and capable of producing more just and liveable worlds, and an opening out to decolonial politics.

- The Anthropocene also poses challenges for development, both in terms of whether it is still desirable and possible in the face of ecological catastrophe and whether it remains a concern in world in which growth has become the dominant economic driver.
- Degrowth theories and activism present alternatives to the dominant notion of development and other alternative futures are being theorized in the global South. To retain relevance, development studies needs to remain attuned to the fact that China and other Asian, African and Latin American countries will undoubtedly play significant and possibly quite divergent roles in shaping these new futures.

## Discussion questions

1 In what ways is the Anthropocene a Eurocentric concept?
2 What challenges does the idea of the Anthropocene present to postcolonial theory and how might the latter respond?
3 What challenges does the idea of the Anthropocene present to the idea of development and what alternatives are being proposed to generate progressive change?
4 What is the relationship between postcolonialism, decolonialism and alternative notions of future development and why does this matter to development studies?

## Further Reading

Baldwin, W.A. (2017) 'Postcolonial Futures: Climate, Race, and the Yet-to-Come' *Interdisciplinary Studies in Literature and Environment* 24, pp. 292–305. An excellent essay that discusses some of the challenges posed by climate change for postcolonial theory and method and suggestion for how they might respond.

Jazeel, T. (2011) 'Spatializing difference beyond cosmopolitanism: rethinking planetary futures' *Theory, Culture & Society* 28, 5, pp. 75–97. A critical engagement with cosmopolitanism that draws on Spivak's postcolonial re-reading of planetarity to engage with difference.

Mbembe, A. (2013) 'Africa and the future: An interview with Achille Mbembe' by Thomas Blaser, 20 November, https://africasacountry.com/2013/11/africa-and-the-future-an-interview-with-achille-mbembe/ (accessed 12/02/18). Read with the two articles below, these are short, accessible accounts from a foremost scholar of Afro-futurism.

Mbembe, A. (2015) 'Discussing African Futures' an interview with Damola Durosomo, 9 November, www.okayafrica.com/achille-mbembe-african-futures-interview/ (accessed 12/02/18).

Mbembe, A. (2016) 'Africa in the New Century', 29 June, https://africasacountry.com/2016/06/africa-in-the-new-century/ (accessed 12/02/18).

Roy, A. (2016) 'When is Asia?' *The Professional Geographer* 68, 2, pp. 313–21. A case for studying Asian futures to allow a broader understanding of the renewal of development in a rearranged South–North world.

Sidaway, J., Woon, C.Y. and Jacobs, J.M. (2014) 'Planetary postcolonialism' *Singapore Journal of Tropical Geography* 35, pp. 4–21. A position paper introducing a special issue on advancing postcolonial geographies, which sets out five pathways for postcolonial geography, including planetary approaches, acknowledging other postcolonialisms and planetary indigeneity.

Weizman, E. and Sheikh, F. (2015) *The Conflict Shoreline* Steidl Books. Collaboration between an architect and a writer/photographer exploring the entanglements of climate change, political conflict and colonialism in the Negev. The book incorporates historical aerial photographs, contemporary remote sensing data, state plans, court testimonies and nineteenth-century travellers' accounts, and is a good counterpoint to the racial-military fantasies in other accounts of climate change and conflict.

# 9  Conclusions

This book has explored the possibilities and implications of, and some of the tensions and contradictions in, bringing together postcolonial approaches with theories and debates in development studies. It has focused specifically on four areas. First, it has provided an account of postcolonial approaches as they relate to development studies, charting the origins of both sets of theories, exploring the ramifications of postcolonialism as a critique of development theory and demonstrating that, despite historical divergences, there is scope for increasing intersection and dialogue between the two bodies of theory. Second, it has provided a comprehensive review of debates about postcolonialism and development, exploring in more detail the key implications of postcolonial critiques on the writing and doing of development studies. Third, it has explored the junctures between postcolonialism and other contemporary re-workings of development theory and practice, such as decoloniality, grassroots and participatory development, feminist critiques, environmental debates, indigenous knowledge and global resistance movements. Finally, it has explored the possibilities for students and researchers to work towards an agenda for development studies that is postcolonial in theory and in practice.

Ultimately, the book has argued that a deeper engagement with each other's theories and practices would be mutually beneficial for both postcolonialism and development. Bringing postcolonial approaches to bear on material issues that determine the conditions of postcoloniality (e.g. biopolitics, poverty, participation) is a means of bringing its abstract ethical and political promise to fruition. Meanwhile, postcolonializing and decolonizing development is important given its origins in Enlightenment notions and colonialism, its continuing significance in shaping global economics and geopolitics, and its role in facilitating interconnections and relationships between North and South – transnational, cross-institutional, intergovernmental and personal. However, this is neither straightforward nor without contradictions. And, as discussed in Chapter 8, responding to the current global shifts in economic power and the emerging ecological crisis requires the rethinking of both postcolonialism and development. The subsequent discussion returns to some the main dilemmas and difficulties of bringing postcolonialism and development into critical dialogue with each other as a means of discussing the radical possibilities of such a move. By way of conclusion, the importance and ramifications of decolonizing development studies are explored in further detail.

## Postcolonial approaches in development: difficulties and dilemmas

Chapter 3 highlighted some of the major criticisms of postcolonialism: that it has become institutionalized, representing the interests of a Northern-based intellectual elite; that greater theoretical sophistication has created greater obfuscation; that postcolonialism is too theoretical and not rooted enough in material concerns, and that emphasis on discourse detracts from an assessment of the material ways in which colonial power relations persist. In response, it might be argued that many of these criticisms are somewhat overstated, particularly given the increased convergence, at least at their margins, of postcolonialism and development studies. As the preceding chapters have demonstrated, theory is itself a practice and discourse also has material effects. Through a critical, but mutually productive, engagement with development studies, postcolonial approaches contribute to debates about globalization and the lived experiences of postcoloniality. They provide new methodologies and new sources for the recovery of these experiences. They thus have the potential to reorient ethics and politics within development studies. Despite this, postcolonialism is still sometimes accused of neglecting urgent life-or-death questions (San Juan 1998), which still inform much of development studies. Given that there is little implicit coherence to postcolonialism, this accusation is perhaps a little unfair. Rather than a single theory or approach, postcolonialism is a plurality of approaches that cohere around particular intellectual concerns with a long heritage in anti-colonial politics (see Chapter 2). And, as Jani (2017: 113) points out, some critics of postcolonialism do not sufficiently acknowledge or appreciate 'the space for progressive and leftwing thought that postcolonial studies – along with ethnic studies, women's studies, and other related fields – have created in a political climate in which the actual Left is weak and fragmented and academia is increasingly corporatized', or that postcolonial scholars 'are committed to social justice, historical truth and other values that postmodernism dismisses'. The onus is on those working within critical development studies to embrace and apply the theoretical and methodological insights of postcolonialism. Adopted in this way, there is no reason why postcolonialism might not inform critical development studies and cast light on questions of contemporary concern, such as inequality of power over and control of resources, human rights, global exploitation of labour, child prostitution, genocide and ecological destruction.

Some critics suggest that postcolonialism cannot easily be translated into action on the ground and that its oppositional stance has not had much impact on the power imbalances between North and South. However, if we think of postcolonialism as a form of anti-colonial politics and as an essential element in the scripting of a political and ethical agenda for development studies, with interconnections with other radical approaches such as decoloniality, post-development and feminism, then there clearly have been notable recent shifts within development studies. In addition, its focus on discourse cannot be thought of as facile in an era when ethnocentric representations continue to disadvantage the South. They also shape global geopolitics, which further entrench political and economic power in the North, and are evident in sources ranging from popular media to the World Bank and government reports.

As we have seen, development texts are both strategic and tactical, promoting and justifying certain interventions and delegitimizing and excluding others (Crush 1995). Power relations are clearly implied in this process; certain forms of knowledge are dominant and others are excluded. The texts of development contain silences and postcolonialism provides a means of interrogating who is silenced, and why. They cast light on the fact that ideas about development are not produced in a social, institutional or literary vacuum, but are assembled within a vast hierarchical apparatus of knowledge, production and consumption – what is referred as the 'development industry' (Crush 1995: 5). This industry is itself implicated in the operation of networks of power and domination that during the twentieth century came to encompass the entire globe and continue to do so in the twenty-first century. Clearly, development discourse promotes and justifies very real interventions with very real consequences. As Chapter 4 illustrates, it is therefore imperative to explore the links between the words, practices and institutional expressions of development, and between the relations of power that order the world and the words and images that represent the world.

Some of the most sustained criticisms of postcolonial approaches have been based around the apparent distance between radical theory and political activism. Debates within gender and development illustrate this extremely well. Shifts towards cultural explanations and concerns with discourse and representation in feminist development theory have been ridiculed by many activists (in the South and in post-communist contexts, but also from within western feminism) as elitist and removed from reality. The problem is often posed as a schism between theory and practice, or the gap between feminist theorizing in the North and the practical needs of women globally. Theoretical preoccupations are not easily translated into direct politics, and are accused of shifting the focus away from the material problems of women's lives. Many critics argue that organizing and obtaining women's human rights cannot be removed from ensuring a better life for men and women in societies characterized by poverty and a lack of freedom and democratic norms. Postcolonialism has been charged with ignoring these issues. Concerns with representation, text and imagery are perceived as too far removed from the exigencies of the daily lives of millions of impoverished people. One response has been a rejection by some feminists of postcolonialism, and an objection to the emphasis on difference and discourse away from material conditions. These objections are based around the notion that 'poverty is real' and not simply discursive.

Jackson (1997: 147) is one such scathing critic of 'postist' feminist understandings of poverty and gender, 'where culture, ideas and symbols are discursively interesting and constitutive of power, whilst materiality is of questionable status, and at least suspect', and where poverty becomes 'largely a state of mind' rather than a matter of material struggle for survival. She argues that 'real' women and the challenges facing them get lost in the morass of text, image and representation. The charge is that the rejection of political economy and the embracing of poststructuralist and postcolonial approaches, therefore, is in danger of 'chucking the baby out with the bathwater' (Udayagiri 1995: 164). In dismissing the universalist assumptions of political economy the material problems of the daily existences of many women are also erased. However, in many ways, such criticisms represent a misreading of postcolonialism, since its various theories have never advocated the rejection of political economy. Perrons (1999) is perhaps more constructive in arguing that political economy still matters, but that rather than rejecting 'postist'[1]Jackson is referring specifically to postmodernism and poststructuralism, but since postcolonialism is part of the intellectual currents which deny universalist ideas of progress and well-being it is also a target of her critique. approaches political economy ought to respond to the challenges posed by their critiques. She argues that political economy and postcolonial approaches should work in tandem to critique the dominant order, particularly through the conceptual integration of production and consumption in their class and gender dimensions (see also Fagan 1999). As we have seen, this is made possible at the intersections of postcolonial and development studies. The importance of continuing to challenge the western-centrism at the heart of development was apparent in December 2017 when plans were made by the UN to appoint Wonder Woman as ambassador for empowering women and girls. The suggestion of using a hypersexualized, scantily clad, white superhero as a role model in challenging female stereotypes and fighting discrimination and violence against women and girls provoked an angry backlash and protests by UN staff.

Similarly, Chakrabarty's (2000: 42) attempts to explain that what 'provincializing' is not about has broader relevance for correcting some of the prevailing misunderstandings about postcolonialism. It is not about 'a simplistic, out-of-hand rejection of modernity, liberal values, universals, science, reason, grand narratives, totalizing explanations, and so on'. In other words, there has never been an out-of-hand rejection of political economy or materiality within postcolonialism. Rather, postcolonialism examines the histories of struggles through which particular universals 'win'. Why is it, for example, that Northern notions of modernity and development provide the model for modernities and development elsewhere, and the conceptual lens through which these other modernities and patterns of development are understood, evaluated and judged? Postcolonial theorists (from Said to Spivak to Chakrabarty) remind us that as intellectuals operating within academia, we are 'not neutral to these struggles and cannot pretend to situate ourselves outside the knowledge procedures of our institutions' (Chakrabarty 2000: 43).

Postcolonialism has also been accused, again quite incorrectly, of leading down the path of cultural relativism in its attempts to deal with difference. Proponents of cultural relativism suggest that the solution to imperialism and universalism is through respect of difference in a plurality of identity politics. However, this offers little assistance in terms of dealing with some of the complex issues confronting people in the global South. By refusing to theorize cultural dominance, relativists implicitly evaluate all cultural positions as equal. This gives them no basis for making moral judgements about social justice in terms of aspirations to deal with inequality, injustice and power in all its guises, including patriarchal power. It also ignores differences (class, regional, religious, ethnic, gendered) between people within specific cultural locations (Schech and Haggis 2000). In addition, the 'culture' that is preserved through such respect is often a hierarchical and patriarchal one that preserves elite male privilege at the expense of marginalized groups (e.g. women, peasants and youths). At the UN International Conferences, for example, 'official' feminisms, often allied to and representing national governments and their political agendas, have previously used arguments about 'cultural respect' to block more radical, 'unofficial' feminisms that pose a greater threat to the status quo (Goetz 1991: 146).

The underlying problem is that relativist arguments share a view of cultures and identities as bounded, coherent and autonomous. However, postcolonialism has never been a project of cultural relativism. Its challenge to the dominance of Northern knowledge forms does not originate from 'the stance that the reason/science/universals that help define Europe as the modern [or developed] are simply "culture-specific" and therefore only belong to European cultures' (Chakrabarty 2000: 43). Rather, it seeks to question why narratives of modernity and development almost universally point to Europe as the 'primary habitus' (Chakrabarty 2000: 43) of the modern and developed. In other words, it exposes the erasure of alternative modernities that come into being in places outside Europe (e.g. in the South), and that might also inform our understanding of what it means to be modern and/or developed. Moreover, notions of cultural relativism have long been rejected by cultural theorists, not least because they replicate notions of culture informing conservative fundamentalisms in a variety of contexts (Butler 1990; Hall 1995). There is no justification, for example, for the barbarism perpetuated by the so-called Islamic State (IS) in the Middle East and in cities like London, Paris and Nice. However, as Adam Shatz (2015) points out in an essay on IS atrocities in Paris in 2015, 'For all its medieval airs, the caliphate holds up a mirror to the world we have made, not only in Raqqa and Mosul, but in Paris, Moscow and Washington'. Resolving the problem of Islamic fundamentalism requires acknowledging the role played in fuelling it by western (neo-)imperialism in the Middle East and the deep resentment that this has historically perpetuated and continues to fuel. The recent and current violent extremism and its manifestations in both the Middle East and Europe can be argued to have roots in the Sykes-Picot Agreement of 1916. This secret deal between the UK and France (assented to by Russia), set out plans for dealing with the defeat and collapse of the Ottoman Empire during World War 1 and establishing British and French spheres of influence in the Middle East. The deal – a profound betrayal of the Arabs – was made public by Russia after the Revolution, much to the embarrassment of Britain and the delight of Turkey, and has fuelled anti-western sentiment ever since. Violent extremism is also related to the failure of post-colonial Europe to deal with its own racialized inequality, which feeds similar resentment in the banlieues of Paris and impoverished inner cities of Britain (see Balibar 2007) and is 'colonial through and through' (Stoler 2016: 377). As Shatz argues, the causes of the current conflict are deeply rooted in histories of postcolonial rage, which need to be understood if a resolution that ends the cycles of violence is to be found.

It is widely acknowledged within postcolonial theory that replacing universal sameness with cultural difference does not disrupt colonial power relations. Cultural difference can be used to deny any possibility of different groups of people 'becoming the same' (i.e. achieving equality). Women in the South, for example, are always marked by difference, since cultural difference is also racialized (Frankenberg and Mani 1993; Narayan 1998). Postcolonialism thus works towards finding an alternative to false universalisms, which subsume difference under hegemonic western understandings, and to relativism that would abandon any universalist claim in favour of reified and absolute conceptions of difference (see, for example, Grewal and Kaplan's (1994) discussion of how postcolonial approaches to human rights reject both universalist and cultural relativist accounts). As Spivak

(1990) argues, there is still a need for greater sensitivity to the relationship between power, authority, positionality and knowledge. The implications, for example, of scholars in the North writing about people outside their own cultural milieu must be considered in the context of the global hegemony of Northern scholarship – in other words, its domination of the production, publication, distribution and consumption of information and ideas.

Postcolonialism need not force development practice in the direction of relativism and immobilize it by nihilism, since such thinking is not intrinsic to postcolonial perspectives or to discourse analysis (Baaz 2005: 175). An analysis of Spivak's work (and the arguments of several other postcolonial theorists) demonstrates that the way forward is not to revert to a simplistic, particularist and relativist position by dismissing those values and ideas – such as democracy, equality and human rights – over which the North assumes proprietorship and which often underpin notions of what development is and should be. Rather, it is important to engage in a form of critique that questions these claims to propriety and the highly selective application of these values. The solution to the problematic nature of development is not to reject the idea out of hand, but to build a consistent critique of the unevenness of development and to challenge its associated meanings, which are often rooted in Eurocentrism, notions of 'trusteeship' and cultural arrogance. To use Spivak's (1993: 284) famous double negative, it is important to 'engage in a persistent critique of what one cannot not want'. As Wainwright (2008) argues, development is absolutely necessary and absolutely inadequate to its task. This is the irresolvable contradiction at the heart of development; but faced with abject poverty and stark inequality, one cannot not desire development. Thus, rejecting development is impossible; instead it is a 'site of fundamental doubt' (*ibid.* 11) through which it might be engaged with and transformed through critique. Improving the quality of life of the millions of impoverished and marginalized people in the world remains of critical importance, and a focal point for intellectual and political endeavour. However, determining the most appropriate ways forward need no longer be the preserve of Northern knowledge forms and expertise. And, as Chapter 8 demonstrates, alternative thinking from the global South will play a critical role in responding to the challenges of ensuring the survival of humans and non-humans in the face of ecological crisis.

## Decolonizing development studies and provincializing Northern development discourses

Postcolonialism poses various challenges to development studies. This book has argued that 'Development' is one of the dominant discourses that postcolonialism seeks to destabilize as unconsciously ethnocentric, rooted in European cultures and reflective of a dominant western world view. Postcolonialism challenges the experiences of speaking and writing by which dominant development discourses come into being. These practices of naming are not innocent, but are part of the process of 'worlding' (Spivak 1990), or setting apart certain parts of the world from others. Said (1985) has shown how knowledge is a form of power, and by implication violence; it gives authority to the possessor of knowledge and knowledge has been, and to a considerable extent within the field of development still is, controlled and produced in the North. The power to name, represent and theorize is still located here, a fact which postcolonialism and decoloniality seek to disrupt. Postcolonialism invokes an explicit critique of the spatial metaphors and temporality employed in western discourses, insisting that the 'other' world is 'in here' (Chambers 1996: 209) and that the South is integral to 'modernity' and 'progress'. We have seen that the global South contributes directly to the economic wealth of Northern countries through its labour and through historical and contemporary forms of exploitation. In addition, the modalities and aesthetics of the South have partially constituted Northern languages and cultures. Postcolonialism, therefore, attempts to re-write the hegemonic accounting of time (history) and the spatial distribution of knowledge (power) that constructs the South. Finally, postcolonialism attempts to recover the lost historical and contemporary voices of the marginalized, the oppressed and the dominated, through a radical reconstruction of history and knowledge production (Guha 1982). Postcolonial theory has developed this radical edge through the works of political and literary critics such as Spivak, Said and Bhabha who, in several ways, have sought to recover the agency and resistance of peoples subjugated by both colonialism and neo-colonialism. This is also of importance in opening spaces in which alternative development knowledge might be formulated.

In the light of these challenges, there are several ways in which development studies might respond and add to its potentially radical insights and effects: firstly, by exploring the possibilities of a more productive engagement between material and discursive concerns; secondly, by developing the intersections between postcolonial approaches and issues of global inequality and the diverse lived experiences of postcoloniality; thirdly, by developing the political and ethical possibilities of postcolonializing development studies. While postcolonialism is cautious of progress narratives, particularly about avoiding universalizing statements of progress, any kind of politics needs some notion of what progress is (Rorty 1998). As with anti-sexism and anti-racism, for example, we need to be able to ask, and keep asking, what a meaningfully postcolonial development studies might look like. Core to this is the notion of decolonizing development studies and exploring shared ethical and political terrain.

As discussed in Chapter 2, while both postcolonial and decolonial theory can inform the critique of development, they are also in tension with each other. Decolonial theory is in the realm of material decolonization and is articulated by theorists who are positioned politically and personally with decolonial agendas (for example, indigenous theorists in former settler colonies). These personal and political struggles are different to and perhaps incommensurate with the concerns of theorists (for example, those in the former imperial metropoles) who are positioned differently in relation to neocolonial power relations and who work from postcolonial perspectives (Noxolo 2017). However, as we have also seen, the concerns of both bodies of theory work towards similar ends in critiquing the material conditions that continue to entrench colonial and imperial power relations and give rise to profound inequalities and injustices. Both bodies of theory work to decolonize development studies by 'provincializing' Europe and 'the North' (Chapter 7). This is not Europe as a geographical entity, but rather European intellectual traditions that have made it impossible to think of anywhere in the world without invoking particular categories and concepts, 'the genealogies of which go deep into the intellectual and even theological traditions of Europe' (Chakrabarty 2000: 4). These categories and concepts include development, modernity, citizenship, the state, civil society, the public sphere, human rights, equality before the law, democracy, the idea of the subject, social justice, scientific rationality and so on. These concepts entail a universal and secular vision of the human, and, as we have seen in this book, were exported globally through imperialism and colonialism. According to Chakrabarty (2000: 43), provincializing Europe thus involves two moves. First is the recognition that Europe's claiming of the adjective 'modern' or 'developed' for itself is an integral part of the story of European imperialism within global history. Second is an understanding of this equating of a certain version of Europe with 'modernity' and 'development' as not solely the work of Europeans. Despite being anti-imperial, nationalisms in the South, for example, were also based on modernizing ideologies and have been equal partners in this process. This in turn leads to an understanding of the histories (and presents) of North and South as fundamentally intertwined and entangled.

Decolonizing development thus starts with provincializing European development thought, which then opens the possibility of reorienting development studies. As Sylvester (1999: 717) explains:

> Postcolonial studies has the potential to be a new and different location of human development thinking – anathema as such a notion sounds. Not infected (as much) by the know-all history of development studies, but just as embroiled in thinking about the West, it is freer to criticize colonialism and the creeds of progress openly and to call upon the types of 'data' that development studies scorns – imaginative literature, postmodern theory and travel writing. It can also wander in between the colonial and postcolonial spaces of many locations in order to point out the ways in which agents of development have been restructured and penetrated by colonized peoples.

According to Pieterse and Parekh (1995: 3), decolonization involves:

> Not the restoration of a historically continuous and allegedly pure precolonial heritage, but an imaginative creation of a new form of consciousness and way of life . . . It involves an engagement with global times that is no longer premised either on Eurocentrism, modernization theory or other forms of Western ethnocentrism passing for universalism, or on Third Worldism, nativism and parochially anti-Western views.

In other words, since colonization was not a straightforward and uniform process, it stands to reason that decolonization cannot be a straightforward process of denial and reversal. Decolonization does not involve rejecting all 'western' knowledge as bad and everything 'indigenous' as good, since this dichotomy has little meaning in a globalized world. Moreover, it involves both the (former) colonized and colonizers. It is not simply a matter of giving voice to those hitherto silenced and marginalized, but about 'challenging the epistemological basis of hegemonic thought' (Mercer *et al.* 2003: 428).

Decolonizing the field of development involves understanding how development concerns the production of narratives and stories, how these have political and material effects, and exploring the nature of these effects. It involves critiquing whiteness in development studies, exploring new forms of activism and their possibilities as genuine partners in the development process. In addition, it involves questioning why it is acceptable to do development research 'over there' (in Africa, India, South East Asia or Latin America, for example) but not 'here' (in Britain, Europe, Japan, Australasia or the USA) (Jones 2000). In short, there is a need to blur the distinction between 'here' and 'there' and to eradicate these binaries from our understanding of the world. And, rather than perpetuating processes of unequal exchange, there is a real need to develop properly collaborative research (from research design to publication) and to share the outcomes of this research with local institutions and peoples. In this way, the problems associated with one-sided knowledge construction about and extraction from countries in the South might be countered, the paternalistic political interventions that still inform Northern interventions in these countries (e.g. DfID or USAID's strategies towards Africa) might be problematized, and the place given to popular agency within the South might be prioritized and foregrounded.

As discussed in Chapter 7, work in 'development education' is already beginning to engage with these ideas. The UK charity Development Education Association, for example, promotes the idea of development starting at 'home'. It aims to raise awareness and build skills to move the public beyond notions of the South based on compassion and charity, and towards an understanding of interdependence. However, promoting education that encourages and enables people to think critically and to aspire towards a more just and sustainable world requires that this understanding of interdependence also recognizes uneven levels of power (Dobson 2005). Connections between aspects of culture, such as identities, representations, otherness, worth and value and economics, such as the distribution of wealth, access to resources and labour, are also critically important. As Andreotti (2007: 3) argues, development education shares with postcolonialism 'the search for a new globalism that has an ethical relationship to 'difference' and that does not reproduce the universalistic and oppressive claims of cultural superiority that were the basis of colonialism'. Several international initiatives are currently spearheading what might be termed the postcolonializing of development education. For example, 'Open Spaces for Dialogue and Enquiry' (www.osdemethodology.org.uk) and 'Through Other Eyes' (www.throughothereyes.org.uk) promote approaches to global citizenship education that emphasize critical literacy, independent thinking and an ethical relationship to difference.

## The radical potential of postcolonialism

Postcolonial approaches demonstrate how the production of western knowledge forms is inseparable from the exercise of western power (Said 1985; Spivak 1990; Young 1990, 2001). They also attempt to resist the power of western knowledge and reassert the value of alternative experiences and ways of knowing (Bhabha 1994; Fanon 1986; Spivak 1998a; Thiong'o 1986). They articulate some challenging questions about imperialist representations and discourses surrounding lands and peoples in the South and about the institutional practice of Northern-based disciplines. They share a social optimism with other critiques, such as feminism, which have helped generate substantial changes in political practice (Darby 1997: 30). While transforming unequal global relations by a politics of difference and agency alone is seemingly impossible, postcolonialism is a much-needed corrective to the Eurocentrism and conservatism of much of western thought, including much of development studies. The potential of postcolonial approaches within development lies in their abilities to interrogate the interconnections and complex spatialities of postcoloniality, and to give proper attentiveness to dialogue and difference.

There is, of course, an inherent possibility that postcolonialism might become a new colonizing discourse and yet another subjection to foreign formations and epistemologies from the English-speaking centres of global power. This is certainly how many critics including some decolonial scholars in Latin America, for example, have viewed postcolonialism (Klor de Alva 1992; see also Ashcroft 2001). However, as this book has demonstrated, postcolonializing development studies is essentially about postcolonializing development studies *in the North* through an engagement with theory from the South, which also has profound implications for how knowledge is created and disseminated and the ethics and politics of the practice of development. In addition, as Ashcroft (2001: 24) argues, rather than a new hegemonic field, we might see postcolonialism as a way of talking about the political and discursive strategies of formerly colonized societies and peoples. Again development scholars are ideally placed in this regard and in more carefully viewing the various forms of anti-systemic operations of global capitalism.

Postcolonialism has made important contributions in theorizing both power and knowledge and the significance of discourse (Rajan 1993; Rose 1987). Postcolonial approaches demand that we see, responsibly and respectfully, from another's point of view. However, in their engagement with development studies it is important that they also engage with material issues of power, inequality and poverty and resist focusing on text, imagery and representation alone. They have the potential to contribute to other radical strategies within development studies that can make a difference. This involves combining the material with the symbolic and encourages the building of coalitions across differences. It might also involve combining a material analysis 'to point to the consequences and inter-relations of different sites of oppression: class, race, nation and sexuality' (Goetz 1991: 151), with recognition of the partial and situated quality of knowledge claims (Haraway 1991). The challenge is to produce something constructive out of disagreement, and to combine material concerns and emphasis on local knowledge with a postcolonial dismantling of knowledge claims.

Ann Ferguson (1998: 95) theorized this as a new 'ethico-politics'. The problem that many scholars in the North need to confront is that they are located in the very global power relations that they might aspire to change; hence there is a 'danger of colluding with knowledge production that valorizes status quo economic, gender, racial and cultural inequalities' (*ibid.*). There is a need for self-reflexivity, recognition of the negative aspects of one's social identity and devaluation of one's moral superiority to build 'bridge identities' across difference. This allows other knowledge to talk back, and creates solidarity between people that 'must be struggled for rather than automatically received' (*ibid.*: 109). This does not mean generalizations cannot be made, but it puts the emphasis back on how they are made. Such approaches to development are not simply about deconstructing Northern knowledge. Rather they provide a more comprehensive project of re-moulding a conceptual framework 'capable of embracing a global politics of social justice in ways which avoid the "colonizing move"' (Schech and Haggis 2000: 113).

Edward Said (1993) pointed out that imperialism bound disparate societies, peoples and cultures together, but in ways that were profoundly unjust and unequal. Postcolonialism allows us to extend this concern by asking: 'How does international development bind disparate societies, peoples and cultures together in postcolonial times and how might this be unjust (for some)?' (Power 2003: 137). Thus, we can examine the geopolitical trajectories of former colonized societies (Slater 1998) and ground these debates in local contexts – in the everyday realities of peoples and places (Paolini 1997) – to meaningfully examine the relations of domination that persist today. The roots of the Rwandan genocide, for example, can be traced back to Africa's colonial past, but also to its postcolonial present. The disadvantages faced by Latin American indigenous peoples are similarly rooted in a colonial past and postcolonial present. This interaction between past and present is a fundamental concern of postcolonialism. It also suggests that postcolonial contexts in both North and South 'have to deal with [their] past to understand [their] present and confront [their] future' (Ahluwalia 2001: 133).

Postcolonial approaches can contribute to new ways of thinking about development in various contexts across the world, in different geographical spaces, rather than being simply a concern of the global South. Development research in global contexts involves shifting the unit of analysis from local, regional and national cultures to relations and processes across cultures. Grounding analyses in specific, local development praxis is necessary, but understanding the local in relation to larger, cross-national processes is also important (see, for example, Katz 2004). A comparative, relational, postcolonial development praxis would be transnational in its response to, and engagement with, global processes of (neo)colonization. This involves acknowledging, and working through, the productive tension between the 'centrifugal force' of discrepant development histories and the 'promising potential of political organizing across cultural boundaries' (Alexander and Mohanty 1997; Sinha et al. 1999: 1). It also requires working with people at grassroots level in diverse cultural contexts, breaking down hierarchies of knowledge/power that privilege the expert/outsider, undermining Northern universalisms and providing a basis for a new understanding of global diversity (Marchand and Parpart 1995).

Here there are clear parallels with approaches in feminism (Box 9.1), indigenous knowledge, anthropology and models of local hybrid cultures, which challenge the orthodoxies of Northern thinking by bringing local knowledge to the fore in ways that dismantle the *a priori* categories of Northern theories. As we saw in Chapters 5 and 6, such approaches are also reflected in the rise of new social movements in the South that are symptomatic of, and inform, moves towards a new societal and development paradigm. In Latin America and South Africa, for example, the very ideas of democracy, community and development are being reinvented in a remarkable flourishing of grassroots activity (Alvarez and Escobar 1992; Munck 1999) from which genuine alternative development strategies might arise (Friedmann 1994).

---

## Box 9.1 Feminism and postcolonialism in development

The tensions and dilemmas in new ways of conceiving a cross-cultural feminist politics informs, and is informed by, postcolonialism. Criticism from black women and feminists in the South has had a considerable impact on gendered approaches within development studies. They have demonstrated 'why women are important, and why gender is an indispensable concept in the analysis of political-cultural movements, of transition, and of social change' (Moghadam 1994: 17). They also suggest that western researchers and observers should not denigrate alternative modes of thought on issues of human development. African feminists, for example, have argued that scholars and activists in the South can depend upon existing legacies of indigenous systems together with the prevailing knowledges about them to formulate an authentic theory of human development (Mangena, in Amadiume 2000: 176). This more holistic understanding of development puts human survival and non-western philosophies at the centre, producing alternative understandings based on relevant and empowering ideas generated by indigenous cultures. These ideas have philosophical merit in their search for an alternative theory of human development and for the emancipation of women. As Amadiume (2000) suggests, they represent an informed contribution to the global debate on human development and feminist methodology, and they should certainly inform the ways in which outsiders approach research in the contemporary South.

Similar interventions have been made by indigenous women in Latin America. In a panel on 'Citizenship and Political Participation by Indigenous and Afro-Descendant Women' at the Tenth Regional Conference on Women in Quito in 2007, indigenous women presented their manifesto for building a 'plurinational' state. This argued that despite quantitative and qualitative advances midway through the decade devoted to achieving the MDGs and the Second International Decade of the World's Indigenous

Peoples, indigenous peoples faced a critical situation exacerbated by the increasing implementation of macroeconomic policies that ignore their collective rights. It argued that advances in respect for the human rights of indigenous women are tied to the broader struggle to protect, respect and exercise the collective rights of their peoples. It recognized the importance of the MDGs as tools to make progress on strategies for women's sustainable development and human rights, but criticized the targets used to measure progress for not including cultural or ethnic indicators. A recent study by the UN found that women account for nearly 60 per cent of the 50 million indigenous people in Latin America and the Caribbean, and they face triple discrimination: as women, as indigenous people and as poor people. However, the 2007 Regional Conference on Women focused on the contribution of indigenous women to the economy and social protection, particularly through unpaid work, challenging the singular representation of them as victims of oppression and asserting their agency. It also examined the emergence of women leaders in the region and the increasingly autonomous electoral behaviour of women as part of a wider shift towards gender parity.

The requirement for postcolonial feminisms to continue to disrupt the dominance of Northern feminisms, particularly in the current geopolitical climate, is illustrated by Algerian sociologist, Marnia Lazreg (2000: 37–8):

> I remain convinced that only a de-centering of the self will make it possible for foreign audiences to 'receive' Other women's work as reflecting another modality of being human. This requires educating not only students but also educators in redefining their purpose in life, acquiring a genuinely critical perspective on their culture, and relinquishing the intoxicating tendency to world the worlds of others . . .. It is crucial to ask whether 'Western' audiences, feminist or otherwise, should insist on the knowability of these women. The many conferences, seminars and courses devoted to these Other women have not yielded a deeper understanding of them. Generalizations and stereotypes still flourish, and for some societies (such as the Middle East) these stereotypes seem to hold greater sway than they did *before* the advent of academic feminism.

Postcolonial feminisms, therefore, have the potential to contribute to the critical exploration of relationships between cultural power and global economic power. Moreover, they point towards a radical reclaiming of the political that is occurring in the field of development and in the broader arena of societal transformation (Figure 9.1).

**Figure 9.1** *Women activists demonstrating in South Africa*

Source: Author

## The importance of postcolonial development studies

Development is important. It is also necessary, given the scale of poverty and inequality that exists in the world. For example, it is estimated that 795 million people (one in ten of the world's population) are undernourished, with 14 million of these in developed countries (UN 2015); 783 million lack access to safe water (*ibid.*); 2.5 billion lack access to basic sanitation (*ibid.*); 1.6 billion (a quarter of the global population) lack adequate shelter, with 1 billion living in informal settlements (UN-Habitat 2015); 1.6 billion lack access to electricity and 2.6 billion lack access to clean cooking facilities (IEA 2015); 2 billion lack access to essential medication (UN 2015); 785 million (one in five) adults are illiterate and two thirds of these are women (*ibid.*); there are an estimated 168 million child labourers (ILO 2014). The latest available WB data (from 2013) suggests that 767 million people are living in severe poverty (defined as living on less than $1.90 a day). Such severe and widespread poverty persists while there is great and rising affluence elsewhere (Pogge 2008). In 2008, the GNI of the world's richest 20 countries was 37 times that of the poorest 20 having doubled in over 40 years (Wainwright 2008). By 2016, the world's richest countries had 46 times the per capita GDP adjusted for purchasing power parity (IMF 2016). Development clearly works in some contexts (e.g. South Korea, Singapore, Malaysia and the Philippines). It is still a field 'whose money and agendas influence the world'; however, 'that it gives itself few channels through which to generate and deliver the types of help or critiques thereof that many local people may want is its great blind spot' (Sylvester 1999: 718). Development has a long history of oscillating between top–down and bottom–up notions of developmentalism. Postcolonial critique reveals it to be too steeped in Northern bureaucratic authority to generate any radically new ideas. Postcolonial approaches might provide a much-needed reinvigoration, particularly if they are able to turn attention to the single biggest failure of development – poverty alleviation – and the challenges of tackling this without worsening the ecological catastrophe currently facing the planet. Developing postcolonial agendas of material well-being that matter on the ground is a challenge facing the next generation of postcolonial scholars:

> There are plenty of critiques and dreams of development circulating ambivalently in the villages and urban areas of specific Third World locales. The challenge is to reach into those spaces, pull out and analyse the stuff of everyday postcolonial deprivation and desire.
>
> (Sylvester 1999: 719)

Foregrounding the desires and aspirations of people in the South is important in defining what development is and should be. That many can still only aspire to safe drinking water, a roof over their heads and a reliable food supply is both an indictment of development and an indication of the need to retain a sense of responsibility towards both distant and not-so-distant others (Simon 1997). The critical nature of much of development theory also tends to focus attention on the failures of development. The deeply unequal power relations between North and South may not have been altered to any great extent, but some countries have made extraordinary progress since independence in improving living conditions for a majority of their population, particularly in Latin America and South East Asia, but also in African countries such as Nigeria and Botswana. Small, incremental improvements might not seem significant in the fight against inequities of global capitalism, but they are significant to those people whose lives are dramatically improved.

A familiar question for advocates of postcolonialism *in* development studies is: 'how will postcolonialism solve the material problems facing millions of people in the South?' Yet this misses the point – postcolonialism does not set out to solve such problems. Postcolonialism is essentially a critique of Northern discourses and has its origins in cultural and literary criticism. It opposes totalizing, grand theories and thus any claim that it *could* provide solutions on such a scale would be anathema. Rather, as we have seen, the importance of postcolonialism is that it raises questions about how development is theorized, how development 'problems' are identified as such, how solutions are authored and implemented, and what knowledge and agents consequently remain utterly marginalized and excluded. It is fundamentally concerned with the relationship between power and knowledge, of how past relationships of power persist into the present, and of how past inequities remain fundamental to understanding contemporary global relations.

Postcolonialism need not, however, be considered as only antagonistic to development, since it raises the possibilities for reworking development theory and practice. Its ethical and political implications create possibilities for hybrid re-combinations and coalition approaches in producing alternative development agendas. Postcolonialism repositions the place of time at the centre of development. Linear conceptualizations that see development as an ever-unfolding process of progress have been shown to be fundamentally flawed. Postcolonialism allows for alternative notions of time, in which the past inheres in the present. This, in turn, allows for cultural assertiveness for those whose knowledge has been erased or ignored. In advocating the sovereignty of formerly colonized peoples, postcolonialism is thoroughly embedded in a wider politics in the pursuit of social, environmental and economic justice. It thus shares a great deal of common ground with more radical approaches within development, as well as with indigenous politics (see Box 9.2) and social movements, which have become increasingly significant in global politics.

One area where this common ground is significant concerns climate change, and postcolonialism remains relevant in exposing some of the inequities that persist in global agreements. The Paris Climate Accord of December 2015, which aims to reduce greenhouse gas emissions and finance adaptation to and mitigation of climate change, is an example of this inequity because it contains a legal agreement by which poor countries – contributing least to greenhouse gas emissions, but worst effected by global warming – are unable to sue global corporations for climate change damage. In contrast, the proposed Transatlantic Trade and Investment Partnership between the UE and the US would allow companies to sue governments if those governments' policies cause a loss of profits. One of the criticisms of debates about tensions between development, on the one hand, and climate change on the other, is that debates remain anchored in conceptual grammars that reiterate Eurocentric nature-culture distinctions. As discussed in Chapter 8, scholars are beginning to retheorize western humanism to engage with Black scholarship concerning racialism and to reorient postcolonial theory away from western humanism to engage with indigenous perspectives that might generate the new ontologies required to prevent global environmental catastrophe. These ontologies, which see no separation between the world of humans and the world of nature, have long been present in indigenous communities. As Jackson (2014: 84–5) argues:

> New demands and new imperatives from anthropogenic climate change to privatized forests to disappearing peoples and languages ask us to reach deeper than provincializing has thus far allowed to address our implicate becoming within an ever hybridizing pluriverse . . . Today's concerns for postcolonial geographies [and development studies] are matters of concern for the commons, emerging social movements, biodiversity, material property rights, resource depletion, migration, cultural erasure, energy scarcity, inequality and the destructive imperatives of growth. We need to attend to these concerns in radical ways, as diverse, often contradictory and sometimes problematic indigenous perspectives have been doing for millennia.

Postcolonialism also foregrounds the interconnectedness between 'them' and 'us', 'here' and 'there' and in so doing forces a consideration of ethics. This is important because while the scale of severe poverty is enormous in human terms, the world poverty problem is tiny in economic terms. It is estimated that just 1 per cent of the national incomes of countries in the North would suffice to end severe poverty worldwide (see Pogge 2008, especially 202–21, for suggestions on how this might be brought about). However, these countries are unwilling to meet this cost and instead continue to impose a profoundly unjust global institutional order that perpetuates the catastrophic situation. As Pogge (2008) argues, most citizens of affluent countries believe that we are doing nothing wrong, largely because they feel disconnected from massive poverty overseas. Postcolonialism dispels this misconception, revealing the world as profoundly intertwined and interconnected, and laying out conceptual terrain for a new, ethical relationship between North and South. It thus provides spaces in which the will to improve that infuses development can be brought into dialogue with debates about human rights, justice and well-being, as well as with new ontologies that radically rethink relationships between the human and the non-human. These ideas do not always sit comfortably together, but productive solutions to some of the world's current problems may emerge from within the tensions and interstices between them.

In addition, as discussed in the Introduction, viewing the South through the idea of 'development' risks erasing the complexities, diversities and dynamism of lived experiences of people who live in the global South. We should not forget that the global South is not a coherent entity and people here do not perceive it as such. Nor do they necessarily aspire to be like countries in the global North. From Cuba to Ghana to Malaysia notions of morality and traditional values are very often perceived to be superior to those in the North. Moreover, significant alternatives to mainstream development have long evolved in the South, often emanating from moral and philosophical traditions – Bolivar and Guevara's tri-continentalism and buen vivir in Latin America, African socialism, Maoism in China, post-apartheid South Africa and Nehru's national development in India (see Chapter 2). Development does not simply originate in the North and to imagine that it does ignores the long history of knowledge exchange and learning between North and South. Postcolonial theory and decolonial scholarship cohere around a shared injunction to learn from southern knowledge. As Grosfoguel (2012: 97) argues, the philosophy of liberation can only come from cross-cultural dialogue between critical thinkers. Women's liberation, democracy, civil rights, alternative forms of development and economic organization 'can only emerge from the creative responses of local ethico-epistemic projects'. While western understandings of liberation, democracy or development cannot be imposed on non-European peoples, this does not represent a call for fundamentalist or nationalist solutions. Rather it sees epistemic diversity as a means of creating a 'decolonized, transmodern world' (*ibid.*).

---

## Box 9.2 Challenging Eurocentric development: indigenous knowledge in Australia

The relatively recent acknowledgement that indigenous Australians have rights based in precolonial social formations unsettles assumptions that underpin dominant policy and practice in a range of social, economic and environmental fields. As a result, different foundations for weaving social, environmental and economic justice into the social fabric might be encouraged. However, the material and discursive spaces of development remain colonized by Eurocentric ideas that marginalize and trivialize indigenous perspectives, specifically on the relationship between people and place. This colonization limits the transformative possibilities in the new discursive and political spaces that have emerged. One example of this is in the development of new systems of wildlife and natural resource management on indigenous lands. Howitt and Suchet-Pearson (2006) demonstrate how the processes adopted in shaping 'co-management' regimes continue to be dominated by Eurocentric notions of 'management', which ontologically privilege non-indigenous ways of understanding people–environment relations.

Dominant management discourses and practices are often presented as the fundamental building blocks for achieving broad societal goals, such as development and conservation, which are treated as self-evidently positive and of universal relevance. In the context of indigenous Australian experience, Howitt and Suchet-Pearson argue that the discourses and practices of both development and conservation reflect highly problematic assumptions about relationships between people, and between people and their surroundings, which are rooted in Eurocentric ontologies. They argue that failure to challenge these assumptions risks reimposing colonial power relations on groups who make different sense of the world. Instead, they advocate cross-cultural wildlife management discourses through the idea of situated engagement with ontological pluralism – in other words, by bringing different understandings of people–environment relations into dialogue with each other. This implies rejection of the dominant management discourses in favour of more open-ended and constructive engagements.

This is one example of how indigenous politics overlap with postcolonial theory. It critiques the idea of ontological privilege being accorded to arrangements that conform to Eurocentric parameters. However, Howitt and Suchet-Pearson (2006: 324) also resist the problematic tendency to romanticize indigenous knowledge. They argue that it is 'equally inappropriate to displace one set of universal propositions with another, indigenized version of universality or a romanticizing of the local to return to some naive vision of what things "once were"'. As they argue, marginalized, traumatized, dispossessed and often dysfunctional indigenous societies 'are no more a source of universal truth than are the flawed, dehumanized and dysfunctional systems whose smoke-and-mirrors approach to being-in-place has entrenched economic, social and environmental injustice'. Indeed, it is in claims to universalism *per se* that they see the roots of injustice. They argue that creating 'building blocks' to create new concepts, categories and exemplars of what might be is not easy, and requires a rethinking of the concepts, language and images used to describe, analyze and address the processes. This requires both indigenous groups and mainstream or progressive development agencies to work together in plural, ethical and egalitarian ways. Postcolonial methodologies (Chapter 7) make these plural ontologies possible.

Source: adapted from Howitt and Suchet-Pearson (2006)

That development scholars now view postcolonial approaches as constructive is significant, yet no coherent project of postcolonial development studies has emerged. This is not necessarily a problem since the use of postcolonialism in development studies must always be provisional and combined with other radical approaches, such as feminism, decoloniality and post-development, as well as critical understandings of globalization and transnationalism. However, if it ignores the lessons of postcolonialism, development studies risks perpetuating its problematic image, rooted as it is in histories of imperialism and contemporary global geopolitics. The concerns of the ascendant (if not the most powerful) majority of the world's population are very different from those of the powerful minority in the global North. Unless the intellectual pre-occupations within the global North acknowledge this shift, they are likely to become ever more out of step with the political and economic dynamics of much of the world's population. As discussed in Chapter 8, development studies risks becoming redundant if it fails to engage with the multiple experiments in redefining and realizing development already underway, especially in the global South. It also risks becoming irrelevant if it continues to cleave to anachronistic linear notions of development and progress when these are being replaced by newer senses of time and of futures founded on more open narrative models in those countries in Asia, Africa and Latin America that will shape new planetary futures.

## Summary

- Both postcolonial and development studies are enriched by a productive engagement between postcolonial criticisms of how we speak and write about the world and development concerns with the material realities of global inequalities.

- Applying postcolonial critiques to development does not imply anti-development; rather it aims to understand the power of development ideas, knowledge and institutions and their consequences in specific places at particular times.
- Producing development studies that is postcolonial in theory and practice means acknowledging the significance of language and representation, and understanding the power of development discourse and its material effects on the lives of people subject to development policies. Decolonizing development is central to this process.
- Development studies scholars are already beginning to view postcolonial approaches as constructive and instructive.
- Because of its focus on the global South, development studies is attuned to a postcolonial sensibility. Postcolonialism has great potential for inspiring a new agenda in development studies and thus maintaining its vitality and relevance in a rapidly changing world.

## Discussion questions

1   What are the major criticisms of postcolonialism in relation to development studies and how valid are these?
2   What is involved in decolonizing development studies?
3   What is the potential of postcolonialism within development studies?
4   Of what contemporary significance is a reinvigorated development studies?

## Further reading

Chakrabarty D. (2000) *Provincializing Europe: Postcolonial Thought and Historical Difference* Princeton, New Jersey, Princeton University Press. A key text within postcolonialism that explains the importance of decolonizing knowledge by 'provincializing' western thought.

McEwan, C. (2003) 'Material geographies and postcolonialism' *Singapore Journal of Tropical Geography* 24, 3, pp. 340–55. This article addresses some of the criticisms of postcolonialism through an examination of how it might be combined with analysis of the materialities of lived experience.

Sylvester, C. (1999) 'Development studies and postcolonial studies: disparate tales of the "Third World"' *Third World Quarterly*, 20, 4, pp. 703–21. Succinct account of the major differences and divergences between postcolonialism and development studies.

Sylvester, C. (2006) 'Bare life as a development/postcolonial problematic' *The Geographical Journal* 172, 1, pp. 66–77. An example of the kinds of intervention that a postcolonial approach to development studies might make.

## Websites

www.ted.com/talks/chimamanda_adichie_the_danger_of_a_single_story.

A powerful TED talk by Nigerian author Chimamanda Adichie in which she tells the story of how she found her authentic cultural voice and warns that if we hear only a single story about another person or country, we risk a critical misunderstanding.

www.un.org/millenniumgoals/2015_MDG_Report/pdf/MDG%202015%20 rev%20(July%201).pdf

UN Millennium Development Goals Report, which provides an indication of the global development challenges that remain despite decades of international development.

# Bibliography

Abbott, D. (2006) 'Disrupting the "whiteness" of fieldwork in geography' *Singapore Journal of Tropical Geography*, 27, 3, pp. 326–41.

Abrahamsen, R. (2001) *Disciplining Democracy: Development Discourse and Good Governance in Africa* London, Zed.

ActionAid (2005) 'Trade & the WTO: An introduction' 19 October 2005.

Adam, B. (2005) *Timescapes of Modernity* London, Routledge.

Adhikari, M. (2010) 'A total extinction confidently hoped for: the destruction of Cape San society under Dutch colonial rule, 1700–1795' *Journal of Genocide Research* 12, 1–2, pp. 19–44.

Afary, J. (1996) *The Iranian Constitutional Revolution, 1906–1911: Grassroots Democracy, Social Democracy, and the Origins of Feminism* New York, Columbia University Press.

Agamben, G. (2005) *State of Exception* (trans. Attell, K.) Chicago, University of Chicago Press.

Agamben, G. (1998) *Homo Sacer: Sovereign Power and Bare Life* (trans. Heller-Roazen, D.) Stanford, Stanford University Press.

Agee, P. (1975) *Inside the Company: CIA Diary* London, Penguin.

Ahluwalia, P. (2001) *Politics and Post-colonial Theory: African Inflections* London, Routledge.

Ahmad, A. (1992) *In Theory: Classes, Nations, Literatures* London, Verso.

Alcoff, L. (1991) 'The problem of speaking for others' *Cultural Critique* 20, pp. 5–32.

Alexander, T. (1996) *Unravelling Global Apartheid: An Overview of World Politics* Cambridge, Polity.

Alfred, T (2007) 'Why revel in birth of imperial monster?' *The Times Higher* May 11, p. 14.

Allen, T. (1992) 'Taking culture seriously'. In T. Allen and A. Thomas (eds), *Poverty and Development in the 1990s* Oxford, Oxford University Press, pp. 331–46.

Allen, T. and Thomas, A. (eds) (2001) *Poverty and Development into the 21st Century* Oxford, Oxford University Press.

Alloula, M. (1986) *The Colonial Harem* Manchester, Manchester University Press.

Alvarez, S. (1990), *Engendering Democracy in Brazil* Princeton, Princeton University Press.

Alvarez, S. and Escobar, A. (1992) 'Conclusion: theoretical and political horizons of change in contemporary Latin American social movement'. In A. Escobar and S. Alvarez (eds), *The Making of Social Movements in Latin America: Identity, Strategy and Democracy* Boulder, Westview Press, pp. 317–30.

Amadiume, I. (1997) *Reinventing Africa. Matriarchy, Religion and Culture* London, Zed.

Amadiume, I. (2000) *Daughters of the Goddess, Daughters of Imperialism: African Women, Culture, Power and Democracy* London, Zed.

Amin, S. (2006) 'The Millennium Development Goals: A Critique from the South', *The Monthly Review* 57, 10, https://monthlyreview.org/2006/03/01/the-millennium-development-goals-a-critique-from-the-south/.

Amnesty International (2016) *The Ugly Side of the Beautiful Game: Exploitation of Migrant Workers on a Qatar 2022 World Cup Site* March, www.amnesty.org/en/documents/mde22/3548/2016/en/ (accessed 5/1/18).

Amos, V. and Parmar, P. (1984) 'Challenging imperial feminisms' *Feminist Review* 17, pp. 3–19.

Andreasson, S. (2005) 'Orientalism and African development studies: the "reductive repetition" motif in theories of African underdevelopment' *Third World Quarterly* 26, 6, pp. 971–86.

Andreotti, V. (2007) 'The contributions of postcolonial theory to development education' Development Education Association Thinkpieces, www.dea.org.uk/uploads/4453d22a64a184b4f76a113996448fcf/dea_thinkpiece_andreotti.pdf.

Anseeuw, W., Boche, M., Breu, T. *et al.* (2012) 'Transnational Land Deals for Agriculture in the Global South' Analytical Report based on the Land Matrix Database, CDE/CIRAD/GIGA, Bern/Montpellier/Hamburg.

Anzaldua, G. and Moraga, C. (eds) (1981) *This Bridge Called My Back: Writings by Radical Women of Colour* New York, Kitchen Table Women of Colour Press.

Appadurai, A. (2000) 'Grassroots globalization and the research imagination' *Public Culture* 12, 1, pp. 1–9.

Appadurai, A. (2004) 'The capacity to aspire: culture and the terms of recognition' In V. Rao and M. Walton (eds), *Culture and Public Action*, Stanford, CA, Stanford University Press, pp. 59–84.

Apffel-Margelin, F. (1996) 'Introduction: rationality and the world'. In F. Apffel-Margelin, and S. Margelin (eds) *Decolonizing Knowledge: From Development to Dialogue*, Oxford, Clarenden Press, pp. 1–40.

Apter, D.E. (1987) *Rethinking Development: Modernization, Dependency and Postmodern Politics* London, Sage.

Arnold, M. (1978) (ed.) *The Testimony of Steve Biko. Black Consciousness in South Africa* London, Grafton.

Asad, T. (1973) *Anthropology and the Colonial Encounter* London, Ithaca Press.

Ashcroft, B. (2001) *On Postcolonial Futures. Transformations of Colonial Culture* London, Continuum.

Ashcroft, B., Griffiths, G. and Tiffin, H. (eds) (1995), *The Post-Colonial Studies Reader* London, Routledge.

Ashcroft, B., Griffiths, G. and Tiffin, H. (eds) (1998) *Key Concepts in Post-Colonial Studies* London, Routledge.

Asher, K. (2013) 'Latin American decolonial thought, or making the subaltern speak' *Geography Compass* 7, 12, 832–42.

Baaz, M.E. (2005) *The Paternalism of Partnership. A Postcolonial Reading of Identity in Development Aid* London, Zed.

Badran, M. (1995) *Feminists, Islam and Nation: Gender and the Making of Modern Egypt* Princeton University Press.

Baldwin, W.A. (2017) 'Postcolonial futures: climate, race, and the yet-to-come' *Interdisciplinary Studies in Literature and Environment* 24, pp. 292–305.

Balibar, E. (2007) 'Uprisings in the "Banlieues"' *Constellations* 14, 1, pp. 47–71.

Bankoff, G. (2001) 'Rendering the world unsafe: "vulnerability" as western discourse' *Disasters* 25, 1, pp. 19–35.

Barnett, C. (1998a) 'The cultural turn: fashion or progress in human geography' *Antipode* 30, pp. 379–94.

Barnett, C. (1998b) 'Impure and worldly geography. The Africanist discourse of the Royal Geographical Society' *Transactions of the Institute of British Geographers* 23, 2, pp. 239–52.

Baron, B. (1994) *The Women's Awakening in Egypt: Culture, Society and the Press* New Haven, Yale University Press.

Barrios de Chungara, D. (1978) *Let Me Speak!* trans. Victoria Ortiz, Mexico City: Siglo 21.

Beattie, L., Miller, D., Miller, E., *et al.* (1999) 'The media and Africa: images of disaster and rebellion'. In *Message Received: Glasgow Media Group Research 1993–1998* Longman, Harlow, pp. 229–67.

Bebbington, A. and Bebbington, D. (2001) 'Development alternatives: practice, dilemmas and theory' *Area* 33, 1, pp. 7–17.

Beck, U. (2000) *What is Globalization?* Cambridge, Polity Press.

Beckwith, C.I. (1987) *The Tibetan Empire in Central Asia. A History of the Struggle for Great Power among Tibetans, Turks, Arabs, and Chinese during the Early Middle Ages* Princeton, Princeton University Press.

Bek, D., McEwan, C. and Bek, K. (2007) 'Ethical trading and socio-economic transformation: critical reflections on the South African wine industry' *Environment and Planning A*, 39 pp. 301–19.

Bell, D. and Klein, R. (1996) 'Beware: Radical feminists speak, read, write, organise, enjoy life, and never forget'. In D. Bell and R. Klein (eds), *Radically Speaking. Feminism Reclaimed* London, Zed, pp. xvii–xxx.

Bell, M. (1994) 'Images, myths and alternative geographies of the Third World' *Human Geography*, pp. 174–199.

Bell, M. (2002) 'Inquiring minds and postcolonial devices: examining poverty at a distance' *Annals of the Association of American Geographers* 92, pp. 507–23.

Benhabib, S. (1999) 'Sexual difference and collective identities: the new global constellation' *Signs* 24, 2, pp. 335–61.

Berger, M.T. (1994) 'The end of the "Third World"?' *Third World Quarterly* 15, 2, pp. 257–75.

Bezzina, L. (2017) 'Disabled voices in development? Listening to People with Disabilities in Burkina Faso' Unpublished PhD thesis, University of Durham.

Bhabha, H. (2004) 'Foreword: framing Fanon'. In F. Fanon *The Wretched of the Earth*, trans. R. Philcox, New York, Grove Press.

Bhabha, H. (1994) *The Location of Culture* London, Routledge.

Bhabha, H. (1985) 'Signs taken for wonders: questions of ambivalence and authority under a tree outside Delhi, May 1817' *Critical Inquiry* 12, 1, pp. 144–65.

Bhambra, G. (2014) 'Postcolonial and decolonial dialogues' *Postcolonial Studies* 17, 2, 115–21.

Biccum, A. (2005) 'Development and the "new" imperialism: a reinvention of colonial discourse in DFID promotional literature' *Third World Quarterly* 26, 6, pp. 1005–20.

Binns, T. (1992) 'The role of fieldwork in teaching development geography: some African perspectives'. In R. Potter and T. Unwin (eds) *Teaching the Geography of Developing Areas*, London, RGS-IBG, pp. 113–29.

Birch, T. (1996) ' "A land so inviting and still without inhabitants". Erasing Koori culture from (post-)colonial landscapes'. In K. Darian-Smith L. Gunner and S. Nuttall (eds) *Text, Theory, Space*, London, Routledge, pp. 173–188.

Bishop, R., Phillips, J. and Wei, Y.W. (2003) *Postcolonial Urbanism* London, Routledge.

Blaney, D. (1996) 'Reconceptualizing autonomy: the difference dependency theory makes' *Review of International Political Economy* 3, 3, pp. 459–97.

Blaut, J.M. (1993) *The Colonizer's Model of the World: Geographical Diffusionism and Eurocentric History* New York, Guilford Press.

Blunt, A. and Wills, J. (2000) *Dissident Geographies: An Introduction to Radical Ideas and Practice* Harlow, Prentice Hall.

Blunt, A. and McEwan, C. (2002) (eds) *Postcolonial Geographies* London, Continuum.

Bolt, C. (1971) *Victorian Attitudes to Race* London, Routledge & Kegan Paul.

Bond, P. (2002) 'NEPAD' *Znet Magazine*, available at www.ifg.org/wssd/bondZnet.htm (accessed 27/09/07).

Bora, S., Bouët, A. and Roy, D. (2007) 'The marginalization of Africa in world trade' *International Food Policy Research Institute*, July, www.ifpri.org/pubs/ib/rb07.asp (accessed 06/11/07).

Bose (1997) 'Instruments and idioms colonial and national development: India's historical experience in comparative perspective'. In F. Cooper and R. Packard (eds) *International Development and the Social Sciences* Berkeley, University of California, pp. 45–63.

Brennan, T. (2014) 'Subaltern stakes' *New Left Review* 89, 67–87.

Briggs, J. and Sharp, J. (2004) 'Indigenous knowledges and development: a postcolonial caution' *Third World Quarterly* 25, 4, pp. 661–76.

Broks, P. (1990) 'Science, the press and empire: Pearson's publications, 1890–1914'. In J.M. MacKenzie, (ed.) *Imperialism and the Natural World* Manchester, Manchester University Press.

Brown, S. (2006) 'Can remittances spur development? A critical survey' *International Studies Review* 8, pp. 55–75.

Buchanan, K. (1977) 'Reflections on a "dirty word" '. In R. Peet (ed.) *Radical Geography: Alternative Viewpoints on Contemporary Social Issues* London, Methuen, pp. 363–7.

Burich, K. (2016) 'Murder by poverty in Indian country: then and now' *Indian Country Today* March 17, https://indiancountrymedianetwork.com/history/events/murder-by-poverty-in-indian-country-then-and-now/.

Burman, E. (1995) 'The abnormal distribution of development: policies for Southern women and children' *Gender, Place and Culture* 2, 1, pp. 21–36.

Burton, A. (1999) 'Some trajectories of "feminism" and "imperialism" '. In M. Sinha, D. Guy and A. Woollacott (eds), *Feminisms and Internationalism* Oxford, Blackwell, pp. 214–24.

Bushong, A.D. (1984) 'Ellen Churchill Semple 1863–1932'. In T.W. Freeman (ed.) *Geographers: Bio-Bibliographic Studies, 8* London, Marshall Publications, pp. 87–94.

Butchart, A. (1998*) The Anatomy of Power. European Constructions of the African Body* London, Zed.

Butler, J. (1990) *Gender Trouble. Feminism and the Subversion of Identity* London, Routledge.

Butz, D. and Besio, K. (2004) 'The value of autoethnography for field research in transcultural settings' *The Professional Geographer* 56, 3, pp. 350–60.

Cairns, A.C. (1965) *Prelude to Imperialism. British Reactions to Central African Society 1840–1890* London, Routledge & Kegan Paul.

Carby, H. (1983) 'White women listen! Black feminism and the boundaries of sisterhood'. In Centre For Cultural Studies, *The Empire Strikes Back* London, Hutchinson.

Castree, N. (1999) ' "Out there"? "In here"? Domesticating critical geography', *Area* 31, 3, pp. 81–6.

Castro-Gómez, S. (2008) '(Post)coloniality for dummies: Latin American perspectives on modernity, coloniality, and the geopolitics of knowledge'. In M.D. Moraña *et al.* (eds) *Coloniality at Large: Latin America and the Postcolonial Debate* Durham, Duke University Press, pp. 396–416.

Césaire, A. (1997) *Return to My Native Land* Bloodaxe Books.

Césaire, A. (2000; orig. 1950) *Discourse on Colonialism* Monthly Review Press.

Chakrabarty, D. (1992) 'Postcoloniality and the artifice of history: Who speaks for Indian pasts?' *Representations*, 37, pp. 1–24.

Chakrabarty, D. (2000) *Provincializing Europe: Postcolonial Thought and Historical Difference* Princeton, New Jersey, Princeton University Press.

Chakrabarty, D. (2001) 'Adda, Calcutta: dwelling in modernity'. In D.P. Gaonkar (ed.) *Alternative Modernities* Durham, NC, Duke University Press.

Chakrabarty, D. (2002) *Habitations of Modernity: Essays in the Wake of Subaltern Studies* Chicago, University of Chicago Press.

Chakrabarty, D. (2005) 'A small history of subaltern studies'. In H. Schwarz and S. Ray (eds), *A Companion to Postcolonial Studies* London, Blackwell, pp. 467–85.

Chakrabarty, D. (2009) 'The climate of history: four theses' *Critical Inquiry* 35, 2, pp. 197–222.

Chakrabarty, D. (2012) 'Postcolonial studies and the challenge of climate change' *New Literary History* 43, 1, pp. 1–18.

Chambers, I. (1996), 'Waiting on the end of the world?'. In D. Morley and K-H. Chen (eds), *Stuart Hall. Critical Dialogues in Cultural Studies* London, Routledge, 201–11.

Chambers, R. (1983) *Rural Development: Putting the Last First* London, Longman.

Chambers, R. (1994) 'Participatory Rural Appraisal (PRA): analysis and experience' *World Development* 22, 9, pp. 1253–68.

Chambers, R. (1997) *Whose Reality Counts? Putting the First Last* London, Intermediate Technology Publications.

Chambers, R. (2005) *Idea of Development* London, Earthscan.

Chang, H-J (2003) *Kicking Away the Ladder: Development Strategy in Historical Perspective* London, Anthem Press.

Chatterjee, P. (1996) *Nationalist Thought and the Colonial World*, London, Zed.

Chatterjee, P. (1998) *A Possible India* Delhi, Oxford India Paperbacks.

Chibber, V. (2013) *Postcolonial Theory and the Specter of Capital* London, Verso.

Childs, P. and Williams, P. (1997) *An Introductory Guide to Postcolonial Theory* New York, Prentice Hall.

Chow, R. (1990) *Women and Chinese Modernity: The Politics of Reading Between East and West* Oxford, University of Minnesota Press.

Chowdhry, G. and Nair, S. (2002) (eds) *Power, Postcolonialism, and International Relations* London, Routledge.

Chowdhry, G. (2002) 'Postcolonial interrogations of child labor: human rights, carpet trade and Rugmark in India'. In G. Chowdhry and S. Nair (eds) *Power, Postcolonialism, and International Relations* London, Routledge, pp. 225–53.

Chutel, L. (2016) 'Germany finally apologizes for its other genocide – more than a century later' *Quartz Africa* July 16.

CIA (2000) *Global Trends 2015: A Dialogue about the Future with Non-Government Experts* National Intelligence Council, available at www/adci.gov/nic/pubs/2015_files/2015.html.

Clayton, D. (2011) 'Subaltern space'. In J. Agnew and D. Livingstone (eds) *Handbook of Geographical Knowledge* London, Sage, pp. 246–60.

Clayton, D. (2002) 'Critical imperial and colonial geographies'. In K. Anderson *et al.* (eds) *Handbook of Cultural Geography*, London, Sage, pp. 354–68.

Clayton, D. (2000) *Islands of Truth: The Imperial Fashioning of Vancouver Island* Vancouver, UBC Press.

Clémencon, R. (2012) 'Welcome to the Anthropocene: Rio+20 and the meaning of sustainable development' *The Journal of Environment and Development* 21, 3, pp. 311–38.

Cobham, A. (2001) 'Capital account liberalization and poverty'. Working Paper 70, Queen Elizabeth House, University of Oxford, available at www.id21.org/society/.

Collinson, H. (1990) *Women and Revolution in Nicaragua* London, Zed.

Comaroff, J. and Comaroff, J.L. (2012) 'Theory from the South: Or, how Euro-America is evolving toward Africa' *Anthropological Forum* 22, 2, pp. 113–31.

Connell, R. (2015) 'Meeting at the edge of fear: theory on a world scale' *Feminist Theory* 16, 1, pp. 49–66.

Connell, R. (2007) *Southern Theory* Cambridge, Polity.

Cook, I. *et al.* (2000) 'Social sculpture and connective aesthetics: Shelley Sacks's "Exchange Values"', *Ecumene* 7, 3, pp. 337–43.

Cook, I. and Crang, M. (1995) *Doing Ethnographies* Norwich, Environmental Publications.

Cooke, B. and Kothari, U. (eds) (2001) *Participation: The New Tyranny* London, Zed Books.

Corbridge, S. (1992) 'Third World development' *Progress in Human Geography* 16, pp. 584–95.

Corbridge, S. (1993a) 'Colonialism, postcolonialism and the political geography of the Third World'. In P.J. Taylor (ed.) *Political Geography of the Twentieth Century* London, Belhaven Press.

Corbridge, S. (1993b) 'Marxisms, modernities and moralities: development praxis and the claims of distant strangers' *Environment and Planning D: Society and Space* 11, pp. 449–72.

Corbridge, S. (1998) 'Development ethics: distance, difference, plausibility' *Ethics, Place and Environment* 1, pp. 35–53.

Corbridge, S. and Mawdsley, E. (eds) (2003) Special issue on fieldwork in the tropics *Singapore Journal of Tropical Geography* 24, 2.

Cornwall, A. (2006) 'Historical perspectives on participation in development' *Commonwealth and Comparative Politics* 44, 1, pp. 62–83.

Coronil, F. (2008) "'Elephants in the Americas?" Latin American postcolonial studies and global decolonization'. In M.D. Moraña *et al.* (eds) *Coloniality at Large: Latin America and the Postcolonial Debate* Durham, Duke University Press, pp. 396–416.

Cowen, M. and Shenton, R. (2006) *Doctrines of Development* London, Routledge.

Cowen, M. and Shenton, R. (1995) 'The invention of development'. In J. Crush (ed.) *Power of Development* London, Routledge, pp. 27–43.

Crespo, I., Palli, C. and Lalueza, J. (2002) 'Moving communities: a process of negotiation with a gypsy minority for empowerment' *Community, Work and Family* 5, pp. 49–65.

Crush, J. (1994) 'Post-colonialism, decolonization and geography'. In A. Godlewska and N. Smith (eds), *Geography and Empire* Oxford, Blackwell.

Crush, J. (ed.) (1995) *Power of Development* London, Routledge.

Cupples, J. and Glynn, K. (2014) 'Indigenizing and decolonizing higher education on Nicaragua's Atlantic Coast' *Singapore Journal of Tropical Geography* 35, pp. 56–71.

Curtis, M. (2016) 'The New Colonialism: Britain's Scramble for Africa's Mineral and Energy Resources' Report for War on Want, July 12.

Dabashi, N. (1993) 'Historical conditions of Persian Sufism during the Seljuk period'. In L. Lewisohn (ed.) *Classical Persian Sufism: From its Origins to Rumi* London and New York, Khaniqahi Nimatallah Publishers.

Dahlgreen, W. (2014) 'The British Empire is "something to be proud of"' https://yougov.co.uk/news/2014/07/26/britain-proud-its-empire/

Darby, P. (ed.) (1997), *At the Edge of International Relations. Postcolonialism, Gender and Dependency* London: Pinter.

Das, B.L. (2006) *The Current Negotiations in the WTO: Options, Opportunities & Risks for Developing Countries* Zed Books.

Datta, K., McIlwaine, C., Wills, J., *et al.* (2007), 'The new development finance or exploiting migrant labour? Remittance sending among low-paid migrant workers in London' *International Development Planning Review* 29, 1, pp. 43–68.

Davis, A. (1982) *Women, Race and Class* New York, Vintage.

Daya, S. (2007) 'Writing the modern body: discursive constructions of the new Indian woman'. Unpublished PhD thesis, University of Durham.

De Haas, H. (2005), 'International migration, remittances and development: myths and facts' *Third World Quarterly* 26, pp. 1269–284.

*The Deloitte Consumer Review. Africa: A Twenty-First Century View* www2.deloitte.com/content/dam/Deloitte/ng/Documents/consumer-business/the-deloitte-consumer-review-africa-a-21st-century-view.pdf.

Delphy, C. (1984) *Close to Home: A Materialist Analysis of Women's Oppression* London: Hutchinson.

Demaria, F., Schneider, F., Sekulova, F. *et al.* (2013) 'What is degrowth? From an activist slogan to a social movement' *Environmental Values* 22, pp. 191–215.

Dengler, C. and Seebacher, L.M. (forthcoming) 'What about the global South? A proposal for a feminist decolonial degrowth approach' *Ecological Economics*.

Department for International Development/HM Treasury (2015) *UK aid: tackling global challenges in the national interest* London, HM Treasury.

Department for International Development) (2006) 'Moving out of poverty: making migration work better for poor people' (draft policy paper) London, DFID, available at www.dfid.gov.uk/pubs/fi les/migration-policy-paper-draft.pdf.

Department for International Development (DFID) (2000a) *Halving World Poverty by 2015: Economic Growth, Equality and Security* London: DFID.

Department for International Development (DFID) (2000b) *Eliminating World Poverty: Making Globalisation Work for the Poor* Cmd 5006, London: HMSO.

Department for International Development (DFID) (1997) *White Paper on Eliminating World Poverty: A Challenge for the Twenty-first Century* London, HMSO.

Derrida, J. (2000) *Of Hospitality* Stanford, Stanford University Press.

Derrida, J. (2001) *On Cosmopolitanism and Forgiveness* London, Routledge.

Dhawan, N. (2013) 'Coercive cosmopolitanism and impossible solidarities' *Qui Parle: Critical Humanities and Social Sciences* 22, 1, pp. 139–66.

'Introduction: colonialism and culture'. In N. Dirks (ed.) *Colonialism and Culture* Ann Arbor, University of Michigan Press, pp. 1–26.

Dirlik, A. (1994) 'The postcolonial aura: Third World criticism in the age of global capitalism' *Critical Inquiry* Winter, pp. 329–356.

Dirlik, A. (1997) *The Postcolonial Aura: Third World Criticism in the age of Global Capitalism* Boulder, Colorado: Westview.

Di Stefano, C. (1990) 'Dilemmas of difference: feminism, modernity, and [ostmodernism'. In L. Nicholson (ed.), *Feminism/Postmodernism* London, Routledge, pp. 63–82.

Dobson, A (2005) 'Globalisation, cosmopolitanism and the environment' *International Relations* 19 3, pp. 259–273.

Dogra, N. (2014) *Representations of Global Poverty. Aid, Development and International NGOs* London, IB Tauris.

Dolan, C.S. (2005) 'Field of obligation. Rooting ethical sourcing in Kenyan horticulture' *Journal of Consumer Culture* 5, 3, 365–89.

Dorling, D. (2017) 'Short Cuts' *London Review of Books* 39, 22, p. 19, 16 November www.lrb.co.uk/v39/n22/danny-dorling/short-cuts.

Doty, R.L. (1996) 'Repetition and variation: academic discourses on North–South relations'. In R.L. Doty, *Imperial Encounters: The Politics of Representation in North–South Relations* Minneapolis: University of Minnesota Press, pp. 145–62.

Duncan, N. and Sharp, J.P. (1993) 'Confronting representation(s)' *Environment and Planning D: Society and Space* 11, pp. 473–486.

Drakulić, S. (1993) *How We Survived Communism and Even Laughed* London, Vintage.

Dreze, J. (2002) 'On research and action' *Economic and Political Weekly*, 37, 9, p. 817.

Driver, F. (2001) *Geography Militant. Cultures of Exploration and Empire* Oxford, Blackwell.

Driver, F. and Yeoh, B. (eds) (2000) 'Constructing the tropics' *Singapore Journal of Tropical Geography* 21, pp 1–98.

Dubois, E.C. (1978) *Feminism and Suffrage: The Emergence of an Independent Women's Movement in America, 1848–1869* Ithaca, Cornell University Press.

Duden, B. (1992) 'Population'. In W. Sachs (ed.) *The Development Dictionary* London, Zed.

Duffield, M. (2001) 'Governing the borderlands: decoding the power of aid' *Disasters* 25, pp. 308–20.

Duffield, M. (2007) *Development, Security and Unending War* Cambridge, Polity.

Eade, D. and Rowlands, J. (2003) *Development Methods and Approaches* Oxford, Oxfam Publications.

Eagleton, T. (1994) 'Goodbye to the enlightenment' *The Guardian* 5 May.

Ehrenreich, B., Dowie, M. and Minkin, S. (1979) 'The charge: gynocide' www.motherjones.com/news/feature/1979/11/ehrenreich.html.

Einhorn, B. (1993) *Cinderella Goes to Market. Citizenship, Gender and Women's Movements in East Central Europe* London, Verso.

El Saadawi, N. (1997) *The Nawal El Saadawi Reader* London, Zed.

Enaudeau, J. (2013) 'In search of the African middle class' www.theguardian.com/world/2013/may/03/africa-middle-class-search.

Escobar, A. (1984) 'Discourse and power in development: Michel Foucault and the relevance of his work to the Third World' *Alternatives* 10, 10, pp. 377–400.

Escobar, A. (1988) 'Power and visibility: the invention and management of development in the Third World' *Cultural Anthropology* 4, 4, pp. 428–443.

Escobar, A. (1992) 'Imagining a post-development era? Critical thought, development and social movements' *Social Text*, 31/32, 16, 243, pp. 20–56.

Escobar, A. (1995a) *Encountering Development: The Making and Unmaking of the Third World* Princeton: Princeton University Press.

Escobar, A. (1995b) 'Imagining a Post-Development Era'. In J. Crush (ed.), *Power of Development* London, Routledge, pp. 211–27.

Escobar, A. (2001) 'Culture sits in places: reflections on globalism and subaltern strategies of localization' *Political Geography* 20, pp. 139–74.

Escobar, A. (2007) 'Worlds and knowledges otherwise. The Latin American modernity/coloniality research program' *Cultural Studies* 21, 2–3, pp. 179–210.

Escobar, A. (2008) *Territories of Difference: Place, Movements, Life, Redes* Durham: Duke University Press.

Escobar, A. (2010) 'Latin America at a crossroads' *Cultural Studies* 21, 2–3, 179–210.

Escobar, A., Rochelau, D and Kothari, S. (2002) 'Environmental social movements and the politics of place' *Development* 45, pp. 28–36.

Eshun, G. and Madge, C. (2012) '"Now let me share this with you": exploring poetry as a method for postcolonial geography research' *Antipode* 44, 4, pp. 1395–428.

Espelund, G., Strudsholm, J. and Miller, E. (2003) *Reality Bites: An African Decade* Cape Town, Double Storey.

Essen, J., Noxolo, P, Baxter, R. *et al.* (2017) 'The 2017 RGS-IBG chair's theme: decolonising geographical knowledges, or reproducing coloniality?' *Area* 49, 3, pp. 384–8.

Esteva, G. (1987) 'Regenerating people's space' *Alternatives* 12, 1, pp. 125–52.

Esteva, G. and Prakash, M. (eds) (1998) *Grassroots Postmodernism: Remaking the Soil of Cultures* London, Zed.

Fabian, J. (1983) *Time and the Other* New York, Columbia University Press.

Fan, S. and Polman, P. (2014) 'An ambitious development goal: ending hunger and undernutrition by 2025'. In A. Marble and H. Fritschel (eds) *2013 Global Food Policy Report* Washington, D.C.: International Food Policy Research Institute, pp. 15–28.

Fanon, F. (1986) *Black Skin White Masks* London, Pluto (first published 1952).

Fanon, F. (1967) *The Wretched of the Earth* London, Penguin (first published 1961).

Fargues, P. (2017) *Four Decades of Cross-Mediterranean Undocumented Migration to Europe: A Review of the Evidence* Geneva, International Organization for Migration.

Fellows, M.L. and Razack, S. (1998) 'The race to innocence: confronting hierarchical relations among women' *The Journal of Gender, Race and Justice* 1, 4, pp. 335–55.

Felski, R. (1997) 'The doxa of difference' *Signs* 23, 1, pp. 1–21.

Ferguson, A. (1998) 'Resisting the veil of privilege: building bridge identities as an ethico-politics of global feminisms', *Hypatia, Special Issue: Border Crossings: Multicultural and Postcolonial Feminist Challenges to Philosophy*, Part 2, 13, 3, pp. 95–114.

Ferguson, J. (2006) *Global Shadows. Africa in the Neoliberal World Order* Durham, NC and London, Duke University Press.

Ferguson, J. (1999) *Expectations of Modernity: Myths and Meanings of Urban Life on the Zambian Copperbelt*, Berkeley, University of California Press.

Ferguson, J. (1994) *The Anti-Politics Machine. 'Development', Depoliticization, and Bureaucratic Power in Lesotho* Minneapolis and London, University of Minneapolis Press.

Ferguson, M. (1992) *Subject to Others: British Women Writers and Colonial Slavery, 1670–1834* London, Routledge.

Fine, B., Lapavitsas, C. and Pincus, J. (eds) (2001) *Development Policy in the Twenty-First Century: Beyond the Post-Washington Consensus* London, Routledge.

Fisher, J. (1993) *Out of the Shadows. Women, Resistance and Politics in South America* London, Latin American Bureau.

Flew, F., Bagilhole, B., Carabine, J. *et al.* (1999) 'Introduction: local feminisms, global futures' *Women's Studies International Forum* 22, 4, pp. 393–40.

Food and Agriculture Organisation of the UN (2015) *The State of Food Insecurity in the World* Rome, FAO www.fao.org/3/a-i4646e.pdf.

Franklin, S. (2017) 'Staying with the manifesto: an interview with Donna Haraway' *Theory, Culture & Society* 34, 4, pp. 49–63.

Freire, P. (1972) *Pedagogy of the Oppressed* [trans. Bergman Ramos, M.] London, Sheed and Ward.

Friedmann, J. (1994) *Empowerment: The Politics of Alternative Development* Oxford, Blackwell.

Frost, N. and Jones, C. (1998) 'Video for recording and training in participatory development' *Development in Practice*, 8, pp. 90–4.

Funk, N. (1993) 'Feminism east and west'. In N. Funk and M. Mueller (eds), *Gender Politics and Post-Communism. Reflections from Eastern Europe and the Former Soviet Union* London, Routledge, pp. 318–30.

Funk, N. and Mueller, M. (eds) (1993) *Gender Politics and Post-Communism. Reflections from Eastern Europe and the Former Soviet Union* London, Routledge.

Frank, A.G. (1997) 'The Cold War and me' *Bulletin of Concerned Asian Scholars* 29, 4, available at http://csf.colorado.edu/bcas/symmpos/syfrank.htm/.

Frankenberg, R. and Mani, L. (1993) 'Crosscurrents, crosstalk: "race", postcoloniality and the politics of location' *Cultural Studies* 7, 2, pp. 292–310.

Gandhi, L. (1998) *Postcolonial Theory: A Critical Introduction*, Edinburgh, Edinburgh University Press.

Gaonkar, D.P. (ed.) (2001) *Alternative Modernities* Durham, NC, Duke University Press.

Gay, P. (1973) *The Enlightenment: An Interpretation. Volume 2: The Science of Freedom* London, Wildwood House.

George, S. (1999) 'A short history of neoliberalism', Conference on Economic Sovereignty in a Globalizing World, Bangkok, 24–26 March, available at www.globalexchange.org/campaigns/econ101/neoliberalism.html.

Ghosh, A. (2016) *The Great Derangement: Climate Chang and the Unthinkable* Chicago: CUP.

Ghosh, D. (2001) 'Water out of fire: novel women, national fictions and the legacy of Nehruvian developmentalism in India' *Third World Quarterly* 22, 6, pp. 951–67.

Gibson-Graham, J.K. (1994) ' "Stuffed if I know!" Reflections on post-modern feminist social research' *Gender, Place and Culture*, 1, pp. 205–24.

Gibson-Graham, J.K. (2002) 'Poststructural interventions'. In E. Sheppard and T. Barnes (eds), *A Companion to Economic Geography* Oxford, Blackwell, pp. 95–110.

Gilman, S. (1985) *Difference and Pathology: Stereotypes of Sexuality, Race and Madness* London, Cornell University Press.

Gilmore, R.W. (2007) *Golden Gulag: Prisons, Surplus, Crisis, and Opposition in Globalizing California* Berkeley, University of California Press.

Gilroy, P. (2000) *Against Race: Imagining Political Culture Beyond the Color Line* Cambridge MA, Harvard University Press.

Gilroy, P. (2015) 'Offshore Humanism' *RGS-IBG*, edited lecture, The Antipode RGS-IBG Lecture https://antipodefoundation.org/2015/12/10/paul-gilroy-offshore-humanism/ accessed 05 June 2018.

Global Justice Now (2016) *The Privatisation of UK Aid: How Adam Smith International is Profiting from the Aid Budget* www.globaljustice.org.uk/sites/default/files/files/resources/the_privatisation_of_uk_aid.pdf (accessed 15 March 2018).

Godlewska, A. and Smith, N. (eds) (1994) *Geography and Empire* Oxford, Blackwell.

Goetz, A.M. (1991) 'Feminism and the claim to know: contradictions in feminist approaches to women in development'. In R. Grant and K. Newland (eds), *Gender and International Relations* Bloomington, Indiana University Press, pp. 133–57.

Goldman, M. (2005) *Imperial Nature. The World Bank and Struggles for Social Justice in the Age of Globalization* London and New Haven, Yale University Press.

Goldtooth, D. (2015) 'Keystone XL would destroy our native lands. This is why we fight' *The Guardian* 09/01/15.

Golooba-Mutebi, F. (2005) 'When popular participation won't improve service provision: primary health care in Uganda' *Development Policy Review* 23, pp. 165–82.

Gooder, H. and Jacobs, J.M. (2002) 'Belonging and non-belonging. The apology in a reconciling nation'. In A. Blunt and C. McEwan (eds) *Postcolonial Geographies* London: Continuum, 200–213.

Graham-Brown, S. (1988) *Images of Women. The Portrayal of Women in the Photography of the Middle-East 1860–1950* New York, Columbia University Press.

Gray, J. (1960). *Amerika Samoa: A History of American Samoa and Its United States Naval Administration* Annapolis, Md.: United States Naval Institute.

Gregory, D. (2004) *The Colonial Present* Oxford, Blackwell.

Grewal, I. and Kaplan, C. (eds) (1994) *Scattered Hegemonies: Postmodernity and Transnational Feminist Practices* Minneapolis, University of Minnesota Press.

Griffiths, M. (2017) 'From heterogeneous worlds: western privilege, class and positionality in the South' *Area* 49, pp. 2–8.

Grosfoguel, R. (2007) 'The epistemic decolonial turn' *Cultural Studies* 21, 2–3, pp. 211–23.

Grosfoguel, R. (2012) 'Decolonizing western uni-versalisms: Decolonial pluri-versalism from Aimé Césaire to the Zapatistas' *Transmodernity* Spring, pp. 88–103.

Guelke, L. and Shell, R. (1992) 'Landscape of conquest: frontier water alienation and Khoikhoi strategies of survival, 1652–1780' *Journal of Southern African Studies* 18, 4, pp. 803–24.

Guha, R. (ed.) (1982), *Subaltern Studies* New Delhi, Oxford University Press.

Gupta, A. (1998) *Postcolonial Developments* Durham, Duke University Press.

Gutiérrez, N. (1995) 'Miscegenation as nation-building: Indian and immigrant women in Mexico'. In D. Stasiulis and N. Yuval-Davis (eds) *Unsettling Settler Societies* London, Sage, 161–87.

Haddour, A. (2006) 'Foreword: postcolonial Fanonism'. In *The Fanon Reader* London, Pluto Press, pp. vii–xxv.

Hall, C. (2002) *Civilising Subjects: Metropole and Colony in the English Imagination, 1830–1867* Chicago: University of Chicago Press.

Hall, R. (2011) 'Land grabbing in Southern Africa: the many faces of the investor rush' *Review of African Political Economy* 38, 128, pp. 119–214.

Hall, S. (1995) 'New Cultures For Old'. In D. Massey and P. Jess (eds), *A Place in the World? Places, Cultures and Globalization* Oxford, Oxford University Press, pp. 175–213.

Hall, S. (1996) 'What was "the post-colonial"? Thinking at the limit'. In I. Chambers and L. Curti (eds), *The Postcolonial Question: Common Skies, Divided Horizons* London, Routledge, pp. 242–60.

Haraway, D. (1991) *Simians, Cyborgs and Women* London, Free Association Books.

Haraway, D. (2016) *Staying with the Trouble: Making Kin in the Chthulucene* Durham: Duke University Press.

Hardoon, D., Ayele, S. and Fuentes-Nieva, R. (2016) *An Economy for the 1%. How privilege and power in the economy drive extreme inequality and how this can be stopped.* OXFAM Briefing Paper 210, 18 January, Oxford: Oxfam International, http://policy-practice.oxfam. org.uk/publications/an-economy-for-the-1–how-privilege-and-power-in-the-economy-drive-extreme-inequ-592643

Harper, D. (2003) 'Framing photographical ethnography: a case study' *Ethnography* 4, 2, pp. 241–66.

Harris, C. (2002) *Making Native Space* Vancouver, UBC Press.

Hartmann, B. (1999) 'Population, environment and security: a new trinity'. In J. Silliman and Y. King (eds) *Dangerous Intersections: Feminist Perspectives on Population, Environment and Development* Boston, MA, South End Press.

Harvey, D. (1990) *The Condition of Postmodernity* Blackwell, Oxford.

Harvey, D. (1993) 'Class relations, social justice and the politics of difference'. In J. Squires (ed.) *Principled Positions: Postmodernism and the Rediscovery of Value* London, Lawrence and Wishart, pp. 85–120.

Hasmath, R. (ed.) (2015) *Inclusive Growth, Development and Welfare Policy: A Critical Assessment* London, Routledge.

Hayter, R. (2003) 'The war in the woods: post-Fordist restructuring, globalization, and the contested remapping of British Columbia's forest economy' *Annals of the Association of American Geographers* 93, pp. 706–29.

Head, L. (2016) *Hope and Grief in the Anthropocene* London, Routledge.

Henry, N., McEwan, C. and Pollard, J.S. (2002) 'Globalization from below: Birmingham – postcolonial workshop of the world?' *Area* 34, 2, pp. 117–27.

Hettne, B. (1995) *Development Theory and the Three Worlds* London, Longman.

Hickey, S. and Mohan, G. (eds) (2004) *Participation: From Tyranny to Transformation? Exploring New Approaches to Participation in Development* London, Zed.

Hickey, S. and Mohan, G. (2005) 'Relocating participation within a radical politics of development' *Development and Change* 36, 2, 237–62.

Hiddleston, J. (2009) *Understanding Postcolonialism* Durham, UK: Acumen.

HM Treasury/DFID (2015) *UK Aid: Tackling Global Challenges in the National Interest* London: HM Treasury.

Hochschild, A (1998) *King Leopold's Ghost: A Story of Greed, Terror and Heroism in Colonial Africa* New York: Pan Macmillan.

Holland, P. (1992) *What is a Child? Popular Images of Childhood* London, Virago.

Holt-Gimenez, E, Altieri, M and Rosset, P. (2008) 'Ten reasons why the Rockefeller and the Bill and Melinda Gates Foundations' Alliance for another green revolution will not solve the problems of poverty and hunger in Sub-Saharan Africa' http://agris.fao.org/agris-search/search.do?recordID=GB2013203243.

Hoogevelt, A. (1997) *Globalization and the Postcolonial World* London, Macmillan.

Hoogevelt, A. (2001) *Globalization and the Postcolonial World: The New Political Economy of Development* London, Palgrave.

hooks, b. (1981) *Ain't I a Woman. Black Women and Feminism* Boston, South End Press.

hooks, b. (1984) *Feminist Theory from Margin to Centre* Boston, South End Press.

hooks, b. (1992) *Black Looks: Race and Representation* Boston, South End Press.

hooks, b. (1994) *Outlaw Culture: Resisting Representations* New York, Routledge.

Hornborg, A. (2009) 'Zero-sum world' *International Journal of Comparative Sociology* 50, 3–4, pp. 237–62.

Hountondji, P. (2002) *The Struggle for Meaning. Reflections on Philosophy, Culture and Democracy in Africa* Athens OH, Ohio University Press.

Howitt, R. and Suchet-Pearson, S. (2006) 'Rethinking the building blocks: ontological pluralism and the idea of "management"' *Geografiska Annaler* 88 B (3), pp. 323–35.

Hunt, S. (2014) 'Ontologies of Indigeneity: the politics of embodying a concept' *Cultural Geographies* 21, 1, pp. 27–32.

Hursh, D.W and Henderson, J.A. (2011) 'Contesting global neoliberalism and creating alternative futures' *Discourse: Studies in the Cultural Politics of Education* 32, 2, pp. 171–85.

Huseman, J. and Short, D. (2012) ' "A slow industrial genocide": tar sands and the indigenous peoples of northern Alberta' *The International Journal of Human Rights* 16, 1, 2012.

Hutchinson, G. (1997) *The Harlem Renaissance in Black and White* New York, Belknap Press.

IEA (2015) *World Energy Outlook* Paris, IEA www.iea.org/publications/freepublications/publication/WEO2015.pdf.

Ignatieff, M. (1998) *The Warrior's Honor. Ethnic War and the Modern Conscience* New York, Henry Holt and Co.

ILO (2014) *Work in Freedom: Preventing Trafficking of Women and Girls in South Asia and the Middle East – Technical Cooperation Progress Report 2013*, Geneva: ILO.

ILO (2017) *Global Estimates of Modern Slavery: Forced Labour and Forced Marriage*, Geneva, ILO.

IMF (2014) *World Economic Outlook Database: GDP List of Countries* October 2014 www.imf.org/external/pubs/ft/weo/2014/02/weodata/index.aspx.

IMF (2016) *World Economic Outlook Database* October 2016 www.imf.org/external/pubs/ft/weo/2016/02/weodata/index.aspx.

Jackson, C. (1997) 'Post-poverty, gender and development?' *Institute of Development Studies (IDS) Bulletin*, 28, 3, pp. 145–53.

Jackson, M. (2014) 'Composing postcolonial geographies: postconstructivism, ecology and overcoming ontologies of critique' *Singapore Journal of Tropical Geography* 35, pp. 72–87.

Jackson, P. and Jacobs, J. (1996) 'Postcolonialism and the politics of race', *Environment and Planning D: Society and Space*, 14, pp. 1–3.

Jacobs, J.M. (2001) 'Touching pasts', *Antipode* 33, pp. 730–4.

Jacobs, J.M. (1996) *Edge of Empire: Postcolonialism and the City* London, Routledge.

James, C.L.R. (1963) *The Black Jacobins. Toussaint l'Ouverture and the San Domingo Revolution* New York, Vintage.

James, W. (1998) *Holding Aloft the Banner of Ethiopia: Caribbean Radicalism in Early Twentieth-Century America* London, Verso.

Jani, P. (2017) Book review: *The Postcolonial Orient* by Vasant Kaiwar, *Race & Class* 58, 4, pp. 106–19.

Jarosz, L. (1992) 'Constructing the Dark Continent: metaphor as geographic representation of Africa' *Geografiska Annaler* 74B, 2, pp. 105–15.

Jayawardena, K. (1986) *Feminism and Nationalism in the Third World* London, Zed.

Jayawardena, K. (1995) *The White Women's Other Burden: Western Women and South Asia During British Rule* London, Routledge.

Jazeel, T. (2017) 'Mainstreaming geographies' decolonial imperative' *Transactions of the Institute of British Geographers* https://doi.org/10.1111/tran.12200.

Jazeel, T. (2011) 'Spatializing difference beyond cosmopolitanism: rethinking planetary futures' *Theory, Culture & Society* 28, 5, pp. 75–97.

Jones, P. (2000) 'Why is it alright to do development "over there" but not "here"? Changing vocabularies and common strategies of inclusion across the "First" and "Third" Worlds' *Area* 32, 2, pp. 237–42.

Kaiwar, V. (2015) *The Postcolonial Orient. The Politics of Difference and the Project of Provincialising Europe* Chicago, Haymarket Books.

Kandiyoti, D. (1991) *Women, Islam and the State* Philadelphia, Temple University Press.

Kaplan, R. (1994) 'The coming anarchy: how scarcity, crime, overpopulation, tribalism, and disease are rapidly destroying the social fabric of our planet' *The Atlantic Monthly* 273, 2, pp. 44–76.

Kaplan, C. (1998) *Questions of Travel: Postmodern Discourses of Displacement* Durham, NC, Duke University Press.

Kapoor, I. (2004) 'Hyper-self-reflexive development? Spivak on representing the Third World "Other"' *Third World Quarterly* 25, 4, pp. 627–47.

Karagiannis, N. (2004) *Avoiding Responsibility: The Politics and Discourse of European Development Policy* London, Pluto Press.

Katz, C. (1995) 'Major/minor: theory, nature, politics' *Annals of the Association of American Geographers* 85, pp. 164–8.

Katz, C. (2004) *Growing Up Global: Economic Restructuring and Children's Everyday Lives* Minneapolis: University of Minnesota Press.

Keck, M. and Sikkink, K. (1998) *Activists beyond Borders: Advocacy Networks in International Politics* New York, Cornell University Press.

Kerstens, P. (2008) ' "Deliver Us from Original Sin": Belgian Apologies to Rwanda and the Congo'. *The Age of Apology. Facing Up to the Past* Philadelphia: University of Pennsylvania Press, pp. 187–201.

Kesby, M. (2005) 'Retheorizing empowerment-through-participation as a performance in space: beyond tyranny to transformation' *Signs* 30, 4, pp. 2037–65.

Kindon, S. (2003) 'Participatory video in geographical research: a feminist practice of looking?' *Area* 35, 2, pp. 142–53.

Klein, N. (2016) 'Let them drown: the violence of othering in a warming world', Edward Said Lecture, *London Review of Books* 2 June, pp. 11–14.

Klor de Alva, J. (1992) 'Colonialism and postcolonialism as (Latin) American mirages' *Colonial Latin American Review*, 1, 1–2, pp. 3–23.

Kofman, E. and Raghuram, P. (2006) 'Women and global labour migrations: incorporating skilled workers' Antipode 38, 2, pp. 282–303.

Kothari, U. (2006) 'Spatial practices and imaginaries: experiences of colonial officers and development professions' *Singapore Journal of Tropical Geography* 27, 3, pp. 235–53.

Kothari, U. (1996) 'Development studies and postcolonial theory' Institute for Development Policy and Management (IDPM) Discussion Paper Series No. 47, University of Manchester.

Krishnaswamy, R. and Hawley, J. (2008) 'Planetarity and the postcolonial'. In R. Krishnaswamy and J. Hawley (eds) *The Postcolonial and the Global* Minneapolic: University of Minnesota Press, pp. 105–8.

Kusno, A. (2000) *Behind the Postcolonial* London, Routledge.

Kuttner, R. (2002) 'US fueled Argentina's economic collapse' *Boston Globe* 7 January.

Lacquer, T.W. (1989) 'Bodies, details and humanitarian narrative'. In L. Hunt (ed.) *The New Cultural History* Berkeley, University of California Press, pp. 176–204.

Larsen, S.C. (2006) 'The future's past: politics of time and territory among Dakelh First Nations in British Columbia'. *Geografiska Annaler* 88 B (3), pp. 311–21.

Larmer, M. (2016) 'Rhodes Must Fall and the study of African history in Oxford', African History Seminar, History Department, Durham University, 26 April.

Latouche, S. (2004) *Degrowth Economics: Why Less Should Be So Much More* Le Monde Diplomatique.

Latouche, S. (2009) *Farewell to Growth* Cambridge, Polity.

Lazarus, N. (2016) 'Vivek Chibber and the spectre of postcolonial theory' *Race & Class* 57, 3, pp. 88–106.

Lazarus, N. (2011) *The Postcolonial Unconscious* Cambridge, Cambridge University Press.

Lazreg, M (2000) 'The triumphant discourse of global feminism: should other women be known?'. In A. Amireh and L.S. Majaj (eds) *Going Global: The Transnational Reception of Third World Women Writers* New York, Garland Publishing, pp. 29–38.

Legg, S. (2017) 'Decolonialism' *Transactions of the Institute of British Geographers* doi: 10.1111/tran.12203.

Lester, A. (2003) 'Colonial and postcolonial geographies' *Journal of Historical Geography* 29, 2, pp. 277–88.

Lewis, D, Bebbington, A.J. *et al.* (2003) 'Practice, power and meaning: frameworks for studying organizational culture in multi-agency rural development projects' *Journal of International Development* 15, 5, 541–57.

Lewis, D., Rodgers, D. and Woolcock, M. (2012) 'The projection of development: cinematic representation as an(other) source of authoritative knowledge?' Brooks World Poverty Institute Working Paper 176.

Lewis, D., Rodgers, D. and Woolcock, M. (2005) 'The fiction of development: knowledge, authority and representation' London: LSE research online. Available at http://eprints.lse.ac.uk/archive/00000379/

Lewis, D.L. (ed.) (1995) *The Portable Harlem Renaissance Reader* New York, Viking Penguin.

Lewis, D.L. (1997) *When Harlem Was in Vogue* New York, Penguin.

Lewis, R. (1996) *Gendering Orientalism: Race, Femininity and Representation* London, Routledge.

Li, T. Murray (2014) 'What is land? Assembling a resource for global investment' *Transactions of the Institute of British Geographers* 39, pp. 589–602.

Li, T. Murray (2007) *The Will to Improve. Governmentality, Development and the Practice of Politics* Durham, NC., Duke University Press.

Lidchi, H. (1999), 'Finding the right image: British development NGOs and the regulation of imagery'. In T. Skelton and T. Allen (eds) *Culture and Global Change* London, Routledge, pp. 87–101.

Loomba, A. (1998) *Colonialism/Postcolonialism* London, Routledge.

Loomba, A. and Kaul, S. (1994) 'Location, culture, postcoloniality' *Oxford Literary Review*, 16, pp. 3–30.

Lorde, A. 1984: *Sister/Outsider: Essays and Speeches* New York, The Crossing Press.

Lorimer, D.A. (1978) *Colour, Class and the Victorians. English Attitudes to the Negro in the Mid-Nineteenth Century* Leicester, Leicester University Press.

Lummis C.D. (1992). 'Equality'. In W. Sachs (ed.) *The Development Dictionary. A Guide to Knowledge as Power* London, Zed Books.

Lunga V. (2008) 'Postcolonial theory: a language for a critique of globalization' *Perspectives on Global Development and Technology* 7, 3/4, pp. 191–9.

MacKinnon, D. and Derickson, K.D. (2012) 'From resilience to resourcefulness: a critique of resilience policy and activism' *Progress in Human Geography* 37, 2, pp. 253–70.

Mackintosh, M. Raghuram, P. and Henry, L. (2006) 'A perverse subsidy: African trained nurses and doctors in the NHS' *Soundings* 34, pp. 103–13.

McClintock, A. (1992), 'The angel of progress: pitfalls of the term "postcolonialism"', *Social Text*, 31–32, pp. 84–98.

McClintock, A. (1995) *Imperial Leather: Race, Gender and Sexuality in the Colonial Contest* New York, Routledge.

McEwan, C. (2000) *Geography, Gender and Imperialism* Aldershot, Ashgate.

McEwan, C. (2002) 'Postcolonialism'. In V. Desai and R. Potter (eds) *The Companion to Development Studies* London, Arnold, pp. 127–31.

McEwan, C. (2003) 'Material geographies and postcolonialism' *Singapore Journal of Tropical Geography* 24, 3, pp. 340–55.

McEwan, C. (2005) 'New spaces of citizenship? Rethinking gendered participation and empowerment in South Africa' *Political Geography* 24, pp. 969–91.

McEwan, C. (2003) '"Bringing government to the people": Women, local governance and community participation in South Africa' *Geoforum* 34, 4, pp. 469–81.

McEwan, C. (2008a) 'Postcolonialism/postcolonial geographies'. In R. Kitchin and N. Thrift (eds) *International Encyclopedia of Human Geography Volume 1* Oxford, Elsevier, pp. 327–33.

McEwan, C. (2008b) 'Subaltern'. In R. Kitchin and N. Thrift (eds) *International Encyclopedia of Human Geography Volume 11* Oxford, Elsevier, pp. 59–64.

McEwan, C. (2006a) 'Mobilizing culture for social justice and development: South Africa's *Amazwi Abesifazane* memory cloths programme'. In S. Radcliffe (ed.) *Culture and Development in a Globalising World: Geographies, Actors and Paradigms* London, Routledge, pp. 203–27.

McEwan, C. (2006b) 'Using images, films and photography'. In R. Potter and V. Desai (eds) *Doing Development Research* London, Sage, pp. 231–40.

McEwan, C. and Bek, D. (2006) '(Re)Politicising empowerment: lessons from the South Africa wine industry' *Geoforum*, 37, 7, pp. 1021–34.

McEwan, C. and Mawdsley, E. (2012) 'Trilateral development cooperation: power and politics in emerging aid relationships' *Development and Change* 43, 6, pp. 1185–209.

McEwan, C., Pollard J.S., and Henry, N. (2005) 'Re-visioning a "global city": multicultural economic development in Birmingham' *International Journal of Urban and Regional Research*, 29, 4, pp. 916–33.

McFarlane, C. (2006) 'Traditional development networks: bringing development and postcolonial approaches into dialogue' *Geographical Journal* 172, 1, pp. 35–49.

McIlwaine, C. (2002), 'Perspectives on poverty, vulnerability and exclusion'. In C. McIlwaine and K. Willis (eds) *Challenges and Change in Middle America* London, Pearson, pp. 82–109.

McKinnon, K. (2006) 'An orthodoxy of "the local": post-colonialism, participation and professionalism in northern Thailand' *The Geographical Journal* 172, 1, pp. 22–34.

Madge, C. (1993) 'Boundary disputes: comments on Sidaway (1992)' *Area* 25, 3, pp. 294–9.

Madge, C. and Eshun, G. (2016) 'Poetic world-writing in a pluriversal world: a provocation to the creative (re)turn in geography' *Social and Cultural Geography* 17, 6, pp. 778–85.

Malm, A. and Hornborg, A. (2014) 'The geology of mankind? A critique of the Anthropocene narrative' *The Anthropocene Review* 1, 1, pp. 62–9.

Marcus, G.E. (1999), *Critical Anthropology Now: Unexpected Contexts, Shifting Constituencies* Santa Fe, School of American Research Press.

Mains, S. (2004) 'Teaching transnationalism in the Caribbean: towards an understanding of representation and neo-colonialism in human geography' *Journal of Geography in Higher Education*, 28, 2, pp. 317–32.

Mani, L. (1992) 'Multiple mediations: feminist scholarship in the age of multinational reception'. In H. Crowley and S. Himmelweit (eds), *Knowing Women* Milton Keynes, Open University Press, pp. 306–21.

Manktelow, E. (2014) 'Review: *The Colonisation of Time* by Giordano Nanni' Family & Colonialism Research Network, March 30, https://colonialfamilies.wordpress.com/2014/03/30/review-the-colonisation-of-time-by-giordano-nanni/ (accessed 11 December 2014).

Marchand, M. and Parpart, J. (eds) (1995) *Feminism/Postmodernism/Development* London, Routledge.

Marchand, M. (1995) 'Latin American women speak on development. Are we listening yet?' In M. Marchand and J. Parpart (eds) *Feminism/Postmodernism/Development* London, Routledge, pp. 56–72.

Marcuse, P. (2000) 'The language of globalization' *Monthly Review* 52, 3, pp. 1–4.

Marx, K. (1977) *Capital* Volume I, New York, Vintage Books.

Mawdsley, E. (2017) 'National interests and the paradox of foreign aid under austerity: Conservative governments and the domestic politics of international development since 2010' *The Geographical Journal* doi: 10.1111/geoj.12219.

Mawdsley, E. (2012) 'The changing geographies of foreign aid and development cooperation: Contributions from gift theory' *Transactions of the Institute of British Geographers*, 37, pp. 256–72.

Mawdsley, E. (2010) 'Non-DAC donors and the changing landscape of foreign aid: the (in) significance of India's development cooperation with Kenya', *Journal of Eastern African Studies* 4, 2, pp. 361–79.

Mawdsley, E. (2008) 'Fu Manchu versus Dr Livingstone in the dark continent? Triangulating China, Africa and the west in UK broadsheet newspapers' *Political Geography* 27, 5, pp. 509–29.

Mawdsley, E. and Rigg, J. (2003) 'The World Development Reports II: continuity and change in development orthodoxies' *Progress in Development Studies*, 3, 4, pp. 271–86.

Mawdsley, E., Townsend, J., Porter, G. *et al.* (2002) *Knowledge, Power and Development Agendas: NGOs North and South* Oxford, INTRAC.

Maxwell, S. and Riddell, R. (1998) 'Conditionality or contract: perspectives on partnership for development' *Journal of International Development* 10, 2, pp. 257–68.

Mbembe, A.(2016) 'Decolonizing knowledge and the question of the archive' *Arts and Humanities in Higher Education* 15, 1, 29–45.

Mbembe, A. (2001) *On the Postcolony*, Berkeley, University of California Press.

Mbembe, A. (2013) 'Africa and the future: An interview with Achille Mbembe' by Thomas Blaser, 20 November, https://africasacountry.com/2013/11/africa-and-the-future-an-interview-with-achille-mbembe/ (accessed 12/02/18).

Mbembe, A. (2015) 'Discussing African Futures' an interview with Damola Durosomo, 9 November, www.okayafrica.com/achille-mbembe-african-futures-interview/ (accessed 12/02/18).

Mbembe, A. (2016) 'Africa in the New Century', 29 June, https://africasacountry.com/2016/06/africa-in-the-new-century/ (accessed 12/02/18).

Mehta, U.S. (1999) *Liberalism and Empire* Chicago, University of Chicago Press.

Melman, B. (1992) *Women's Orients: Englishwomen and the Middle East 1718–1918* London, Macmillan.

Memmi, A. (1973) 'The impossible life of Frantz Fanon' *Massachusetts Review* (winter) pp. 2–34.

Menchu, R. (1984) *I, Rigoberta Menchu. An Indian Woman in Guatemala* trans. A. Wright, London, Verso.

Mercer, C., Mohan, G. and Power, M. (2003) 'Toward a critical political geography of African development' *Geoforum* 34, pp. 419–36.

Midgley, C. (1992) *Women Against Slavery: The British Campaigns, 1780–1870* London, Routledge.

Mignolo, W. (2007a) 'Introduction' *Cultural Studies* 21, 2–3, 155–67.

Mignolo, W. (2007b) 'Delinking: the rhetoric of modernity, the logic of coloniality and the grammar of de-coloniality' *Cultural Studies* 21, 2–3, 449–514.

Mignolo, W. (2008) 'The geopolitics of knowledge and colonial difference'. In M.D. Moraña *et al.* (eds) *Coloniality at Large: Latin America and the Postcolonial Debate* Durham, Duke University Press, pp. 225–58.

Miyoshi, M. and Harootunian, H. (eds) (2002) *Learning Places. The Afterlives of Area Studies* Durham, NC, Duke University Press.

Miyoshi, M. (1997) 'Sites of resistance in the global economy'. In K. Ansell-Pearson, B. Parry and J. Squires (eds) *Cultural Readings of Imperialism: Edward Said and the Gravity of History* London, Lawrence and Wishart, pp. 49–66.

Milman, O. and Ryan, M. (2016) 'Lives in the balance: climate change and the Marshall Islands' *The Guardian* 15 September www.theguardian.com/environment/2016/sep/15/marshall-islands-climate-change-springdale-arkansas (accessed 21/12/17).

Moghadam, V. (1994) 'Introduction: women and identity politics in theoretical and comparative perspective'. In V. Moghadam (ed.), *Identity Politics and Women. Cultural Reassertions and Feminisms in International Perspective* Boulder, Westview Press, pp. 3–26.

Mohan, G. (2001) 'Beyond participation: strategies for deeper empowerment'. In B. Cooke and U. Kothari (eds) *Participation: The New Tyranny* London, Zed, pp. 153–67.

Mohan, G. (1999), 'Not so distant, not so strange: the personal and the political in participatory research' *Ethics, Place and Environment* 2, 1, pp. 41–54.

Mohan, G. and Stokke, K. (2000) 'Participatory development and empowerment: the dangers of localism' *Third World Quarterly* 21, pp. 247–68.

Mohanty, C.T. (1988) 'Under western eyes: feminist scholarship and colonial discourses' *Feminist Review* 30, pp. 61–88.

Mohanty, C. T. (2002), ' "Under Western eyes" revisited: feminist solidarity through anticapitalist struggles' *Signs* 28, 2, 499–535.

Mollett, S. (2017) 'Irreconcilable differences? A postcolonial intersectional reading of gender, development and Human Rights in Latin America' *Gender, Place & Culture* 24, 1, 1–17.

Monk, J., Manning, P. and Denman, C. (2003) 'Working together: feminist perspectives on collaborative research and action' *Acme* 2, pp. 91–106.

Moore, D. (2000) 'The crucible of cultural politics: reworking "development" in Zimbabwe's eastern highlands' *American Ethnologist* 26, pp. 654–89.

Moore-Gilbert, B. (1997) *Postcolonial Theory. Contexts, Practices, Politics* London, Verso.

Moraña, M.D., Dussel, E. and Jáuregui, C.A. (eds) (2008) *Coloniality at Large: Latin America and the Postcolonial Debate* Durham, Duke University Press.

Morgan, R. (1984) *Sisterhood is Global: The International Women's Movement Anthology* New York, Anchor.

Morris, R. (2010) 'Introduction'. In R. Morris (ed.) *Can the Subaltern Speak? Reflections on the History of an Idea* New York: Colombia University Press, pp. 1–18.

Mosse, D. (1991) ' "People's knowledge", participation and patronage: operations and representations in rural development'. In B. Cooke and U. Kothari (eds) *Participation:The New Tyranny?* London, Zed, pp. 16–35.

Murphy, D. (2001) *Imagining Alternatives in Film and Fiction – Sembene* Oxford: Africa World Press.

Mufti, A. (2005) 'Global comparativism' *Critical Inquiry* 31 (Winter), pp. 472–89.

Munck, R. (1999) 'Deconstructing development discourse: of impasses, alternatives and politics' in R. Munck and D. O'Hearn (eds) *Critical Development Theory: Contributions to a New Paradigm* London, Zed Books, pp. 189–210.

Myers, G. (2006) 'The unauthorized city: late colonial Lusaka and postcolonial geography' *Singapore Journal of Tropical Geography* 27, 3, pp. 289–308.

Nagar, R. and Ali, F. (2003) 'Collaboration across borders: moving beyond positionality' *Singapore Journal of Tropical Geography* 24, 3, pp. 356–72.

Nain, G.T. (1991) 'Black women, sexism and racism: Black or anti-racist feminism?' *Feminist Review* 37, pp. 1–22.

Nanni, G. (2012) *The Colonisation of Time* Manchester, Manchester University Press.

Narayan, D. (2000) *Voices of the Poor: Can Anyone Hear Us?* World Bank report, available at www1.worldbank.org/prem/poverty/voices/reports/canany/vol1.pdf, (accessed 27/09/07).

Narayan, U. (1997) *Dislocating Cultures. Identities, Traditions and Third World Feminism* New York, Routledge.

Narayan, U. (1998) 'Essence of culture and a sense of history: a feminist critique of cultural essentialism' *Hypatia, Special Issue: Border Crossings: Multicultural and Feminist Challenges to Philosophy* Part 1, 13, 2, pp. 86–107.

Nash, C. (2002) 'Cultural geography: postcolonial cultural geographies' *Progress in Human Geography* 26, 2, pp. 219–30.

Newfarmer, R. (ed.) (2006) *Trade, Doha and Development: A Window on the Issues* New York, World Bank.

Norris, C. (1993) *The Truth about Postmodernism* Oxford, Blackwell.

Nowrojee, B. (1996) *Shattered Lives. Sexual Violence during the Rwandan Genocide and its Aftermath* Human Rights Watch/Africa, available at www.hrw.org/reports/1996/Rwanda. htm (accessed 27/09/07).

Noxolo, P, Raghuram, P. and Madge, C. (2012) 'Unsettling responsibility: postcolonial interventions' *Transactions of the Institute of British Geographers* 37, 3, pp. 418–29.

Noxolo, P. (2017) 'Decolonial theory in a time of re-colonisation of UK research' *Transactions of the Institute of British Geographers* doi: 10.1111/tran.12202.

Noxolo, P. (2006) 'Claims: a postcolonial geographical critique of "partnership" in Britain's development discourse' *Singapore Journal of Tropical Geography* 27, 3, pp. 254–69.

OECD Observer (2014) *OECD in Figures, 2013–14 Supplement* OECD Publishing.

O'Kane, D and Hepner, T.R. (2011) (eds) *Biopolitics, Militarism, and Development: Eritrea in the Twenty-First Century* New York, Berghahn.

Omvedt, G. (2006) *Dalit Visions: The Anti-caste Movement and the Construction of an Indian Identity* New Delhi, Orient Longman (2nd edition).

Omvedt, G. (2005) 'Farmer's movements and the debate on poverty and economic reforms in India'. In R. Ray and M.F. Katzenstein (eds), *Social Movements in India: Poverty, Power, and Politics* New Delhi, Oxford University Press, 179–202.

Omvedt, G. (2004) 'Struggle against dams or struggle for water? Environmental movements and the state'. In R. Vora and S. Palshikar (eds) *Indian Democracy: Meanings and Practices* New Delhi, Sage, pp. 410–31.

Ong, A. (1988) 'Colonialism and modernity: feminist representations of women in non-western societies' *Inscriptions*, 3, 4, pp. 79–104.

Ong, A. (1999) *Flexible Citizenship: The Cultural Logics of Transnationality* Durham, N.C: Duke University Press.

Ong, A. (2006) *Neoliberalism as Exception* Durham, NC., Duke University Press.

Onibokun, A.G. and Faniran, A. (1995) *Community Based Organisations in Nigerian Urban Cities*, Nigeria Centre for African Settlement Studies and Development.

Otzen, U. *et al.*, (1999) *Integrated Development Planning. A New Task for Local Government in South Africa. Participatory planning for socio-economic development in two municipalities in Mpumalanga* Bramfontein, GTZ.

Panos Institute (2006a) *Signed and Sealed? Time to Raise the Debate on International Trade Talks* July, www.panos.org.uk/PDF/reports/prsptoolkit2.pdf.

Panos Institute (2006b) *Making or missing the links? The politics of trade reform and poverty reduction* August, www.panos.org.uk/PDF/reports/prsptoolkit3.pdf.

Paolini, A. (1997) 'The place of Africa in discourse about the postcolonial, the global and the modern' *New Formations* 31, pp. 83–106.

Paral, R. (2002) 'Mexican workers and the US economy: an increasingly vital role' *AILF Immigration Policy Focus* 1, 2, pp. 1–15, available at www.ailf.org/ipc/ipf0902.pdf.

Parpart, J. (1995a) 'Deconstructing the development "expert": gender, development and the "vulnerable groups"'. In M. Marchand and J. Parpart (eds), *Feminism/Postmodernism/ Development* London, Routledge, pp. 221–43.

Parpart, J. (1995b) 'Post-modernism, gender and development'. In J. Crush (ed.) *Power of Development* London, Routledge, pp. 253–65.

Parry, B. (2002) 'Directions and dead ends in postcolonial studies'. In D.T. Goldberg and A. Quayson (eds), *Relocating Postcolonialism* Oxford, Blackwell, pp. 66–81.

Patel, R. and McMichael, P. (2004) 'Third Worldism and the lineages of global fascism: the regrouping of the global South in the neoliberal era' *Third World Quarterly* 25, 1, pp. 231–54.

Peake, L. and Trotz, A. (2002) 'Feminists and feminist issues in the South'. In V. Desai and R. Potter (eds) *The Companion to Development Studies* London, Arnold, pp. 34–7.

Peet, R. (1985) 'The social origins of environmental determinism' *Annals of the Association of American Geographers* 75, pp. 309–33.

Peet, R. and Watts. M. (eds) (1996) *Liberation Ecologies: Environment, Development and Social Movements* London, Routledge.

Perrons, D. (1999) 'Reintegrating production and consumption, or why political economy still matters'. In R. Munck and D. O'Hearn (eds), *Critical Development Theory* London, Zed, pp. 91–112.

Perry, S. and Schenck, C. (eds) (2001) *Eye-to-Eye: Women Practising Development Across Cultures* London, Zed.

Phillips, J. and Potter, R. (2006) ' "Black skins-white masks": postcolonial reflections on "race", gender and second generation return migration to the Caribbean' *Singapore Journal of Tropical Geography* 27, 3, pp. 309–25.

Philo, C. (2000) 'More words, more worlds: reflections on the "cultural turn" and human geography', in I. Cook, D. Crouch, S. Naylor *et al.* (eds), *Cultural Turns/Geographical Turns: Perspectives on Cultural Geography* Harlow, Longman, pp. 26–53.

Pieterse, J.N. (1991) 'Dilemmas of development discourse: the crisis of developmentalism and comparative method' *Development and Change* 22, pp. 5–29.

Pieterse, J.N. (1992) *White on Black: Images of Africa and Blacks in Western Popular Culture* London, Yale University Press.

Pieterse, J.N (2001a) *Globalization and Social Movements* London, Macmillan.

Pieterse, J.N. (2001b) *Development Theory: Deconstructions/Reconstructions* London, Sage.

Pieterse, J.N. (2004) *Globalization and Culture: Global Mélange* Lanham, MD, Rowman and Littlefield.

Pieterse, J.N. and Parekh, B. (1995) (eds) *The Decolonization of the Imagination* London, Zed Books.

Pinney, C. and Peterson, N. (2003) *Photography's Other Histories* Durham, NC and London, Duke University Press.

Pletsch, C. (1981) 'The three worlds or the division of social scientific labour 1950–1975' *Comparative Studies in Society and History* 23, pp. 565–90.

Pogge, T. and Sengupta, M. (2016) 'Assessing the sustainable development goals from a human rights perspective' *Journal of International and Comparative Social Policy* 32, 2, pp. 83–97.

Pogge, T. (2008) *World Poverty and Human Rights* Cambridge, Polity.

Pollard, J., McEwan, C. and Hughes, A. *Postcolonial Economies* London, Zed.

Porter, D., Allen, B. and Thompson, G. (1991) *Development in Practice: Paved with Good Intentions* London, Routledge.

Potter, R.B. (2001) 'Progress, development and change' *Progress in Development Studies*, I, pp. 1–4.

Povinelli, E. interviewed by Mat Coleman and Kathryn Yusoff (2014) 'On biopolitics and the Anthropocene' *Society & Space* March 6, http://societyandspace.org/2014/03/06/on-biopolitics-the-anthropocene-and-neoliberalism/

Povinelli, E. (2011a) *Economies of Abandonment* Durham: Duke University Press.

Povinelli, E. (2011b) 'The woman on the other side of the wall: archiving the otherwise in postcolonial digital archives' *Differences* 22, 1, pp. 146–71.

Power, M., Mohan, G. and Mercer, C. (2006) 'Postcolonial geographies of development' *Singapore Journal of Tropical Geography* 27, 3, pp. 231–4.

Power, M. (2003) *Rethinking Development Geographies* London, Routledge.

Prabhu, A. (2003) 'Mariama Bâ's So Longa Letter: women, culture and development from a francophone/postcolonial perspective'. In K.-K. Bhavnani, J. Foran and P. Kurian (eds), *Feminist Futures: Re-imagining Women, Culture and Development* London, Zed, pp. 239–55.

Prashad, V. (2012) *The Poorer Nations: A Possible History of the Global South* New York, Verso.

Pratt, G. (2012) *Families Apart* Minneapolis, MN, University of Minnesota Press.

Pratt, M.L. (1992) *Imperial Eyes: Travel Writing and Transculturation* London, Routledge.

Pritchett, L. and Woolcock, M. (2004) 'Solutions when the solution is the problem: arraying the disarray in development' *World Development* 32, pp. 191–212.

Prunier, G. (1995) *The Rwanda Crisis 1959–1994: History of a Genocide* London, Hurst.

Quayson, A. (2000) *Postcolonialism: Theory, Practice or Process?* Cambridge, Polity.

Quayson, A. and Goldberg, D.T. (2002) 'Introduction: scale and sensibility' in D.T. Goldberg and A. Quayson (eds), *Relocating Postcolonialism* Oxford, Blackwell, pp. xi–xxii.

Radcliffe, S. (1994) '(Representing) post-colonial women: authority, difference and feminisms' *Area* 26, 1, pp. 25–32.

Radcliffe, S. (1999) 'Re-thinking development'. In P. Cloke, P. Crang and M. Goodwin (eds) *Introducing Human Geographies* London, Arnold, pp. 84–91.

Radcliffe, S. (2005) 'Development and geography II: towards a postcolonial development geography?' Progress in Human Geography, 29, 3, pp. 291–8.

Radcliffe, S. (ed.) (2006) *Culture and Development in a Globalising World: Geographies, Actors and Paradigms* London, Routledge.

Radcliffe, S. (2015) *Dilemmas of Difference. Indigenous Women and the Limits of Postcolonial Development Policy* Durham, NC, Duke University Press.

Radcliffe, S. (2017) 'Decolonising geographical knowledges' *Transactions of the Institute of British Geographers* doi: 10.1111/tran.12195.

Rao, V. and Walton, M. (eds) (2004) *Culture and Public Action* Stanford, University Press.

Raghuram, P., Noxolo, P. and Madge, C. (2014) 'Rising Asia and postcolonial geography' *Singapore Journal of Tropical Geography* 35,1, pp. 119–35.

Raghuram, P., Madge, C. and Noxolo, P. (2009), 'Rethinking responsibility and care for a post-colonial world' *Geoforum*, 40, 1, pp. 5–13.

Raghuram, P. and Madge, C. (2006) 'Towards a method for postcolonial development geography? Possibilities and challenges' *Singapore Journal of Tropical Geography*, 27, 3, pp. 270–88.

Rahnema, M. and Bawtree, V. (eds) (1997) *The Post-Development Reader* London, Zed.

Rajan, R.S. (1993) *Real and Imagined Women. Gender, Culture and Postcolonialism* London, Routledge.

Raju, S. (2002) 'We are different but can we talk?' *Gender, Place and Culture* 9, pp. 173–7.

Ramamurthy, P. (2003) 'Material consumers, fabricating subjects: perplexity, global connectivity discourses, and transnational feminist research' *Cultural Anthropology* 18, 4, pp. 524–50.

Ranta, E. (2016) 'Toward a decolonial alternative to development? The emergence and short-comings of Vivir Bien as Bolivian state policy in the era of globalization' *Globalizations* 13, pp. 425–39.

Rattansi, A. (1997) 'Postcolonialism and its discontents' *Economy and Society* 26, 4, pp. 480–500.

Reddy, S. and Hogge, T. (2002) 'How *not* to count the poor' unpublished paper, Department of Economics, University of Columbia, available at www.socialanalysis.org/.

Reiff, D. (1991) *Los Angeles: Capital of the Third World* New York, Simon Schuster.

Rich, A. (1986) *Blood, Bread and Poetry: Selected Prose, 1979–1985* New York, Norton.

Richards, P. (2016) *Ebola: How a People's Science Helped End an Epidemic* London, Zed.

Rigg, J. (2007) *An Everyday Geography if the Global South* London, Routledge.

Rigg, J. (1997) *Southeast Asia. The Human Landscape of Modernization and Development* London, Routledge.

Rivera Cusicanqui, S. (2012) '*Ch'ixinakax utxiwa*: a reflection on the practices and discourses of decolonisation' *The South Atlantic Quarterly* 111, pp. 95–109.

Roberts, C. and Connell, R. (2016) 'Feminist theory and the global South' *Feminist Theory* 17, 2, pp. 135–40.

Robinson, J. (1994) 'White women researching/representing "others": from anti-apartheid to postcolonialism?' In A. Blunt and G. Rose (eds), *Writing Women and Space* New York: Guilford, pp. 197–226.

Robinson, J. (2003a) 'Postcolonialising geography: tactics and pitfalls' *Singapore Journal of Tropical Geography* 24, 3, 273–89.

Robinson, J. (2003b) 'Political geography in a postcolonial context' *Political Geography* 22, 6, pp. 647–51.

Robinson, N. (2017) 'A quick reminder of why colonialism was bad' *Current Affairs* September 14, www.currentaffairs.org/2017/09/a-quick-reminder-of-why-colonialism-was-bad.

Robson, E. (2002) '"An unbelievable academic and personal experience": issues around teaching undergraduate field courses' *Journal of Geography in Higher Education* 26, 3, pp. 327–44.

Robson, E. and Willis, K. (eds) (1997) *Postgraduate Fieldwork in Developing Areas: A Rough Guide*. Monograph #9, Developing Areas Research Group, London, RGS-IBG.

Rorty, R. (1998) *Achieving our Country: Leftist Thought in Twentieth Century America* Cambridge, MA, Harvard University Press.

Rostow, W.W. (1960) *The Stages of Economic Growth: A Non-Communist Manifesto* Cambridge University Press.

Roy, A. (2010) *Poverty Capital: Microfinance and the Making of Development* New York, Routledge.

Roy, A. (2014) 'Slum-free cities of the Asian century: Postcolonial government and the project of inclusive growth' *Singapore Journal of Tropical Geography* 35, pp. 136–50.

Roy, A. (2016) 'When is Asia?' *The Professional Geographer* 68, 2, pp. 313–21.

Routledge, P. (2003) 'Convergence space: process geographies of grassroots globalization networks' *Transactions of the Institute of British Geographers* 28, 3, pp. 333–49.

Rose, J. (1987) 'The state of the subject (II): the institution of feminism', *Critical Inquiry*, 29, 4, pp. 9–15.

Rothenberg, T. (1994) 'Voyeurs of imperialism: The National Geographic Magazine before World War II'. In A. Godlewska and N. Smith (eds), *Geography and Empire* Oxford, Blackwell, pp. 155–72.

Routledge, P. (2003) 'Convergence space: process geographies of grassroots globalization networks' *Transactions of the Institute of British Geographers* 28, pp. 333–49.

Rist, G. (1997) *History of Development. From Western Origins to Global Faith* London, Routledge.

Rist, G. (2014) *History of Development. From Western Origins to Global Faith* Fourth Edition London, Routledge.

Russett, C.E. (1989) *Sexual Science. The Victorian Construction of Womanhood* Cambridge, Mass., Harvard University Press.

Russo, L. (n.d.) 'Place-relation ecopoetics: A collective glossary' *Jacket2 https://jacket2.org/ commentary/place-relation-ecopoetics-collective-glossary* (accessed 11/10/17).

Sachs, J. (2005) *The End of Poverty* London, Penguin.

Said, E. (2004) *Humanism and Democratic Criticism* New York Colombia University Press.

Said, E. (1999) *Out of Place* London, Granta.

Said, E. (1994) *Representations of the Intellectual* London, Vintage.

Said, E. (1993) *Culture and Imperialism* London, Vintage.

Said, E. (1985), *Orientalism* Harmondsworth, Penguin (first publ. 1978).

Said, E. (1984) *The World, the Text and the Critic* London, Faber and Faber.

Said, E. (1980) 'Islam through western eyes' *The Nation* April 26 www.thenation.com/ doc/19800426/19800426said

Sangari, K. and Vaid, S. (1989) *Recasting Women: Essays in Colonial History* New Delhi, Kali for Women.

San Juan, E. (1998) *Beyond Postcolonial Theory* London, Palgrave.

Sardar, Z. (1996) 'Beyond development: an Islamic perspective' *European Journal of Development Research* 8, 2, pp. 36–55.

Sardar, Z. (1999) 'Development and the locations of Eurocentrism'. In R. Munck and D. O'Hearn (eds), *Critical Development Theory* London, Zed, pp. 44–62.

Satre, L. (2005) *Chocolate on Trial: Slavery, Politics and the Ethics of Business* Athens, Ohio University Press.

Sayer, A. (2001) 'For a critical cultural political economy' *Antipode* 33, pp. 687–708.

Schech, S. and Haggis, J. (2000) *Culture and Development. A Critical Introduction* Oxford, Blackwell.

Schneider, F., Kallis, G. and Martinez-Alier, J. (2010) 'Crisis or opportunity? Economic degrowth for social equity and ecological sustainability. Introduction to this special issue' *Journal of Cleaner Production* 18, 6, pp. 511–18.

Schroeder, R.A. (1999) *Shady Practices? Agro-forestry and gender politics in the Gambia* Berkeley, University of California Press.

Schuurman, F. (ed.) (2001) *Globalisation and Development: Challenges for the 21st Century* London, Sage.

Seery, E. and Caistor Arendar, A. (2014) 'Even it up: time to end extreme inequality' Oxfam, www.oxfam.org/sites/www.oxfam.org/files/file_attachments/cr-even-it-up-extreme-inequality-291014–en.pdf.

Sen, A. (1999) *Development as Freedom* Oxford, Oxford University Press.

Senghor, L.S. (1998), *The Collected Poetry* University of Virginia Press.

Senier, S. (2014) 'Digitizing indigenous history: trends and challenges' *Journal of Victorian Culture* 19, 3, pp. 396–402.

Shahidian, H. (1995) 'Islam, politics, and the problems of writing women's history in Iran' *Journal of Women's History* 7, pp. 113–44.

Sharp, J. (2005) 'Geography and gender: feminist methodologies in collaboration in the field' *Progress in Human Geography* 29, pp. 304–9.

Sharp, J. and Briggs, J. (2006) 'Postcolonialism and development: new dialogues?' *The Geographical Journal* 172, 1, pp. 6–9.

Shatz, A. (2015) 'Magical thinking about ISIS' *London Review of Books*, November 20 www.lrb.co.uk/v37/n23/adam-shatz/magical-thinking-about-isis.

Shiva, V. (2016) *The Violence of the Green Revolution* Lexington, The University Press of Kentucky.

Shiva, V. (1989) *Staying Alive. Women, Ecology and Development* London, Zed.

Sidaway, J., Woon, C.Y. and Jacobs, J.M. (2014) 'Planetary postcolonialism' *Singapore Journal of Tropical Geography* 35, pp. 4–21.

Sidaway, J. (2012) 'Geographies of development: new maps, new visions' *The Professional Geographer* 64, 1, pp. 49–62.

Sidaway, J. (2002) 'Post-development'. In V. Desai and R. Potter (eds) (2002) *The Companion to Development Studies* London: Arnold.

Sidaway, J. (2000) 'Postcolonial geographies: an exploratory essay' *Progress in Human Geography* 24, pp. 591–612.

Simon, D. (2006) 'Separated by common ground? Bringing (post)development and (post)colonialism together' *The Geographical Journal* 172, 1, pp. 10–21.

Simon, D. (1998) 'Rethinking (post)modernism, postcolonialism, and post-traditionalism: South–North perspectives' *Environment and Planning D: Society and Space* 16, pp. 219–45.

Simon, D. (1997) 'Development reconsidered: new directions in development thinking' *Geografiska Annaler* 79B, 4, pp. 183–201.

Simon, D., McGregor, D., Nsiah-Gyabaah, K., *et al.* (2003) 'Poverty elimination, North–South research collaboration, and the politics of participatory development' *Development in Practice* 13, pp. 40–56.

Sinha, M., Guy, D. and Woollacott, A. (1999) 'Introduction: why feminisms and internation-alism?'. In M. Sinha, D. Guy and A. Woollacott (eds), *Feminisms and Internationalism* Oxford: Blackwell, pp. 1–13.

Sittirak, S. (1998) *The Daughters of Development* London, Zed.

Silvey, R. and Rankin, K. (2011) 'Development geography: critical development studies and political geographic imaginaries' *Progress in Human Geography* 35, 3, pp. 1–8.

Slater, D. (1998) 'Post-colonial questions for global times' *Review of International Political Economy* 5, pp. 647–78.

Slater, D. (1997) 'Geopolitical imaginations across the North–South divide: issues of differ-ence, development and power' *Political Geography* 16, 8, pp. 631–53.

Slater, D. (1992a) 'Theories of development and the politics of the postmodern' *Development and Change* 23, pp. 283–319.

Slater, D. (1992b) 'On the borders of social theory: learning from other regions' *Society and Space* 10, pp. 307–27.

Slater, D. and Bell, M. (2002) 'Aid and the geopolitics of the postcolonial: critical reflections on New Labour's overseas development strategy' *Development and Change* 33, 2, pp. 335–60.

Sleman, S. (1991) 'Modernism's last post'. In I. Adam and H. Tiffin (eds), *Past the Last Post: Theorizing Post-colonialism and Post-modernism* Hemel Hempstead: Harvester Wheat-sheaf, pp. 1–11.

Smith, D. (2002) 'Responsibility to distant others'. In V. Desai and R. Potter (eds) *The Com-panion to Development Studies* London, Arnold, pp. 131–4.

Smith, N. (1997) 'The Satanic geographies of globalization: uneven development in the 1990s' *Public Culture* 10, 1, pp. 169–92.

Sogge, D. (2002) *Give and Take: What's the Matter with Foreign Aid?* London, Zed.

Sotomayor, A.M., Rodríguez, J.C., Vélez Martínez, S.I. *et al.* (2017) 'The cruelest storm: 200+ academics speak out for Puerto Rico' *Common Dreams*, September 30, www.com-mondreams.org/views/2017/09/30/cruelest-storm-200–academics-speak-out-puerto-rico accessed March 1 2018.

Southard, B. (1995) *The Women's Movement and Colonial Politics in Bengal, 1921–1936* New Delhi, Manohar.

Spivak, G. (1985a) 'The Rani of Sirmur: an essay in reading the archives' *History and Theory* 24, pp. 247–72.

Spivak, G. (1985b) 'Subaltern Studies: Deconstructing Historiography' in R. Guha (ed.), *Subal-tern Studies IV* New Delhi, Oxford University Press, pp. 330–63.

Spivak, G. (1987) *In Other Worlds: Essays in Cultural Politics* London, Routledge.

Spivak, G. (1988) 'Can the subaltern speak?'. In C. Nelson and L. Grossberg (eds) *Marxism and Interpretation of Culture*, Chicago, University of Illinois Press, pp. 271–313.

Spivak, G. (1990) *The Postcolonial Critic: Interviews, Strategies, Dialogue* London, Routledge.

Spivak, G. (1993) *Outside in the Teaching Machine* London, Routledge.

Spivak, G. (1999) *A Critique of Postcolonial Reason; Toward a History of the Vanishing Present* Cambridge MA, Harvard University Press.

Spivak, G. (2003a) 'A conversation with Gayatri Chakravorty Spivak: politics and the imagina-tion, interview by Jenny Sharpe' *Signs: Journal of Women in Culture and Society* 28, 2, pp. 609–24.

Spivak, G. (2003b) *Death of a Discipline* New York: Colombia University Press.

Spivak, G. (2008) *Other Asias* Oxford, Blackwell.

Spivak, G. (2012) *An Aesthetic Education in An Era of Globalization* Cambridge MA, Harvard University Press.

Spivak, G. (2014) 'Humanities and Development' Durham University Castle Lecture Series, 22 January https://youtube/PX031X4–bmc.

Stengers, I. (2015) *In Catastrophic Times: Resisting the Coming Barbarism* trans. A. Goffey, Ann Arbor: Open Humanities Press.

Stephen, M. (2012) 'Rising regional powers and international institutions: the foreign policy orientations of India, Brazil and South Africa' *Journal of Interdisciplinary International Relations* 26, 3, pp. 289–309.

Stiglitz, J. (2006) 'Development in defiance of the Washington consensus' *The Guardian* 13 April, available at www.guardian.co.uk/china/story/0,,1752851,00.html (accessed 19/11/07).

Stiglitz, J. & Charlton, A. (2005) *Fair Trade for All: How Trade Can Promote Development* Oxford, Oxford University Press.

Stoler, A.L. (1995), *Race and the Education of Desire: Foucault's History of Sexuality and the Colonial Order of Things* Durham, NC, Duke University Press.

Stoler, A.L. (2016) *Duress. Imperial Durabilities in Our Times* Durham, NC, Duke University Press.

Storper, M. (2001) 'The poverty of radical theory today: from the false promises of Marxism to the mirage of the Cultural Turn' *International Journal of Urban and Regional Research* 25, 1, pp. 155–79.

Streeten, P. (1995) *Thinking About Development*, Cambridge, Cambridge University Press.

Sumner, A. (2012) 'Global poverty and the "New Bottom Billion" revisited: exploring the paradox that most of the world's extreme poor no longer live in the world's poorest countries' *IDS Working Paper* www.ids.ac.uk/project/the-new-bottom-billion.

Sundberg, J. (2014) 'Decolonizing posthumanist geographies' *Cultural Geographies* 21, 1, pp. 33–47.

Sylvester, C. (2011) 'Postcolonial takes on biopolitics and economy'. In J. Pollard, C. McEwan and A. Hughes *Postcolonial Economies* London, Zed, 185–204.

Sylvester, C. (2006) 'Bare life as a development/postcolonial problematic' *The Geographical Journal* 172, 1, pp. 66–77.

Sylvester, C. (1999) 'Development studies and postcolonial studies: disparate tales of the "Third World"' *Third World Quarterly*, 20, 4, pp. 703–21.

Silvey, R. and Rankin, K. (2011) 'Development geography: critical development studies and political geographic imaginaries' *Progress in Human Geography* 35, 5, pp. 696–704.

Talbot, C. (2002) 'Blair's neo-colonialist vision for Africa' *World Socialist Web Site* www.wsws.org/articles/2002/feb2002/afri-f16.shtml (accessed 26/09/07).

Thiong'o, N. (1986) *Decolonising the Mind* London, James Curry.

Thiong'o, N. (2013) 'My encounters with Chinua Achebe' *Journal of Asian and African Studies* 48, pp. 760–62.

Thomas (1994) *Colonialism's Culture: Anthropology, Travel and Government* Princeton, Princeton University Press.

Thompson, C.B. (2006) 'Africa: Green Revolution or rainbow evolution' *Review of African Political Economy* 113, pp. 562–5.

Thompson, E.P. (1967) 'Time, work-discipline, and industrial capitalism' *Past & Present* 38, pp. 56–97.

Ticktin, M. (2011) 'The gendered human of humanitarianism: medicalising and politicising sexual violence' *Gender & History* 23, pp. 250–65.

Tlostanova, M., Thapar-Björkert, S. and Koonak, R. (2016) 'Border thinking and disidentification: Postcolonial and postsocialist feminist dialogues' *Feminist Theory* 17, 2, pp. 211–28.

Todd, Z. (2015) 'Indigenizing the Anthropocene'. In H. Davis and E. Turpin (eds) *Art in the Anthropocene: Encounters Among Aesthetics, Politics, Environment and Epistemology* Open Humanities Press, pp. 241–54.

Townsend, J (1995). In collaboration with U. Arrevillaga, J. Bain, S. Cancino, *et al.* (1995) *Women's Voices From the Rainforest* London, Routledge.

Trinh, T. Minh-Ha (1989) 'Difference: "A Special Third World Women's Issue"'. In Trinh, T. Minh-Ha, *Woman, Native, Other. Writing Post-Coloniality and Feminism* Bloomington, Indiana University Press.

Tsing, A. (2015) *The Mushroom at the End of the World: On the Possibility of Life in Capitalist Ruins* Durham: Duke University Press.

Tuck, E. and Yang, W. (2012) 'Decolonization is not a metaphor' *Decolonization: Indigeneity, Education, & Society* 1, 1, pp. 1–40.

Tuhiwai Smith, L (1999) *Decolonizing Methodologies: Research and Indigenous Peoples* London, Zed.

Tuhiwai Smith, L (2012) *Decolonizing Methodologies: Research and Indigenous Peoples* London, Zed (2nd edition).

Udayagiri, M. (1995) 'Challenging modernisation: gender and development, postmodern feminism and activism'. In M. Marchand and J. Parpart (eds), *Feminism/Postmodernism/ Development* London, Routledge, pp. 159–77.

UNICEF (2000) *Progress of Nations* New York, UNICEF

UNICEF (2006) *1946–2006 Sixty Years For Children* New York, UNICEF.

UNDP (2013) *Human Development Report* New York: UNDP. Available at: http://hdr.undp.org/en/media/HDR_2013_EN_complete.pdf.

United Nations Development Programme (1996) *Human Development Report 1996* New York, Oxford University Press.

United Nations Development Programme (UNDP) (2003), *Human Development Report 2003* New York, Oxford University Press.

United Nations Development Programme (UNDP) (2004), *Human Development Report 2004: Cultural Liberty in Today's Diverse World* New York, Oxford University Press http:// hdr.undp.org/reports/global/2004/?CFID=6693733&CFTOKEN=41cb38335de70ce4– 247ABBB9–1321–0B50–354A7618719CB99E&jsessionid=e6301af3b16458598382 (accessed 13 June 2007).

United Nations Development Programme (2016) *Human Development Report 2016* http://hdr. undp.org/sites/default/files/2016_human_development_report.pdf (accessed 14 March 2017).

UN-Habitat (2015) *Habitat III Issue Paper 22 on Informal Settlements* Nairobi, United Nations.

United Nations (2012) *The Global Partnership for Development: Making Rhetoric a Reality* www.un.org/millenniumgoals/2012_Gap_Report/MDG_2012Gap_Task_Force_ report.pdf.

United Nations (2015) *The Millennium Development Goals Report 2015* New York, United Nations, available at www.un.org/millenniumgoals/2015_MDG_Report/pdf/MDG%20 2015%20rev%20(July%201).pdf.

Urdang, S. (1979) *Fighting Two Colonialisms: Women in Guinea-Bissau* London, Monthly Review Press.

VMSDFI (2001) 'Meet the Philippines Homeless People's Federation' *Environment and Urbanization* 13, 2, pp. 73–84.

Van Ausdal, S. (2001) 'Development and discourse among the Maya of Southern Belize' *Development and Change* 32, pp. 577–606.

Van der Post, L. (1955) *Flamingo Feather. A Story of Africa* London, Hogarth.

Van der Post, L. (1958) *The Lost World of the Kalahari* London, Hogarth.

Verhelst, T. (1990) *No Life Without Roots: Culture and Development* London, Zed.

Vizenor, G. (2001) *Fugitive Poses* Lincoln, University of Nebraska Press.

Wainwright, J. (2008) *Decolonizing Development* Oxford: Blackwell.

Wainaina, B. (2005) 'How to Write about Africa' *Granta* 92, winter.

Wade, R. (2001) 'Winners and losers' *The Economist* April, pp. 79–82.

Walsh, C. (2010) 'Development as *Buen Vivir*: institutional arrangements and decolonial entanglements' *Development* 53, 1, pp. 15–21.

Walsh, C. (2007) 'Shifting the geopolitics of critical knowledge: decolonial thought and cultural studies "others" in the Andes' *Cultural Studies* 21, 2–3, pp. 224–39.

Wang, Y. (2008) 'Public diplomacy and the rise of Chinese soft power' *Annals of the American Academy of Political and Social Science* 616, 1, pp. 257–73.

War on Want (2004) *Profiting from Poverty: privatizations consultants, DfID and public services* London www.waronwant.org/download.php?id=254 (accessed 7 June 2007).

War on Want (2006) *Coca-Cola. The Alternative Report* London http://wow.webbler.org/downloads/cocacola.pdf (accessed 6 November 2001).

Ware, V. (1994) *Beyond the Pale: White Women, Racism and History* London, Verso.

Warner, M. (2016) "Under the shadow of Rhodes", Letters, *London Review of Books* 21 April: p. 4.

Watson, E. (2003) 'Examining the potential of indigenous institutions for development: a perspective from Borana, Ethiopia' *Development and Change* 34, pp. 287–309.

Watson, M. (2014) 'Derrida, Stengers, Latour and subalternist cosmopolitics' *Theory, Culture & Society* 31, pp.75–98.

Watts, M. (1995) ' "A New Deal in Emotions". Theory and practice and the crisis of development'. In J. Crush (ed.) *Power of Development* London, Routledge, pp. 44–62.

Watts, M. (2003) 'Alternative modern – development as cultural geography'. In K. Anderson, M. Domosh, S. Pile *et al.* (eds) *Handbook of Cultural Geography*, London, Sage, pp. 433–53.

Weisskopf (1964) 'Economic growth and human well-being' *Quarterly Review of Economics and Business*, 4, 2.

Werbner, P. (2002) 'Introduction: postcolonial subjectivities: the personal, the political and the moral'. In Werbner, R. (ed.) *Postcolonial Subjectivities in Africa* London, Zed, pp. 1–21.

Weizman, E. and Sheikh, F. (2015) *The Conflict Shoreline* Steidl Books.

Wilder, G. (2015) *Freedom Time: Negritude, Decolonization, and the Future of the World* Durham NC, Duke University Press.

Wilkinson, S. and Kitzinger, C. (eds) (1996) *Representing the Other: A Feminism & Psychology Reader* London, Sage.

Williams, G., Meth, P. and Willis, K. (2014) *Geographies of Developing Areas: The Global South in a Changing World* London, Routledge (2nd edition).

Williams, G. (1995) 'Modernizing Malthus. The World Bank, population control and the African environment'. In J. Crush (ed.) *Power of Development* London, Routledge, pp. 158–75.

Willis, K. (2005) *Theories and Practices of Development* London, Routledge.

Wimaladharma, J., Pearce, D. and Stanton, D. (2004) 'Remittances: the new development finance?' *Small Enterprise Development*, 15, pp. 12–20.

Woehrer, V. (2016) 'Gender Studies as a multi-centred field? Centres and peripheries in academic gender research' *Feminist Theory* 17, 3, pp. 323–43.

Wolfensohn, J. (2002) 'Remarks at the signing ceremony to welcome China as a founding member of the Development Gateway Foundation' http://web.worldbank.org/WBSITE/EXTERNAL/NEWS/0,,contentMDK:20049134~menuPK:34474~pagePK:34370~piPK:34424,00.html.

Woods, N. (2007) *The Globalizers: The IMF, the World Bank, and Their Borrowers* Ithaca, NY, Cornell University Press.

World Bank (2011) *Global Development Horizons* Washington DC, World Bank.

World Bank (1997) *World Development Report 1997: The State in a Changing World* New York, Oxford University Press.

World Bank (2006) *Global Economic Prospects: Economic Implications of Remittances and Migration* Washington DC, World Bank.

World Bank (2015) *World Development Report: Mind, Society and Behaviour* Washington DC, World Bank.

World Bank (2016) *Migration and Remittances Factbook 2016* (3rd edition) Washington D.C., World Bank Group.

World Economic Forum (2015) *Africa Competitiveness Report 2015* http://reports.weforum.org/africa-competitiveness-report-2015/overview/?doing_wp_cron=1522149514.9336779117584228515625#view/fn-12 (accessed 27/03/18).

Yapa, L. (2002) 'How the discipline of geography exacerbates poverty in the Third World' *Futures* 34, pp. 33–46.

Yeboah, I. (2006) 'Subaltern strategies and development practice: urban water privatization in Ghana' *The Geographical Journal* 172, 1, pp. 50–65.

Yeoh, B.S.A. (1991) 'Postcolonial Cities' *Progress in Human Geography* 25, 3, pp. 456–68.

Yeoh, B.S.A. and Willis, K. (1999) ' "Heart" and "wing", nation and diaspora: gendered discourses in Singapore's regionalisation process' *Gender, Place and Culture* 6, 4, pp. 355–72.

Young, R. (2001) *Postcolonialism: An Historical Introduction* Oxford, Blackwell.

Young, R. (1990) *White Mythologies: Writing History and the West* London, Routledge.

Yusoff, K. (forthcoming) *Geologic Life.*

Yuval-Davies, N. (1997) *Gender and Nation* London, Sage.

# Index